Control Theory

Control Theory

Multivariable and Nonlinear Methods

Torkel Glad and Lennart Ljung

CRC Press
Taylor & Francis Group
Boca Raton London New York

CRC Press is an imprint of the
Taylor & Francis Group, an informa business

Reprinted 2010 by CRC Press

First published 2000 by Taylor & Francis
11 New Fetter Lane, London EC4P 4EE

Simultaneously published in the USA and Canada
by Taylor & Francis
29 West 35th Street, New York, NY 10001

Taylor & Francis is an imprint of the Taylor & Francis Group

© 2000 Torkel Glad and Lennart Ljung

Translated by the authors from the Swedish *Reglerteori: flervariabla och olinjära metoder* by Torkel Glad and Lennart Ljung. Published by Studenlitteratur 1997.

Every effort has been made to ensure that the advice and information in this book is true and accurate at the time of going to press. However, neither the publisher nor the authors can accept any legal responsibility or liability for any errors or omissions that may be made. In the case of drug administration, any medical procedure or the use of technical equipment mentioned within this book, you are strongly advised to consult the manufacturer's guidelines.

Publisher's Note

This book has been prepared from camera-ready copy provided by the authors

British Library Cataloguing in Publication Data
A catalogue record for this book is available from the British Library

Library of Congress Cataloging in Publication Data
A catalogue record for this book has been requested

ISBN 0-7484-0878-9 (pbk)
ISBN 0-7484-0877-0 (hbk)

Contents

Preface

This book is intended for someone with basic knowledge of control theory who wants only *one* additional book on control in the bookshelf.

Control theory is subject to rapid development and many new methods have been introduced in recent years. There are now also many new control design packages that implement advanced control synthesis techniques. One objective with this book is to provide sufficient background and understanding so that such packages can be used in an insightful manner. Another objective is to describe and discuss basic aspects of control design, the approaches, the key quantities to consider and the fundamental limitations inherent in the designs.

The book can be used for self study and as a complement to program manuals, but is primarily intended as a text book for one or several undergraduate or graduate control courses. It takes off where a typical undergraduate control course ends. This means that we assume the reader to be familiar with continuous time, single-input–single-output systems, with frequency functions and elementary state space theory. This corresponds to the material in e.g.,

Franklin, G. F., Powell, J. D. & Amami-Naeini, A. (1994), *Feedback Control of Dynamic Systems*, 3rd edn, Addison-Wesley, Reading, MA.

or any equivalent text book.

The material can be studied in several different ways. It is, for example, possible to leave out the parallel treatment of discrete time systems, since the basic properties are described in detail for continuous time systems. Part III, on non-linear systems can be used separately, etc. Please visit the home page of the book

`http://www.control.isy.liu.se/controlbook`

for ideas on course curricula, and for exercises and software.

Many people have helped us in our work. Professor Dan Rivera taught us IMC and MPC and also helped us with the examples on the distillation column. Professor Anders Helmersson taught us H_∞-methods and helped us with Example 10.5. Dr. Lars Rundqwist, SAAB AB, supplied us with the Gripen aircraft dynamics and "approved" our design. Professor Per-Olof Gutman shared his knowledge of QFT. Professor Keith Glover has given valuable comments on Chapter 10.

Many students in Linköping have supplied important comments on the material during its development. The book has also been used at the Royal Institute of Technology in Stockholm and Professor Alf Isaksson and Dr. Anders Hansson have shared their experiences in a very useful way.

A number of colleagues have commented the manuscript. We want to mention especially Dr. Bo Bernhardsson, Dr. Claes Bretiholz, Professor Bo Egardt, Dr. Svante Gunnarsson, Professor Per Hagander, Dr. Kjell Nordström, M.Sc. Per Spångeus and Professor Torsten Söderström.

Mrs. Ulla Salaneck has helped us with the translation of the book and given us invaluable support with many other things in the preparation of the material.

We thank all these people for their help.

Notation

Operators, Signals, and Matrices

Notation	Name	Def in eq
p	Differentiation operator	(2.5)
q	Shift operator	(2.24)
s	Laplace variable, replaces p	
z	Z-transform variable, replaces q; also the controlled variable	
r	Reference signal (setpoint)	
e	Control error $r - z$	
u	Control input	
y	Measured output	
w	System disturbance	(6.2)
n	Measurement disturbance	(6.2)
v_1	Noise source (white noise) affecting the states	(5.55)
v_2	Noise source (white noise) affecting measured output	(5.55)
R_1, R_2, R_{12}	Intensities for these sources	Thm 5.4
A, B, C D, N, M	State space matrices	(5.55)
K	Observer gain	(5.64)
L	State feedback	(8.25)

Notation	Name	Def. in eq.
F_y	Feedback controller	(6.1)
F_r	Reference prefilter	(6.1)
G_c	Closed loop system	(6.8)
S	Sensitivity function	(6.9)
T	Complementary sensitivity function also the sampling interval	(6.10)
R_u	"Covariance matrix" for a signal u	(5.15c)
Φ_u	Spectrum for a signal u	Sec. 6.2
$f_x(x,t)$	$\frac{\partial}{\partial x} f(x,t)$	Chap. 18

Norms

We will also work with various vector, matrix, and signal norms. In the table below, z denotes a column vector, A a matrix, and f a function, e.g., of time (i.e., a signal) or of frequency (i.e., a frequency function). A^* denotes matrix adjoint, i.e., the matrix obtained by taking the transpose of the matrix and the complex conjugates of its elements. $\mathrm{tr}(A)$ denotes the trace of the square matrix A, i.e., the sum of its diagonal elements.

Notation	Definition	Name
$\lvert z \rvert_2 = \lvert z \rvert$	$\lvert z \rvert_2^2 = z^* z$	Euclidian vector norm
$\lvert z \rvert_P$	$\lvert z \rvert_P^2 = z^* P z$	Weighted vector norm
$\lvert A \rvert$	$\sup_{z \neq 0} \frac{\lvert Az \rvert_2}{\lvert z \rvert_2}$	(Operator) matrix norm
$\overline{\sigma}(A)$	$\overline{\sigma}^2(A) = \max \lambda(A^* A)$	Largest singular value of A
	Note that $\lvert A \rvert = \overline{\sigma}(A)$	($\lambda(A)$ are the eigenvalues of A) A
$\lvert A \rvert_2$	$\lvert A \rvert_2^2 = \mathrm{tr}(A^* A)$	2-norm
$\lvert A \rvert_P$	$\lvert A \rvert_P^2 = \mathrm{tr}(A^* P A)$	Weighted 2-norm
$\lVert f \rVert_2$	$\lVert f \rVert_2^2 = \int \lvert f(x) \rvert_2^2 dx$	$\mathcal{H}_2$-norm
	$\lim_{N \to \infty} \frac{1}{N} \int_0^N \lvert f(x) \rvert_2^2 dx$	Alternative interpretation
$\lVert f \rVert_P$	$\lVert f \rVert_P^2 = \int \lvert f(x) \rvert_P^2 dx$	Weighted $\mathcal{H}_2$-norm
$\lVert f \rVert_\infty$	$\sup_x \lvert f(x) \rvert$	$\mathcal{H}_\infty$-norm

Chapter 1

Introduction

Control theory deals with analysis of dynamical systems and methodologies to construct controllers. The object to be controlled can be modeled in several ways. Basic concepts, like feedback, performance, stability, sensitivity, etc., are quite general, but techniques and tools vary with the type of models used. A typical first course in control covers continuous time systems described by linear systems with one input and one output (Single-Input–Single-Output, SISO, systems). In this book we assume the reader to be familiar with this basic theory, including frequency domain analysis and simple state-space representations. The aim of the book is to extend this theory to cover

- Multivariable systems (multiple inputs and multiple outputs, MIMO)

- Discrete time theory

- Nonlinear systems

For SISO systems we will also convey a deeper understanding of the fundamental performance limitations that may be at hand. To motivate the ensuing treatment, we start by considering two examples, where the behavior cannot be understood from linear SISO theory. Later in this introductory chapter we will treat some general issues in control, like the role of discrete and continuous time models (Section 1.3), list the basic concepts and notations (Section 1.4), and discuss general aspects of stability and signal sizes (Sections 1.5 and 1.6).

1

1.1   Multivariable Systems

A straightforward way to control a multivariable system is to couple each input to an output and then control each pair as a SISO system. The following examples show the possible pitfalls of such an approach.

Example 1.1:   A Multivariable System

Consider a system with two inputs, u_1 and u_2, and two outputs, y_1 and y_2, whose dynamics is given by [1]

$$y_1 = \frac{2}{s+1}u_1 + \frac{3}{s+2}u_2$$
$$y_2 = \frac{1}{s+1}u_1 + \frac{1}{s+1}u_2 \tag{1.1}$$

Suppose that r_1 is the setpoint (reference signal) for y_1 and r_2 is the setpoint for y_2. If we let u_1 control y_1, we could use a PI-regulator

$$u_1 = \frac{K_1(s+1)}{s}(r_1 - y_1) \tag{1.2}$$

giving the transfer function

$$\frac{2K_1}{s+2K_1} \tag{1.3}$$

from r_1 to y_1 if $u_2 = 0$. For $K_1 = 1$ we obtain the step response of Figure 1.1a. Similarly, if u_2 controls y_2 with the PI-regulator

$$u_2 = \frac{K_2(s+1)}{s}(r_2 - y_2) \tag{1.4}$$

we have the transfer function

$$\frac{K_2}{s+K_2} \tag{1.5}$$

from r_2 to y_2 if $u_1 = 0$. With $K_2 = 2$ we obtain the same step response as from r_1 to y_1, i.e., as in Figure 1.1a. What happens if both controllers (1.2) and (1.4) operate simultaneously? The results are shown in Figure 1.1b for $K_1 = 1$ and $K_2 = 2$ and a step in r_1 with $r_2 = 0$. The control is successful in the sense that y_1 tends to 1 and y_2 tends to zero. The response is however much slower than in case (a). Trying to speed up the response by increasing the gains to $K_1 = 4$ and $K_2 = 8$ gives an unstable system, as illustrated in Figure 1.1c. Notice that this happens despite the fact that each regulator by itself, (1.3) and (1.5), gives stability for all positive gains.

---

[1]In these introductory examples we assume the reader to be familiar with some standard notation in control theory. Formal definitions will be given in Chapter 2.

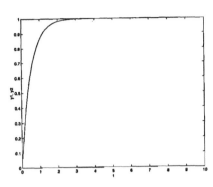

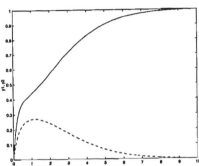

(a) Step response from r_1 to y_1 when $u_2 = 0$ and u_1 is controlled from y_1 by a PI-regulator (1.2) with $K_1 = 1$. The same response is obtained from r_2 to y_2 when $u_1 = 0$ and u_2 is controlled from y_2 by a PI-regulator (1.4) with $K_2 = 2$.

(b) y_1 (solid) and y_2 (dashed) for a step in r_1 when u_1 and u_2 are controlled simultaneously by PI-regulators from y_1 and y_2. $K_1 = 1$, $K_2 = 2$.

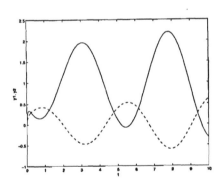

(c) y_1 (solid) and y_2 (dashed) for a step in r_1 when u_1 and u_2 are controlled simultaneously by PI-regulators from y_1 and y_2. $K_1 = 4$, $K_2 = 8$.

Figure 1.1: Step responses for a multivariable system. Various control strategies.

Of course, nothing forces us to use u_1 to control y_1 and u_2 to control y_2. We could equally well do the reverse and use, for example

$$u_1 = K_1 \frac{s+1}{s}(r_2 - y_2), \quad u_2 = K_2 \frac{s+2}{s}(r_1 - y_1) \tag{1.6}$$

This leads to even greater difficulties. Inserting the controllers (1.6) into the system (1.1) gives the following transfer function from r_1 to y_1

$$\frac{K_2(3s^2 + (3 + K_1)s - K_1)}{s^3 + (1 + K_1 + 3K_2)s^2 + (K_1 + 3K_2 + K_1 K_2)s - K_1 K_2} \tag{1.7}$$

A necessary condition for stability is that all coefficients in the denominator are positive. This means that either K_1 or K_2 must be chosen as negative. This in turn implies that one of the gains must have "the wrong sign". The system will then be unstable if it is operated with this controller alone. The result is a vulnerable system where one of the regulators cannot be disconnected. Moreover, it is difficult to tune the controller parameters for good performance. Figure 1.2 shows step responses for $K_1 = -0.75$, $K_2 = 5$.

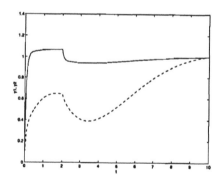

Figure 1.2: y_1 (solid) and y_2 (dashed) for a step in r_1 ($t = 0$) and r_2 ($t = 2$) when u_2 is chosen as PI-control from y_1 and u_1 is controlled from y_2.

After discussing the general theory for multivariable systems, we shall return to this example in Chapter 7. We will show that the fundamental difficulty is that the multivariable transfer function has a zero in the right half plane (the system is non-minimum phase), despite the fact that none of the individual SISO transfer functions in (1.1) have any zeros at all.

1.2   Nonlinear Systems

We know from experience that many physical systems can be described well by linear theory under certain operating conditions, while other conditions may show strong nonlinear effects. An example of this is when a signal reaches a certain physical limit – it is "saturated".

Example 1.2:   Current-controlled DC-motor with Saturated Current

A DC-motor with the current as input, u, and angular position as output, y can be described approximately as two integrations

$$\ddot{y}(t) = u(t) \tag{1.8}$$

The input is always bounded. Assume for simplicity that the limits are ± 1

$$|u(t)| \leq 1 \tag{1.9}$$

Consider a servo with a linear controller of the form

$$u = F(r - y) \tag{1.10}$$

where r is the setpoint for y. Let F be a phase advance (lead) link with transfer function

$$F(s) = \frac{30(s+1)}{s+10} \tag{1.11}$$

The total loop gain for the system will then be

$$\frac{30(s+1)}{s^2(s+10)} \tag{1.12}$$

with crossover frequency 3 rad/s and phase margin 55°, which should give a reasonable step response. When the controller has computed a control input outside the bounds (1.9), we simply truncate it to the corresponding bound. This gives step responses according to Figure 1.3. Here, the output has been scaled with the setpoint amplitude, so that it tends to 1 in both cases. Figure 1.3a shows the response when the step amplitude is so small the the input hits its limit during just a short interval. This case therefore shows good agreement with what linear theory predicts. The rise time is about 0.3 s and there is just a small overshoot. When the step amplitude is increased to 20, the response is as in Figure 1.3b. The rise time here is much longer, as a natural consequence of the limitation of the input amplitude. In addition, we see that the qualitative behavior of the step response is much worse with large over- and undershoots. The reason why linear theory fails to describe this behavior is clear from the input signal. It is at its bounds practically all the time, and never equal to the value computed by the controller.

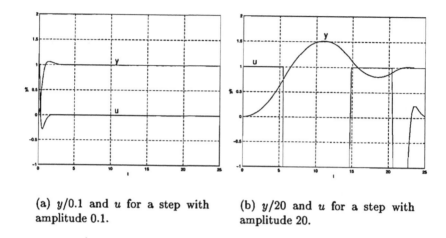

(a) $y/0.1$ and u for a step with   (b) $y/20$ and u for a step with
amplitude 0.1.   amplitude 20.

Figure 1.3: Step responses for the double integrator servo with constrained input amplitude and different setpoint steps.

The example shows that it is difficult to predict performance from linear theory when there are essential nonlinearities (in this case control signal saturation). Part III deals with analysis and design of systems with nonlinearities. In particular, we will see more sophisticated ways of dealing with input saturation than just truncating the computed control signal. This includes internal model control (Section 15.2), parametric optimization (Section 15.3), model predictive control (Chapter 16) and optimal control (Chapter 18).

1.3   Discrete and Continuous Time Models and Controllers

In modern control systems, the controllers are almost exclusively digitally implemented in computers, signal processors or dedicated hardware. This means that the controller will operate in discrete time, while the controlled physical system naturally is described in continuous time. There are two different ways to handle this.

I. Do the Calculations as if the Controller Were Time Continuous

With today's fast computers the sampling in the controller can be chosen very fast, compared to the time constants of the controlled system. Then the controller approximately can be considered as a time continuous system. The design can be carried out in continuous time, and the discrete time theory is needed only in the final implementation phase, when the controller is transformed to discrete time. This transformation typically will deteriorate the performance somewhat, and this fact must be considered at the robustness analysis.

II. Do the Calculations as if Both Controller and the Physical System Were Time Discrete

The controlled system can always approximately be described by a time discrete model. For a linear system whose input is constant between the sampling instants (those instants when the control input is changed and the output is measured) a linear, discrete time model can be derived that exactly gives the relationships between the variables at the sampling instants. Then it is possible to do the controller design entirely by discrete time calculations. An advantage with this is that the time discretization effects are properly taken care of. A disadvantage is that the model only describes what happens at the sampling instants. The method is complicated for nonlinear systems, since a continuous time model of a nonlinear system typically has no simple, exact, discrete time counterpart.

If only a finite time horizon is considered, the method has the advantage that only a finite number of input values have to be determined (in contrast to continuous time, which in principle requires an uncountable number of values). This leads to discrete time controllers that have no continuous time counterpart, e.g., model predictive controllers. See Chapter 16.

The Treatment of Continuous and Discrete Time in This Book

We will not make a definite choice between the two methods above. It is quite dependent on the application and available design and implementation tools. Since the theory for linear systems is very similar in continuous and discrete time, we shall present the theory for

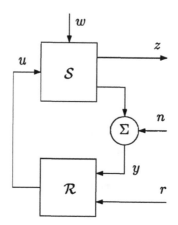

Figure 1.4: A controlled system.

continuous time systems, and then, at the end of each chapter, briefly
discuss the corresponding discrete time results. The considerations
about continuous and discrete time that are relevant for the choice
between methods I and II are discussed in Chapter 4 and Section 6.7.
For nonlinear systems we focus entirely on continuous time, except in
Chapter 16 on model predictive control.

1.4   Some Basic Concepts in Control

The control problem can in general terms be formulated as follows:

The Control Problem
Given a system S, with measured signals y, determine a
feasible control input u, so that a controlled variable z as
closely as possible follows a reference signal (or setpoint)
r, despite influence of disturbances w, measurement
errors n, and variations in the system.

The problem is typically solved by letting u be automatically generated
from y and r by a controller (or regulator) (R). See Figure 1.4.

The control problem as formulated leads to a number of issues. One
is to describe the properties of the system (S) and the disturbances (w).
Another is to construct methods to calculate the regulator (R) for wide

classes of system descriptions. In this book we shall treat the cases where S may be multivariable and/or nonlinear. Mathematically, this means that S is described by linear or nonlinear differential equations. In discrete time, the counterpart is difference equations. Some basic principles for the control design will be common to all cases, but the actual computation of the controllers will depend strongly on the system description.

In the remainder of this chapter we shall deal with some aspects of the control problem that do not depend on the system description.

System

By a *system* we mean an object that is driven by a number of inputs (external signals) $u(t)$, $-\infty < t < \infty$ and as a response produces a number of outputs $y(t)$ $-\infty < t < \infty$. See Figure 1.5.

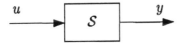

Figure 1.5: A system S.

From a mathematical point of view, a system is a mapping (a function) from the input u to the output y

$$u \xrightarrow{S} y$$

or

$$y = S(u) \tag{1.13}$$

Note that the mapping (1.13) is from the entire input $u(t)$, $-\infty < t < \infty$ to the entire output $y(t)$, $-\infty < t < \infty$. The value of the output at time t_1, i.e., $y(t_1)$ could thus very well depend on the values of the signal u at all time points t, $-\infty < t < \infty$.[2] Starting

---

[2] For a dynamical system, one must normally define an initial state for the output to be uniquely defined. Since the initial state in our formalism is defined at $-\infty$, its influence will have "died out" at any finite time point for a wide class of systems. If this is not the case, we will let different initial states correspond to different systems. Systems with initial state at a finite time point t_0 can in this formalism often be handled by considering an input over $-\infty < t \le t_0$ that leads to the desired state at time t_0.

from the general description (1.13) we shall define a number of useful concepts for systems. The system S is

1.
- *causal* if for every time point t_1, $y(t_1)$ only depends on $u(t)$, $-\infty < t \le t_1$
- *non-causal* otherwise

2.
- *static* if for every time point t_1, $y(t_1)$ only depends on $u(t)$ for $t = t_1$.
- *dynamic* otherwise

3.
- *time discrete* if u and y are defined only for a countable number of time points, kept strictly apart by a smallest, positive time distance:

$$(y(t), u(t)) \qquad t = t_k, \ k = 0, \pm 1, \pm 2, \ldots$$

- *time continuous* if u and y are defined for all real t over an interval or the whole of the time axis

4.
- *SISO (single-input–single-output)* if, for every time point t_1, $u(t_1)$ and $y(t_1)$ are scalars (real numbers)
- *multivariable* otherwise

5.
- *time invariant* if the mapping (1.13) does not depend on absolute time (i.e., if u is shifted by τ time units, then the corresponding output also will be shifted by τ time units).
- *time varying* otherwise

6.
- *linear* if S is a linear mapping, i.e.,

$$S(\alpha_1 u_1 + \alpha_2 u_2) = \alpha_1 S(u_1) + \alpha_2 S(u_2) \tag{1.14}$$

- *nonlinear* otherwise

Example 1.3: Scalar Differential Equation

Let the relationship between u and y be given by the differential equation

$$\frac{d}{dt}y(t) + y(t) = u(t) \tag{1.15}$$

Solving this equation gives

$$y(t) = \int_{-\infty}^{t} e^{-(t-\tau)} u(\tau) d\tau.$$

It is now easy to verify that (1.15) is a SISO, linear, causal, time continuous, and time invariant system.

Example 1.4: Nonlinearity

The system

$$y(t) = u^3(t), \qquad t = 1, 2, ...$$

is static, nonlinear, time invariant, time discrete, and SISO.

Let us consider some typical systems that are part of the control problem. See also Figure 1.4.

- *The Control Object* ("the plant") is the system to be controlled. It has

    - Inputs: A *control input* (*u*) which is used by us to affect the system, and a *disturbance* (*w*) that also affects the system. See the following section.

    - Outputs: A *controlled variable* (*z*) which is that variable that should be controlled to desired values, and a *measured output* (*y*) which contains all measured signals from the plant. The signal *y* may very well include some disturbances.

- *The Controller or Regulator* is also a system with

    - Inputs: *The Reference signal or Setpoint* (*r*), which is the desired value of the controlled variable *z*. The measured output from the controlled object, *y*, is also an input to the controller.

    - Outputs: The controller output is the control signal *u* to the controlled object.

- *The Closed Loop System* is the system obtained when the control loop is closed. This system has

    - Inputs: Reference signal and disturbance signal.

– Outputs: *The Control Error* ($e = r - z$), the difference between the desired and actual values of the controlled variable. Moreover, it is natural to consider also the control signal u from the controller as an output from the closed loop system, since it may be necessary to study its properties as a function of the inputs.

Disturbances

The external signals that affect a controlled object are of two different kinds: (1) The signals that we use to actively control the object. These are the *control inputs*, and in the sequel we reserve the notation u for these. (2) Signals that we cannot choose ourselves, but still affect the system. These are called *disturbance signals* and will generally be denoted by w.

Example 1.5:   Room Temperature Control

Consider heating of a room. The controlled variable is the room temperature, and it is affected by the external signals

1. temperature and flow of water into the radiators
2. outdoor temperature
3. number of persons in the room (each person radiates some 100W)
4. draft
5. solar radiation into the room.

Of these external signals, 1. corresponds to control inputs and the rest are disturbances.

From a control point of view it is important to distinguish between

- measured/measurable disturbances w_m

- non measured/measurable disturbances w_s

Measurable disturbances shall of course be used in the controller as feedforward. In the example above, the outdoor temperature is a typical measurable disturbance that is often used in room temperature control. The other disturbances are usually considered to be unmeasurable, but this is a concept that also depends on the sensor budget.

The non-measurable disturbances in turn are of two fundamentally different kinds. To understand this, it is important to distinguish between *the controlled variable* $z(t)$ and *the measured output* $y(t)$, cf. Figure 1.1. The controlled variable $z(t)$ should follow the reference signal. The measured output $y(t)$ contains the measured information about how the system behaves. Often there is the simple relationship

$$y(t) = z(t) + n(t) \tag{1.16}$$

where $n(t)$ is the *measurement error* or the *measurement disturbance*. In, e.g., the paper industry, it is of interest to control the paper thickness to a certain setpoint. The thickness is typically measured by a traversing sensor with radioactive or X-ray equipment, and these measurements always have some errors. This is the situation (1.16).

The measurement disturbances thus do not affect the controlled variable $z(t)$. By *system disturbances* we mean such disturbances that also affect $z(t)$. In the paper machine example, system disturbances include, e.g., variations in the raw material.

Let us summarize the classification of external signals for a controlled object as follows:

- $u(t)$: Control input. Signals that are actively used to affect the controlled variable z.

- $w(t)$ System disturbance. Affects the controlled variable z, but cannot be chosen by the user.

  - $w_m(t)$: Measured/measurable system disturbance. Should be used in the controller.

  - $w_s(t)$: Non measured/measurable system disturbance.

- $n(t)$: Measurement disturbance. Affects the measured output y, but not the controlled signal z.

The basic relationship is

$$z = S(u, w_m, w_s), \qquad y = z + n \tag{1.17}$$

See also Figure 1.6.

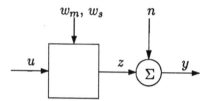

Figure 1.6: A control system with measured disturbance w_m, measurement disturbance n and system disturbance w_s. The controlled variable is z and the measured output is y.

Measurement and System Disturbances

It is important to understand the fundamental distinction between system and measurement disturbances. The influence of the former should (ideally) be cancelled by the control input, while the latter should have as little impact as possible on the control input. Normally measurement and system disturbances are of quite different physical origins. Consider, e.g., an aircraft, whose outputs are position and attitude, and whose control inputs are commanded rudder and aileron angles. The system disturbances are signals that actually affect the aircraft, like wind forces and turbulence. The measurement disturbances, however, are deficiencies and disturbances in the gyros and accelerometers that measure the motion of the aircraft, typically friction, misalignment, and electronic noise.

The simple relation (1.16) between controlled variable and measured output does not always hold. Sometimes the relationship can be quite complicated. In various combustion processes, like car engines, some of the controlled variables are concentrations of polluting substances in the exhaust fumes. These concentrations can seldom be measured directly. Instead, other variables, like temperatures and flows, are measured and from those measurements estimates of the variables of interest are formed. This gives two different relationships

$$z(t) = S_1(u, w)$$
$$y(t) = S_2(u, w) + n(t) \qquad (1.18)$$

where $z(t)$ is the controlled variable, e.g., NOX concentration, and $y(t)$ is the measured output. $S_2(u, w)$ then are "the true values" of the measured variables, temperatures, flows, etc.

1.5 Gain and Signal Size

Scalar Measures of Size; Signal Norms

The general control problem as formulated in Section 1.4 contains indirect characterizations of signal size ("as closely as possible", "feasible control inputs"). In Chapter 5 we shall discuss in more detail how to measure and change the "size" of a signal. Here we shall introduce some basic concepts.

Consider a signal $z(t)$. In the time continuous case t assumes all real values, while we let t assume integer values in the discrete time case. If $z(t)$ is a (column) vector of dimension n, its size at time t is measured by the usual vector norm

$$|z(t)|^2 = \sum_{j=1}^{n} z_j^2(t) = z^T(t)z(t) \tag{1.19}$$

Typical candidates for measuring the size of the entire signal z will then be

$$\|z\|_\infty = \sup_t |z(t)| \quad \text{(infinity-norm, } \mathcal{H}_\infty\text{-norm)} \tag{1.20}$$

$$\|z\|_2^2 = \int_{-\infty}^{\infty} |z(t)|^2 dt \quad \text{(2-norm, } \mathcal{H}_2\text{-norm)} \tag{1.21}$$

or, in discrete time

$$\|z\|_2^2 = \sum_{t=-\infty}^{\infty} |z(t)|^2 \tag{1.22}$$

Of these size measures, the 2-norm in (1.21) or (1.22) is the most natural one from a mathematical point of view. We shall use this norm for formal definitions and calculations. The drawback is that it requires the "energy" of the signal to be bounded. Often control systems are subject to disturbances, for which this is not a reasonable assumption. We shall discuss this in more detail in Section 5.2.

Remark 1. The $\mathcal{H}$ in the norm notation refers to a function space, the Hardy space, that is normally used for causal signals and systems.

Remark 2. For practical reasons we reserve the "double-bar" notation for norms of functions and operators. Norms for vectors and matrices in finite dimensional spaces are marked with "single-bars".

Gain

Linear mappings can be written in matrix form as follows

$$y = Ax \tag{1.23}$$

The norm of the mapping A – the (operator) norm of the matrix A – is then defined as how much larger the norm of y can be, compared to the norm of x:

$$|A| = \sup_{x \neq 0} \frac{|y|}{|x|} = \sup_{x \neq 0} \frac{|Ax|}{|x|} \tag{1.24}$$

We may interpret $|A|$ as the "gain" of the mapping.

Based on this interpretation of the operator norm, we define the concept of *gain* also for general systems. For a system

$$y = S(u) \tag{1.25}$$

we introduce

Definition 1.1 (Gain) *With the **gain** of the system S we mean*

$$\|S\| = \sup_{u} \frac{\|y\|_2}{\|u\|_2} = \sup_{u} \frac{\|S(u)\|_2}{\|u\|_2} \tag{1.26}$$

The supremum (maximum) in the expression above is to be taken over all well defined inputs u, i.e., those with a finite 2-norm, which give well defined outputs y (which need not have finite 2-norm). The gain may very well be infinite.

Remark. For this definition of gain to be meaningful, the zero levels of the signals should be defined so that $u \equiv 0$ corresponds to $y \equiv 0$.

When cascading systems, we have the natural relation

$$\|S_1(S_2)\| \leq \|S_1\| \cdot \|S_2\| \tag{1.27}$$

Example 1.6:   The Gain of a Static Nonlinear System

Consider a static system given by

$$y(t) = f(u(t))$$

where the function f is subject to

$$|f(x)| \leq K \cdot |x|$$

and where equality holds for at least one x, say x^*.

This gives

$$\|y\|_2^2 = \int_{-\infty}^{\infty} (f(u(t)))^2 dt \leq \int_{-\infty}^{\infty} K^2(u(t))^2 dt = K^2\|u\|_2^2$$

The gain is therefore less than or equal to K. On the other hand, we can use the input $u(t) \equiv x^*$, which gives equality in the above expression. The gain of the system $y(t) = f(u(t))$ is therefore equal to K.

Example 1.7: Integrator

Consider a system that is an integrator:

$$y(t) = \int_0^t u(\tau)d\tau \tag{1.28}$$

Take, e.g., the input

$$u(t) = \begin{cases} 1 & \text{for} \quad 0 < t < 1 \\ 0 & \text{otherwise} \end{cases}$$

with 2-norm equal to 1. The output is identically 1 for $t \geq 1$. The output has infinite 2-norm and *the gain of an integrator is thus infinite*.

Example 1.8: A SISO Linear System

Consider a linear, time-continuous system with input u, output y, and impulse response $g(\tau)$, such that $\int |g(\tau)|d\tau < \infty$. Then its Fourier transform (i.e., the frequency function, see Section 3.5) $G(i\omega)$ is well defined. Limiting ourselves to inputs with well defined Fourier transforms $U(i\omega)$ gives

$$Y(i\omega) = G(i\omega)U(i\omega)$$

where $Y(i\omega)$ is the output Fourier transform. Parseval's equality then tells us that

$$\|y\|_2^2 = \int_{-\infty}^{\infty} y^2(t)dt = \frac{1}{2\pi} \int_{-\infty}^{\infty} |Y(i\omega)|^2 d\omega$$

Suppose that

$$|G(i\omega)| \leq K$$

and that there is equality for some ω^*. Then

$$\|y\|_2^2 = \frac{1}{2\pi} \int_{-\infty}^{\infty} |Y(i\omega)|^2 d\omega = \frac{1}{2\pi} \int_{-\infty}^{\infty} |G(i\omega)|^2 |U(i\omega)|^2 d\omega$$
$$\leq K^2\|u\|_2^2$$

The gain of the system is consequently at most K. Moreover, we realize that we can come arbitrarily close to equality in the above expression by concentrating the input energy more and more closely around the frequency ω^*. *The gain of the system thus is*

$$\|G\| = \sup_\omega |G(i\omega)|$$

This gain can be seen as a norm of the function $G(i\omega)$ and is then written $\|G\|_\infty$. The operator norm for a linear system thus coincides with the norm (1.20) of its frequency function $G(i\omega)$.

1.6   Stability and the Small Gain Theorem

Stability is a fundamental concept for control systems. The price for the good properties that feedback gives, is the risk of making the closed loop system unstable. In this book we will therefore show considerable interest in conditions that guarantee stability of the closed loop system under varying conditions.

Input–Output Stability

There are a number of different stability concepts. A simple one is *input–output stability* for general systems. With this is meant that an input with bounded norm must lead to an output with bounded norm. In terms of the general concept of *gain* in (1.26) the definition is simply:

Definition 1.2 (Input-Output Stability) *A system is* input–outpu stable *if it has finite gain.*

The Small Gain Theorem

A very simple, but useful result on stability of general closed loop systems is given by the following theorem.

Theorem 1.1 (Small Gain Theorem) *Consider two stable systems S_1 and S_2 in a feedback loop according to Figure 1.7. The closed loop system, thus defined with r_1, r_2 as inputs and e_1, e_2, y_1, y_2 as outputs is input-output stable if the product of the gains is less than 1:*

$$\|S_2\| \cdot \|S_1\| < 1 \tag{1.29}$$

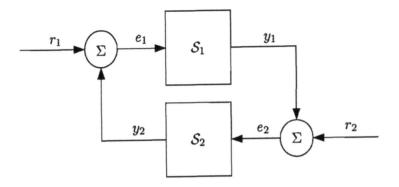

Figure 1.7: A closed loop system.

Corollary 1.1 *If the systems S_1 and S_2 are linear, the condition (1.29) can be replaced by the weaker one*

$$\|S_1 S_2\| < 1 \qquad (1.30)$$

Proof: (Sketch.) The loop gives the following relations

$$e_1 = r_1 + S_2(r_2 + y_1)$$
$$y_1 = S_1(e_1)$$

Take the norm of the signals and use the triangle inequality and the definition of gain:

$$\|e_1\| \le \|r_1\| + \|S_2\|(\|r_2\| + \|S_1\|\|e_1\|)$$

which gives

$$\|e_1\| \le \frac{\|r_1\| + \|S_2\|\|r_2\|}{1 - \|S_2\|\|S_1\|}$$

The gain from r_1 and r_2 to e_1 is consequently finite. Similar calculations show that the gains to the other outputs are also finite. (The reason why these calculations are not a formal proof, is that the norms of the output must exist for the calculations to be valid. We should have truncated the outputs at some time point and shown that these truncated outputs have an upper bound for their norms.) To show the corollary, we realize that a linear, stable system can be moved around the loop without affecting the stability. We then view the loop as the system $S_1 S_2$ in feedback with the system 1. $\square$

Note that it does not matter if we define the feedback as positive or negative (S_2 has the same gain as $-S_2$).

Example 1.9:   Nonlinear Static Feedback

Consider the system in Figure 1.8. The input of the linear system

$$G(s) = \frac{0.4}{s+1}$$

is determined by feedback from a static nonlinearity $f(\cdot)$ shown in Figure 1.9. This means that $e(t) = r(t) - f(y(t))$. According to Example 1.8 the linear system has a gain of 0.4, while Example 1.6 shows that the gain of the nonlinearity is 2. The loop gain is thereby 0.8, and Theorem 1.1 guarantees the stability of the closed loop system, regardless of the details of the nonlinear function in Figure 1.9.

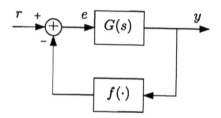

Figure 1.8: A linear system with static, nonlinear feedback.

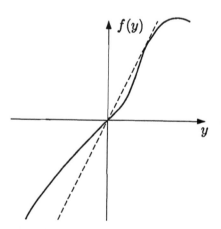

Figure 1.9: The static nonlinearity (solid) and a straight line with slope 2 (dashed).

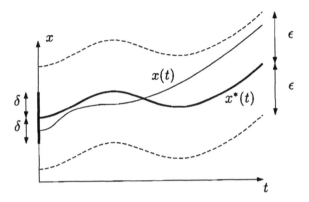

Figure 1.10: Definition of stability: For all $\epsilon > 0$ there is a δ so that the figure holds.

Stability of Solutions

Stability of a certain solution to a differential or difference equation concerns the question whether small changes in initial conditions give small changes in the solutions. The concepts can be summarized by the following definitions. We state them for the case where the system is given in state space form, i.e.,

$$\dot{x}(t) = f(x(t)) \tag{1.31}$$

where the dot denotes differentiation with respect to time t. (See also Chapter 11.) The initial values then correspond to the initial state $x(0)$.

Definition 1.3 *Let $x^*(t)$ be a solution of the differential equation (1.31) corresponding to the initial state $x^*(0)$. This solution is said to be **stable** if for each $\epsilon > 0$ there is a δ such that $|x^*(0) - x(0)| < \delta$ implies that $|x^*(t) - x(t)| < \epsilon$ for all $t > 0$. ($x(t)$ is the solution that corresponds to the initial state $x(0)$.) It is said to be **asymptotically stable** if it is stable and there exists a δ such that $|x^*(0) - x(0)| < \delta$ implies that $|x^*(t) - x(t)| \to 0$ as $t \to \infty$.*

The definition is illustrated in Figure 1.10. Here $x^*(t)$ is the nominal solution and the distance δ around $x^*(0)$ shows where to start to remain in an ϵ-environment around $x^*(t)$. This definition of stability of solutions to differential equations is also called *Lyapunov-stability*.

In Section 3.4 and Chapter 12 we shall return to stability of linear and nonlinear systems, respectively.

1.7   Comments

Main Points of the Chapter

In this chapter we have

- seen an example of a multivariable system, whose behavior is difficult to understand from SISO theory (Example 1.1)

- seen an example of a nonlinear system whose behavior cannot be predicted from linear theory (Example 1.2)

- described two ways of handling discrete time controllers (methods I and II in Section 1.3)

- presented the general structure for a control system (Figure 1.4)

- discussed various types of disturbances (Figure 1.6)

- defined signal norms $\|z\|$, system gain $\|\mathcal{S}\|$, and stability

- shown the small gain theorem: that a closed loop system is stable if the product of the gains in the loop is less than 1.

Literature

Mathematical foundations. Control theory builds on a mathematical foundation and we shall use concepts and results from a variety of mathematical subfields. The prerequisites necessary to follow this book are contained in normal undergraduate engineering courses. Here we shall give examples of books that can be used to refresh or deepen relevant mathematical knowledge. For *Fourier transforms*, Bracewell (1978) is a classical and thorough treatment, which goes far beyond what is required here. We also use a fair amount of *matrix theory and linear algebra*. For a user oriented account and for numerical applications Golub & Loan (1996) is a standard reference, while Horn & Johnson (1985) gives more theoretical background. A treatment of *analytical functions and complex analysis* is given in Churchill & Brown (1984). For *ordinary differential equations* we can point to Verhulst (1990). In this book we also treat stochastic disturbance models and *stochastic processes.* A introductory account of this is given in Papoulis (1977).

The basic knowledge of control theory that we assume the reader to have, can be found in e.g., any of Friedland (1988), Franklin, Powell & Amami-Naeini (1994), or Dorf & Bishop (1995). Moreover, Ljung & Glad (1994) gives complementary material on models of signals and systems. For deeper studies, we point in each chapter to suitable literature. Among books that cover broad aspects of the subject, we may mention Maciejowski (1989), Zhou, Doyle & Glover (1996) and Skogestad & Postlethwaite (1996) for multivariable theory and robustness, and Åström & Wittenmark (1997) and Franklin, Powell & Workman (1990) for detailed treatments of sampled data control. The mathematical theory of nonlinear systems is presented in Isidori (1995) and Nijmeijer & van der Schaft (1990).

Software

To use the tools that we treat in this book almost exclusively requires good, interactive software. There are now many commercially available software packages for control design. Many of these are directly aimed at certain application areas, like the aerospace industry, etc. The most widely spread packages belong to the MATLAB-family with a large number of toolboxes that cover most of today's control design methods, including simulation and implementation. Another general product is the MATRIX-X-family with design support, simulation and code generation for real time applications. The third family is MATHEMATICA which now has a package for control design, CONTROL SYSTEM PROFESSIONAL. It handles both symbolic and numeric calculations.

PART I:
LINEAR SYSTEMS

This part describes general properties of linear systems and signals. It has a special focus on multivariable systems. Other than that, the material is of general character; corresponding descriptions of signals and systems and their properties can be found in circuit theory, signal processing and basic control theory.

Chapter 2

Representation of Linear Systems

In the general classification of systems in Chapter 1, we defined linear, time invariant and causal systems. Systems with these three properties constitute, by far, the most common models of control objects. There are two reasons for this: (1) many physical objects can be described well by linear, time invariant models, and (2) such models give good possibilities for analysis and systematic control design.

In this chapter we will discuss different ways to represent linear, time invariant and causal systems. The *multivariable* aspects of the system descriptions will be a special focus. The output at time t, i.e., $y(t)$, is a p-dimensional column vector and the input $u(t)$ is an m-dimensional column vector. In this chapter we will not distinguish between control signals and disturbance signals. The vector $u(t)$ can contain signals of both kinds. This does not influence the forms of representation.

The representation of continuous time and discrete time systems is quite similar. The main part of this chapter deals with continuous time systems. In Section 2.6 the corresponding expressions for the discrete time case will be presented.

2.1 Impulse Response and Weighting Function

The output from a linear system is the weighted sum of input values at all times. For a causal system only old values of the input signal are included, and for a time invariant system the contributions are weighted by a factor dependent only on the time distance (not absolute time).

All this means that the output can be written as

$$y(t) = \int_0^\infty g(\tau)u(t-\tau)d\tau \tag{2.1}$$

The function $g(\tau)$ is called the *weighting function*, and shows how much the value of the input τ time units ago, now influences the output. It is also the *impulse response* for the system, i.e., the output that is obtained when the input is an impulse $\delta(t)$ (Dirac's delta function).

The impulse response $g(\tau)$ is for every τ a $p \times m$ matrix. Its (k,j)-element shows how output number k (k:th component of the column vector y) depends on the j:th input signal.

2.2   Transfer Function Matrices

The impulse response representation (2.1) gives the output as a convolution between the impulse response and the input.   Let us introduce the Laplace transforms

$$G(s) = \mathcal{L}(g) = \int_0^\infty e^{-st}g(t)dt$$
$$U(s) = \mathcal{L}(u), \qquad Y(s) = \mathcal{L}(y) \tag{2.2}$$

That $g(t)$ is a matrix and $y(t)$ and $u(t)$ are column vectors only means that the integration is carried out element-wise. The Laplace transform of a matrix function is thus a matrix obtained by Laplace transforming its elements.

Equation (2.1) can now be written as:

$$Y(s) = G(s)U(s) \tag{2.3}$$

provided u is zero up to time 0. The product $G(s)U(s)$ is computed in the normal way as a matrix times a vector, and the resulting elements are thus Laplace transforms of the corresponding output. The matrix $G(s)$ is called the *transfer function matrix* for the system. It has the dimension $p \times m$. Its (k,j)-element, $G_{kj}(s)$, describes how the input j influences the output k. If we only vary this input (and keep the others equal to zero) and only observe the output k, we obtain a single-input–single-output (SISO) system with the transfer function $G_{kj}(s)$. The transfer function matrix is thus a direct extension of the transfer function concept to a multivariable system.

A transfer function is called *proper* if the degree of the numerator does not exceed that of the denominator. It is said to be *strictly proper* if the degree of the denominator is larger than that of the numerator. In the multivariable case we say that a transfer function matrix is (strictly) proper if all its elements are (strictly) proper transfer functions. For a strictly proper transfer function matrix $G(s)$

$$\lim_{s \to \infty} G(s) = 0$$

while the limit is bounded if G is proper.

Example 2.1: A System With One Output and Two Inputs

Consider the system with the transfer function matrix

$$G(s) = [1/(s+1) \quad 2/(s+5)]$$

Assume that input number 1 is a unit step, that input number 2 is an impulse, and that the system is at rest for $t \le 0$. We then have

$$U(s) = \begin{bmatrix} 1/s \\ 1 \end{bmatrix}$$

The Laplace transform of the output is then

$$Y(s) = G(s)U(s) = \frac{1}{s(s+1)} + \frac{2}{s+5} = \frac{1}{s} - \frac{1}{s+1} + \frac{2}{s+5}$$

and the output is

$$y(t) = 1 - e^{-t} + 2e^{-5t}$$

We will not use explicit time functions to any great extent in this book, transform tables are therefore not particularly important here. However, there might be reason to recall the *final value theorem*

$$\lim_{t \to \infty} y(t) = \lim_{s \to 0} sY(s) \qquad (2.4)$$

if the left hand side exists.

2.3　Transfer Operator

We will use the symbol p for the *differentiation operator,*

$$py(t) = \frac{d}{dt}y(t) \tag{2.5}$$

With the symbol $e^{p\tau}$ we refer to the operator obtained by formal series expansion of the exponential function:

$$e^{p\tau} = 1 + p\tau + \frac{1}{2}p^2\tau^2 + \dots \frac{1}{k!}p^k\tau^k + \dots$$

This thus means that

$$e^{p\tau}u(t) = u(t) + \tau\frac{d}{dt}u(t) + \dots + \frac{\tau^k}{k!}\frac{d^k}{dt^k}u(t) + \dots$$

which we (for an infinitely differentiable function u) recognize as the Taylor expansion of $u(t+\tau)$.

We can write $u(t-\tau) = e^{-p\tau}u(t)$ and inserted into (2.1) we have

$$y(t) = \int_0^\infty g(\tau)u(t-\tau)d\tau = \int_0^\infty g(\tau)e^{-p\tau}d\tau\, u(t) = G(p)u(t)$$

In the last step we define the *transfer operator* $G(p)$ as the integral of $g(\tau)e^{-p\tau}$. Note the formal resemblance to the definition of the transfer function matrix as a Laplace transform of the impulse response: We have only exchanged s for p. The transfer operator has the advantage that we can work directly with the time functions. The link between the Laplace variable s and the differentiation operator is of course also well known from transform theory.

2.4　Input–Output Equations

In the SISO case, most often the transfer function is rational in s:

$$G(s) = \frac{B(s)}{A(s)} \tag{2.6}$$

where

$$A(s) = s^n + a_1 s^{n-1} + \dots + a_{n-1}s + a_n$$
$$B(s) = b_1 s^{n-1} + b_2 s^{n-2} + \dots + b_{n-1}s + b_n$$

With the transfer operator we have

$$y(t) = \frac{B(p)}{A(p)}u(t) \quad \text{or} \quad A(p)y(t) = B(p)u(t)$$

which corresponds to the differential equation

$$\frac{d^n}{dt^n}y(t) + a_1\frac{d^{n-1}}{dt^{n-1}}y(t) + \ldots + a_{n-1}\frac{d}{dt}y(t) + a_ny(t)$$

$$= b_1\frac{d^{n-1}}{dt^{n-1}}u(t) + \ldots + b_nu(t) \tag{2.7}$$

This is a differential equation of order n, only containing input and output. Modeling can often lead directly to such a representation of the system.

In the *multivariable case* the counterpart is a *vector-differential equation* (2.7), where the coefficients a_i and b_i are $p \times p$ and $p \times m$ matrices, respectively. The counterpart of the rational expression (2.6) becomes

$$G(s) = A^{-1}(s)B(s) \quad \text{or} \quad G(s) = B(s)A^{-1}(s) \tag{2.8}$$

where $A(s)$ and $B(s)$ are matrix polynomials, defined as above but with the coefficients a_i and b_i as matrices. The representation (2.8) of multivariable systems is called *Matrix Fraction Description, MFD*.

2.5 State Space Form

The *state* of a system at time t represents the amount of information necessary to unambiguously determine future outputs if future inputs are known. If necessary, see Franklin et al. (1994) for a review.

It is well known that for continuous time systems, the state space form is obtained, if the relationship between input and output is written as a system of first order differential equations:

$$\dot{x}(t) = Ax(t) + Bu(t)$$
$$y(t) = Cx(t) + Du(t) \tag{2.9}$$

Here $x(t)$ is the state vector, a column vector of dimension n, and A, B, C, and D are matrices of compatible dimensions.

A state space representation of a linear system has many advantages, and will be our main alternative in this book. It is an advantage that

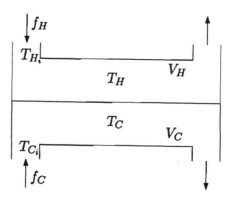

Figure 2.1: A simple heat exchanger. f_C and f_H are the flows of cold and hot water. T_H, T_C, T_{H_i}, T_{C_i} denote the temperatures at the marked points, and V_C and V_H are the volumes of cold and hot water.

SISO and multivariable systems formally are treated in exactly the same way. Another advantage is that state space models usually are obtained as a result of physical modeling.

Example 2.2:   Heat Exchanger

Consider the heat exchanger in Figure 2.1. Its behavior can be described in the following simplified way. The lower part is the "cold part" into which water flows with temperature T_{C_i}. The flow is f_C (m^3/min). The upper part is the "hot part" with input water temperature T_{H_i} and flow f_H. When the flows "meet" (in separate pipes), the hot water heats the cold water to temperature T_C. At the same time it is itself cooled to the temperature T_H. Let us make the simplifying approximation that the water in the different parts is perfectly mixed, so that it has the same temperatures as the outlet ones, i.e., T_H and T_C. By setting up the heat balance in the cold part we find that the temperature changes according to

$$V_C \frac{dT_C}{dt} = f_C(T_{C_i} - T_C) + \beta(T_H - T_C) \tag{2.10}$$

The first term in the right hand side then represents the cooling due to the inflow of cold water (normally, $T_{C_i} \leq T_C$, so this will give the decrease of temperature). The other term corresponds to the heat transfer from the hot to the cold part of the heat exchanger. It is proportional to the difference in temperature, and the constant of proportionality, β, depends on the heat transfer coefficient, the heat capacity of the fluids, etc. (In more accurate models the fact that β depends on flows and temperatures has to be taken into account, but here we fix it to a constant.)

Correspondingly, we have for the hot part

$$V_H \frac{dT_H}{dt} = f_H(T_{H_i} - T_H) - \beta(T_H - T_C) \tag{2.11}$$

We now assume that the flows f_H and f_C are constant and equal to f and view the inflow temperatures as inputs: $u_1 = T_{C_i}$ and $u_2 = T_{H_i}$. With the state variables $x_1 = T_C$ and $x_2 = T_H$ we then have the state space representation

$$\dot{x} = \begin{bmatrix} -(f+\beta)/V_C & \beta/V_C \\ \beta/V_H & -(f+\beta)/V_H \end{bmatrix} x + \begin{bmatrix} f/V_C & 0 \\ 0 & f/V_H \end{bmatrix} u \tag{2.12}$$

$$y = x$$

We will use the numeric values $f = 0.01$ (m³/min), $\beta = 0.2$ (m³/min) and $V_H = V_C = 1$ (m³), which gives

$$\dot{x} = \begin{bmatrix} -0.21 & 0.2 \\ 0.2 & -0.21 \end{bmatrix} x + \begin{bmatrix} 0.01 & 0 \\ 0 & 0.01 \end{bmatrix} u \tag{2.13}$$

Example 2.3: The Gripen Aircraft

In a couple of examples we will study the dynamics of the Swedish military aircraft Gripen, manufactured at SAAB AB in Linköping. It is a good approximation in normal flight to disregard the coupling between pitch angle and turning. In this example we are going to study the latter, and consider the aircraft in the horizontal plane. The course of the plane can be influenced in two ways. There is a rudder on the tail fin that can give a direct turning torque to the plane. There are also ailerons on the wings. Their main effect is to let the aircraft roll and thus change its course.

We introduce angles according to Figure 2.2.

Here are

- ψ = the course angle (angle between aircraft body and given course, typically north)
- v_y = velocity across the aircraft body in y-direction
- ϕ = roll angle (the rolling angle around the speed vector)
- δ_a = aileron deflection
- δ_r = rudder deflection

We can produce a linear state model for the system that describes the dynamics for small angles rather well. The following state variables are chosen

$x_1 = v_y$,

$x_2 = p =$ roll angular rate in a body-fixed system,

$x_3 = r =$ turning angular rate in a body-fixed system,

$x_4 = \phi$, $x_5 = \psi$, $x_6 = \delta_a$, $x_7 = \delta_r$

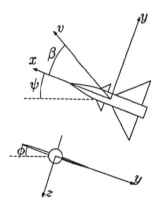

Figure 2.2: Angles for description of the motion of the aircraft. The xyz-axes mark a fixed body coordinate system. The velocity vector of the aircraft is v. The dashed lines denote north and a horizontal plane, respectively. β illustrates the so called side slip angle, the precise definition of which is technically complex. For small angles $v_y \approx \beta v_0$, where v_0 is the velocity.

Note that p and r are the angular rates around x-axis, and z-axis, respectively in Figure 2.2. They differ slightly from $\dot{\phi}$ and $\dot{\psi}$ since the body-fixed coordinate system xyz is not fixed in the horizontal plane. As inputs we choose the commanded aileron and rudder angles

$$u_1 = \delta_a^{cmd}, \quad u_2 = \delta_r^{cmd}$$

which are different from the real rudder angles because of the dynamics in the hydraulic servos. All angles are measured in radians and the time unit is seconds.

For the Gripen aircraft the following model applies in a certain case (attack missiles mounted on wing tips, Mach 0.6, altitude 500 m, angle of attack, 0.04 rad)

$$\dot{x} = Ax + Bu$$

where

$$A = \begin{bmatrix} -0.292 & 8.13 & -201 & 9.77 & 0 & -12.5 & 17.1 \\ -0.152 & -2.54 & 0.561 & -0.0004 & 0 & 107 & 7.68 \\ 0.0364 & -0.0678 & -0.481 & 0.0012 & 0 & 4.67 & -7.98 \\ 0 & 1 & 0.0401 & 0 & 0 & 0 & 0 \\ 0 & 0 & 1 & 0 & 0 & 0 & 0 \\ 0 & 0 & 0 & 0 & 0 & -20 & 0 \\ 0 & 0 & 0 & 0 & 0 & 0 & -20 \end{bmatrix}$$

$$B = \begin{bmatrix} 0 & -2.15 \\ -31.7 & 0.0274 \\ 0 & 1.48 \\ 0 & 0 \\ 0 & 0 \\ 20 & 0 \\ 0 & 20 \end{bmatrix}$$

Transformation Between State Space Form and Transfer Operator

The state space model (2.9) is in operator form

$$\dot{x} = px = Ax + Bu$$
$$y = Cx + Du \tag{2.14}$$

This gives $(pI - A)x(t) = Bu(t)$ or $x(t) = (pI - A)^{-1}Bu(t)$ which leads to

$$y(t) = Cx(t) + Du(t) = (C(pI - A)^{-1}B + D)u(t) = G(p)u(t)$$

with

$$G(p) = C(pI - A)^{-1}B + D \tag{2.15}$$

This expression gives the relationship between state representation and the transfer operator (or transfer function matrix: exchange p for s).

Example 2.4:   The Transfer Function Matrix for the Heat Exchanger

From (2.13) we have

$$G(s) = I \begin{bmatrix} s+0.21 & -0.2 \\ -0.2 & s+0.21 \end{bmatrix}^{-1} \begin{bmatrix} 0.01 & 0 \\ 0 & 0.01 \end{bmatrix}$$
$$= \frac{0.01}{(s+0.01)(s+0.41)} \begin{bmatrix} s+0.21 & 0.2 \\ 0.2 & s+0.21 \end{bmatrix} \tag{2.16}$$

The step from state space form to transfer operator is straightforward matrix algebra. It can be more difficult to move in the other direction. From basic state space theory we know that the transfer function for SISO systems,

$$G(p) = \frac{b_1 p^{n-1} + b_2 p^{n-2} + \ldots + b_{n-1} p + b_n}{p^n + a_1 p^{n-1} + \ldots + a_{n-1} p + a_n}$$

can be represented easily either by *controller canonical form*

$$\dot{x}(t) = \begin{bmatrix} -a_1 & -a_2 & \cdots & -a_{n-1} & -a_n \\ 1 & 0 & \cdots & 0 & 0 \\ 0 & 1 & \cdots & 0 & 0 \\ \vdots & \vdots & & \vdots & \vdots \\ 0 & 0 & \cdots & 1 & 0 \end{bmatrix} x(t) + \begin{bmatrix} 1 \\ 0 \\ 0 \\ \vdots \\ 0 \end{bmatrix} u(t)$$

$$y(t) = \begin{bmatrix} b_1 & b_2 & \cdots & b_{n-1} & b_n \end{bmatrix} x(t)$$

$$(2.17)$$

or by *observer canonical form*

$$\dot{x}(t) = \begin{bmatrix} -a_1 & 1 & 0 & \cdots & 0 \\ -a_2 & 0 & 1 & \cdots & 0 \\ \vdots & \vdots & \vdots & & \vdots \\ -a_{n-1} & 0 & 0 & \cdots & 1 \\ -a_n & 0 & 0 & \cdots & 0 \end{bmatrix} x(t) + \begin{bmatrix} b_1 \\ b_2 \\ \vdots \\ b_{n-1} \\ b_n \end{bmatrix} u(t)$$

$$y(t) = \begin{bmatrix} 1 & 0 & 0 & \cdots & 0 \end{bmatrix} x(t)$$

$$(2.18)$$

Using this starting point it is also easy to represent a transfer function matrix with one output and several inputs in observer canonical form. This is illustrated in the following example.

Example 2.5: System with Several Inputs

Consider the system

$$y(t) = \frac{p+2}{p^2 + 2p + 1} u_1(t) + \frac{1}{p^2 + 3p + 2} u_2(t)$$

We start out by describing the system using a common denominator:

$$y(t) = \frac{1}{p^3 + 4p^2 + 5p + 2} \begin{bmatrix} p^2 + 4p + 4 & p + 1 \end{bmatrix} \begin{bmatrix} u_1(t) \\ u_2(t) \end{bmatrix}$$

Assume now that we realize the system from input 1 to the SISO output in an observer canonical from. Assume then that we realize the SISO system from input 2 in the same way. The realizations will have the same A and C matrices. The actual output is the sum of the parts originating from the two input signals, and we can simply directly add the effects as follows:

$$\dot{x} = \begin{bmatrix} -4 & 1 & 0 \\ -5 & 0 & 1 \\ -2 & 0 & 0 \end{bmatrix} x + \begin{bmatrix} 1 & 0 \\ 4 & 1 \\ 4 & 1 \end{bmatrix} u$$
$$y = \begin{bmatrix} 1 & 0 & 0 \end{bmatrix} x$$

The example, with obvious modifications, can also be used to realize a transfer function with one input signal and several output signals in controller canonical form. There is, however, no simple formula for transforming to state form from a transfer function matrix with several inputs as well as several outputs. It is best to introduce states via direct interpretation, either physical or related to the different scalar transfer functions. See Examples 2.3 and 3.1.

2.6 Discrete Time Systems

By definition, discrete time systems only consider inputs and outputs at an enumerable number of time points. There are several situations when it is necessary, or natural, to describe a system by a discrete time model. A typical case is that the signal values only are available for measurement or manipulation at certain times, for example, because a continuous time system is sampled at certain times. In Chapter 4 we will treat, in more detail, how discrete time systems can be obtained from continuous time systems. Another case is when the system naturally is described directly in discrete form. This is typical for financial systems where, for example, sales figures and inventory levels are naturally associated with a certain day or week.

In this section we will consider how to describe discrete time systems. The signals are only defined at discrete time points, and to simplify matters we enumerate these times using integers. Inputs and outputs are thus defined as

$$u(t), y(t), \quad t = \dots, -2, -1, 0, 1, 2, \dots$$

Impulse Response

A general linear representation from the input sequence to the output at a given time t can be written

$$y(t) = \sum_{\ell=-\infty}^{\infty} g_t(\ell)u(t-\ell) \tag{2.19}$$

If we limit ourselves to time invariant systems, g does not depend on t and for a causal system no influence on $y(t)$ from future u can occur. We then have

$$y(t) = \sum_{\ell=0}^{\infty} g(\ell)u(t-\ell) \tag{2.20}$$

The interpretation is completely analogous to the continuous time case. The weighting function $g(\ell)$ is here also the (im)pulse response to the system, i.e., the system's response when the input is a pulse:

$$u(t) = \begin{cases} 1 & \text{when } t=0 \\ 0 & \text{when } t \neq 0 \end{cases}$$

Transfer Function Matrices

For discrete time signals we use the $\mathcal{Z}$-transform, that is

$$G(z) = \mathcal{Z}(g) = \sum_{t=0}^{\infty} g(t)z^{-t}$$

$$U(z) = \sum_{t=0}^{\infty} u(t)z^{-t}, \qquad Y(z) = \sum_{t=0}^{\infty} y(t)z^{-t} \tag{2.21}$$

If values of u, prior to $t=0$ are zero, the relationship (2.20) can be written as

$$Y(z) = G(z)U(z) \tag{2.22}$$

Here $G(z)$ is the transfer function matrix for the discrete time system, analogously to the continuous time case (2.3). Everything concerning continuous systems also applies to discrete time systems if the Laplace table is exchanged for the $\mathcal{Z}$-transform table. Recall the *final value result:*

$$\lim_{t \to \infty} y(t) = \lim_{z \to 1}(z-1)Y(z) \tag{2.23}$$

if the left hand side exists.

Transfer Operators

The transfer operator is considerably more straightforward in discrete time than in continuous time. We introduce the *shift operator q* as

$$qu(t) = u(t+1) \tag{2.24}$$

Its inverse, the *backwards shift operator*, is denoted by q^{-1}

$$q^{-1}u(t) = u(t-1) \tag{2.25}$$

Using the shift operator and the impulse response (2.20) we define the transfer operator as

$$G(q) = \sum_{\ell=0}^{\infty} g(\ell)q^{-\ell} \tag{2.26}$$

We then have

$$y(t) = \sum_{\ell=0}^{\infty} g(\ell)u(t-\ell) = \sum_{\ell=0}^{\infty} g(\ell)q^{-\ell}u(t) = G(q)u(t)$$

The transfer operator has, formally, been obtained by exchanging z for q in the definition of the transfer function matrix. Other than that, the same comments apply as for the continuous time case.

Difference Equations

A rational transfer operator $G(q) = B(q)/A(q)$, with

$$B(q) = b_1 q^{n-1} + b_2 q^{n-2} + \ldots + b_n$$
$$A(q) = q^n + a_1 q^{n-1} + \ldots + a_n$$

leads to a difference equation for inputs and outputs as follows:

$$y(t) = \frac{B(q)}{A(q)}u(t) \Rightarrow A(q)y(t) = B(q)u(t) \tag{2.27}$$
$$\Rightarrow y(t+n) + a_1 y(t+n-1) + \ldots + a_{n-1}y(t+1) + a_n y(t)$$
$$= b_1 u(t+n-1) + b_2 u(t+n-2) + \ldots$$
$$+ b_{n-1}u(t+1) + b_n u(t)$$

In the above expression we assumed the degree of the numerator polynomial $B(q)$ to be lower than that of the denominator polynomial.

In (2.27), this means that we have a time delay between input and output, i.e., a change in input at time t does not influence the output until at time $t + 1$. It is easy to realize that if the degree of the denominator is n as above, while the degree of the numerator is $n - k$, there is a time delay of k steps between input and output.

If the A and B polynomials have the same degree, it means that there is a direct influence from $u(t)$ to $y(t)$. This corresponds to the fact that there is a *direct term*, i.e., a D-matrix that is not zero in the state space representation (2.29).

Example 2.6:　Inventory Control

Consider a simple inventory model. Let $y(t)$ be the level of the inventory at time (day) t. Let $u_1(t)$ be orders for day t. Ordered goods is, however, not delivered until two days later. Let $u_2(t)$ be the sales for day t. The inventory level thus varies as

$$y(t) = y(t-1) + u_1(t-2) - u_2(t-1) \qquad (2.28)$$

This is a difference equation of the form (2.27), with one output and two inputs, $n = 2$, and with the coefficients

$$a_1 = -1, a_2 = 0, b_0 = \begin{bmatrix} 0 & 0 \end{bmatrix}, b_1 = \begin{bmatrix} 0 & -1 \end{bmatrix}, b_2 = \begin{bmatrix} 1 & 0 \end{bmatrix}$$

Here $u(t)$ is a column vector

$$u(t) = \begin{bmatrix} u_1(t) \\ u_2(t) \end{bmatrix}$$

We then have

$$y(t) - y(t-1) = \begin{bmatrix} 0 & -1 \end{bmatrix} u(t-1) + \begin{bmatrix} 1 & 0 \end{bmatrix} u(t-2)$$

The transfer function matrix for this system is

$$G(z) = \begin{bmatrix} \dfrac{1}{z^2 - z} & \dfrac{-1}{z - 1} \end{bmatrix}$$

State Space Form

The discrete time variant of the state form is a system of first order difference equations, written as

$$\begin{aligned} x(t+1) &= Ax(t) + Bu(t) \\ y(t) &= Cx(t) + Du(t) \end{aligned} \qquad (2.29)$$

with a n-dimensional state vector $x(t)$.

Example 2.7:   Inventory Control

Let us return to Example 2.6. What is a reasonable state vector for this system? What information do we need for day t in order to know enough about how the inventory situation will evolve? Obviously we need to know the inventory level for the day in question, i.e., $y(t)$. Further, we need to know yesterday's orders, i.e., $u_1(t-1)$, since they are still undelivered. We choose

$$x(t) = \begin{bmatrix} y(t) \\ u_1(t-1) \end{bmatrix}$$

as the state and then have the state space model

$$x(t+1) = \begin{bmatrix} 1 & 1 \\ 0 & 0 \end{bmatrix} x(t) + \begin{bmatrix} 0 & -1 \\ 1 & 0 \end{bmatrix} u(t) \tag{2.30}$$

$$y(t) = \begin{bmatrix} 1 & 0 \end{bmatrix} x(t)$$

In operator form (2.29) becomes

$$qx = Ax + Bu$$
$$y = Cx + Du$$

analogously to (2.14). Just exchange p for q. The algebra for the relationship between state space representations and other forms of representation for a linear system is therefore the same for time continuous and discrete time models. All results of the type (2.15), (2.17), (2.18) and Example 2.5 thus apply also for discrete time, by letting q replace p.

2.7   Comments

The Main Points of the Chapter

We have shown how linear systems can be represented in different forms. One observation is that the multivariable case in this respect is a straightforward generalization of the SISO case: it is "just" a question of interpreting certain variables such as matrices and vectors. In addition, the formal similarity between continuous time and discrete time systems is obvious: the differentiation operator p plays the same role as the shift operator q. In the same way, the Laplace variable s corresponds to the $\mathcal{Z}$-transform variable z. With these interpretations all relationships between the forms of representation look alike in discrete and continuous time.

Literature

The treatment of continuous time systems – linear differential equations – with Laplace-technique, input–output forms and state representation goes back to the pioneers of the 19th century. The discrete time formalism, as described here, was developed in the 1950s and is summarized in classical textbooks such as Jury (1958), Ragazzini & Franklin (1958) and Tsypkin (1958).

There are many modern textbooks that treat the material in this chapter. A thorough study of multivariable systems is given in Kailath (1980). Both state space form and matrix fraction description are treated in detail. Descriptions with emphasis on discrete time systems are given in, for example, Åström & Wittenmark (1997) and Franklin et al. (1990).

Software for System Representation

Analysis and design of control systems is now predominantly done using software packages. Especially, multivariable systems require access to efficient software. We will in the different chapters comment on the character of this computer support. As a starting point we have the CONTROL SYSTEMS TOOLBOX (CSTB), used together with MATLAB. This is probably the most widely spread program package for analysis and control design of linear systems.

In MATLAB-5 the representation of systems is object oriented and thereby continuous time and discrete time systems are treated in parallel in a transparent manner. For the methodology described in this chapter we have the following MATLAB commands.

ss, tf Create a system with state form or transfer function form. Transformation between these forms.

ss2tf Carry out the transformation (2.15) "state-space to transfer function"

tf2ss Do state realizations for systems with one input

impulse Impulse response

step Step response

Chapter 3

Properties of Linear Systems

Explicit solutions of the system equations are not widely used in control theory analysis. The focus is, instead, on other concepts describing the properties of the solutions: poles, zeros, frequency function, etc. In this chapter we will discuss these concepts.

3.1 Solving the System Equations

Despite what was said above, it is useful to have the expressions for the general solution of the linear system equations. For the system

$$\dot{x} = Ax + Bu$$
$$y = Cx + Du \tag{3.1}$$

it is well known that the solution can be written

$$x(t) = e^{A(t-t_0)}x(t_0) + \int_{t_0}^{t} e^{A(t-\tau)}Bu(\tau)d\tau \tag{3.2}$$

For a diagonal A-matrix

$$A = \begin{bmatrix} \lambda_1 & 0 & \cdots & 0 \\ 0 & \lambda_2 & \cdots & 0 \\ \vdots & \vdots & \vdots & \vdots \\ 0 & 0 & \cdots & \lambda_n \end{bmatrix}$$

we have

$$
e^{At} = \begin{bmatrix} e^{\lambda_1 t} & 0 & \cdots & 0 \\ 0 & e^{\lambda_2 t} & \cdots & 0 \\ \vdots & & \vdots & \vdots \\ 0 & \cdots & \cdots & e^{\lambda_n t} \end{bmatrix}
$$

The ith component of x becomes

$$
x_i(t) = e^{\lambda_i(t-t_0)}x_i(t_0) + \int_{t_0}^t e^{\lambda_i(t-\tau)}B_i u(\tau)d\tau \tag{3.3}
$$

Here B_i is the ith row of the matrix B. The component x_i thus develops independently of the other state variables. x_i is said to correspond to a *mode* of the system with the eigenvalue λ_i.

If we denote the ith column of the matrix C by C_i, the output becomes

$$
y(t) = C_1 x_1(t) + \ldots + C_n x_n(t) + Du(t) \tag{3.4}
$$

It is thus a weighted sum of the modes of the system.

In general, if the A-matrix is diagonalizable, we may consider a change of variables in the state space $\xi = Tx$:

$$
\begin{aligned}
\dot{\xi} &= TAT^{-1}\xi + TBu \\
y &= CT^{-1}\xi + Du
\end{aligned} \tag{3.5}
$$

so that TAT^{-1} is diagonal. The modes of the system are the components $\xi_i = T_i x$ (T_i is the ith row of T) which are certain linear combinations of the original state vector x, corresponding to A's eigenvectors. The mode is generally named for its corresponding eigenvalue ("the mode λ_i").

From the general result (3.2) an explicit expression for the impulse response for the system follows:

$$
g(\tau) = D\delta(\tau) + Ce^{A\tau}B \tag{3.6}
$$

3.2   Controllability and Observability

Definitions

The concepts of controllability and observability describe how state vectors in the state space are influenced by inputs and how they show up in the output. They are also important for understanding what happens when factors are cancelled in the transfer operator as well as for the control synthesis with observer and state feedback.

Definition 3.1 (Controllability) *The state x^* is said to be* **controllable** *if there is an input that in finite time gives the state x^* from the initial state $x(0) = 0$. The system is said to be* **controllable** *if all states are controllable.*

Definition 3.2 (Observability) *The state $x^* \neq 0$ is said to be* **unobservable** *if, when $u(t) = 0, t \geq 0$ and $x(0) = x^*$, the output is $y(t) \equiv 0, t \geq 0$. The system is said to be* **observable** *if it lacks ubobservable states.*

Criteria

From a basic control course the criteria for controllability and observability are well known, see for example, Franklin et al. (1994), Appendix D.

Theorem 3.1 *The controllable states of the system (3.1) form a linear subspace, viz, the range of the matrix (the controllability matrix)*

$$S(A, B) = \begin{bmatrix} B & AB & A^2B & \cdots & A^{n-1}B \end{bmatrix} \tag{3.7}$$

where n is the order of the system. The system is thus controllable if and only if S has full rank.

A proof of this theorem in the discrete time case is given below (Theorem 3.11). Note that the theorem is valid also if the system has several inputs ($u(t)$ is a vector). If u is scalar, S is a square matrix and has full rank exactly when $\det S \neq 0$.

Theorem 3.2 *The unobservable states constitute a linear subspace, viz, the null space of the matrix (the observability matrix)*

$$\mathcal{O}(A, C) = \begin{bmatrix} C \\ CA \\ \vdots \\ CA^{n-1} \end{bmatrix} \tag{3.8}$$

The system is thus observable if and only if $\mathcal{O}$ has full rank.

A proof of this theorem in the discrete time case is given below (Theorem 3.11).

Recall especially that:

- A system in controller canonical form is always controllable

- A system in observer canonical form is always observable

- A system with one input and one output is both controllable and observable exactly when no cancellations can be made when the transfer operator (2.15) is computed.

A state space representation that is both controllable and observable is said to be a *minimal realization* of the system. There is then no other state representation with lower dimension of the state vector that realizes the same input–output relationship.

Controllable and Observable Modes

Let the system be given *in diagonal form* with distinct eigenvalues and consider (3.3) and (3.4). We see that a mode x_i with corresponding row vector $B_i = 0$ is not influenced at all by the input. If any row in B is zero for a system given in diagonal form, the system is thus not controllable. We then speak of the *uncontrollable modes* as the state variables x_i, with associated eigenvalue λ_i, corresponding to the rows with $B_i = 0$. The other modes are called *controllable modes*. In the same way, a column $C_i = 0$ results in a *unobservable mode* while columns in the C-matrix that are not identically equal to zero correspond to *observable modes*. It is easy to verify that the eigenvectors of the controllable modes span the subspace of controllabe states, and similary for the uncontrollable modes.

If the system is not in diagonal form, but the A-matrix can be diagonalized, the modes are associated with the eigenvectors of the A-matrix; they are thus linear combinations of the state variables. These modes are controllable, observable, etc., depending on the corresponding row (column) in TB (CT^{-1}) in (3.5).

If the system cannot be diagonalized, the mode concept becomes more complicated.

The PBH-Test

There is an equivalent test to Theorems 3.1 and 3.2:

Theorem 3.3 *The system* (A, B, C, D) *in (3.1) is controllable if and only if*

$$\begin{bmatrix} A - \lambda I & B \end{bmatrix} \text{ has full rank } \forall \lambda \qquad (3.9)$$

It is observable if and only if

$$\begin{bmatrix} A - \lambda I \\ C \end{bmatrix} \text{ has full rank } \forall \lambda \qquad (3.10)$$

Proof: The proof is simple in the "if"-direction: If (3.9) does not have full rank for some λ there is a non-zero vector v, such that $v \begin{bmatrix} A - \lambda I & B \end{bmatrix} = 0$, i.e., $vA = v\lambda$ and $vB = 0$. It is easy to verify that $vS(A, B) = 0$, with S defined by (3.7), so the system is not controllable. To prove the "only if" direction, the system can be transformed to a triangular block form that will yield the result. See, e.g., Kailath (1980), Section 2.4.3.   □

The criterion is called the *PBH-test* for the originators Popov, Belevitch and Hautus. Note that the matrix in (3.9) is $n \times (n + m)$, and that it automatically is of full rank for all λ that are not eigenvalues of A. The requirement is thus that B "fills out" the null space of $A - \lambda I$ when λ is an eigenvalue of A. If this is not the case for some eigenvalue λ_i, the corresponding mode is not controllable.

Modifying the Eigenvalues

An important usage of the controllability and observability concepts is to describe when the eigenvalues of the A-matrix can be modified using feedback. If we have the system $\dot{x} = Ax + Bu$ and use the state feedback $u = -Lx$ the closed loop system will be $\dot{x} = (A - BL)x$. In later chapters we will have reasons to study how the eigenvalues of $A - BL$ can be modified using the matrix L. Here we give an important formal result on matrices, which shows the connection to controllability.

Theorem 3.4 *Let A be an $n \times n$ matrix and B an $n \times m$ matrix. The $m \times n$ matrix L can be selected so that $A - BL$ attains arbitrary, predetermined eigenvalues if and only if*

$$S(A, B) = \begin{bmatrix} B & AB & A^2B & \cdots & A^{n-1}B \end{bmatrix} \qquad (3.11)$$

has full rank. (Complex eigenvalues must, however, occur in conjugated pairs.)

Equivalently formulated: Let C be a $p \times n$ matrix. Then the $n \times p$ matrix K can be selected so that $A - KC$ attains arbitrary, predetermined eigenvalues if and only if

$$\mathcal{O}(A,C) = \begin{bmatrix} C \\ CA \\ CA^2 \\ \vdots \\ CA^{n-1} \end{bmatrix} \qquad (3.12)$$

has full rank.

The proof is given in Appendix 3A.

Stabilizability and Detectability

Theorem 3.4 can be extended to show that if a system is not controllable, the controllable modes (eigenvalues) can be modified, but the uncontrollable cannot. If a certain mode is unstable (the definition of stability will be discussed in Section 3.4) it is particularly interesting to know if it is controllable, and thereby can be modified. The following concepts therefore are useful.

Definition 3.3 (Stabilizability, Detectability) *A system (A, B, C) is said to be **stabilizable** if there exists a matrix L, so that $A - BL$ is stable (i.e., it has all the eigenvalues in the stability region). It is said to be **detectable** if there exists a matrix K, so that $A - KC$ is stable.*

(The stability region will be defined in Definitions 3.6 and 3.11.) A controllable system is obviously always stabilizable according to Theorem 3.4. Likewise, an observable system is always detectable. The definitions are of great importance in Sections 5.7 and 9.2 on optimal state reconstruction and state feedback.

3.3   Poles and Zeros

Poles and zeros are important for characterization of the properties of a system. In this section we will discuss how they are defined and calculated for multivariable systems.

Poles: Definition

The eigenvalues of the state matrix A, corresponding to controllable and observable modes describe according to (3.3) the general time response of the system. In particular we have seen in (3.3)–(3.6) how the impulse responses are built by $e^{p_i t}$ where p_i are the eigenvalues of the system. These are thus essential for the behavior of the system.

Analogously to the SISO case we introduce the term *poles* for these eigenvalues:

Definition 3.4 (Poles) *By* **poles** *of a system we mean the eigenvalues (counted with multiplicity) of the system matrix A in a minimal state space realization of the system. By the* **pole polynomial** *we mean the characteristic polynomial for the A-matrix: $\det(\lambda I - A)$.*

The order of the system is thus equal to the number of poles (counted with multiplicity).

Calculation of Poles From the Transfer Function Matrix

To realize a multivariable system, $G(s)$, in minimal state space form may be laborious. Therefore, it would be convenient if the poles could be calculated directly from $G(s)$. In the SISO case this is simple: The poles are the zeros of the denominator polynomial (after all cancellations between numerator and denominator). In the multivariable case it is somewhat more difficult. We have from (2.15) that

$$G(s) = C(sI - A)^{-1}B + D$$
$$= \frac{1}{\det(sI - A)}(C(sI - A)^a B + D\det(sI - A)). \tag{3.13}$$

In the second representation we have used Kramer's rule for matrix inversion: F^a is the adjunct matrix to F, obtained by forming determinants of submatrices (minors, see below) of F. The point here is that the second factor of the right hand side of (3.13) is a matrix with elements that are polynomials in s. From this we realize that all denominator polynomials in $G(s)$ must be divisors to the pole polynomial $\det(sI - A)$. This consequently has to be "at least" the least common denominator for all denominator polynomials in $G(s)$.

In order to be able to calculate poles and to define zeros for multivariable system we must use the notion of *minors*. A minor of a matrix A is the determinant of a square sub-matrix of A, obtained by deleting rows and columns in A. For example, the matrix

$$\begin{bmatrix} 1 & 2 & 3 \\ 4 & 5 & 6 \end{bmatrix}$$

has the 9 minors

$$\begin{vmatrix} 1 & 2 \\ 4 & 5 \end{vmatrix} = -3, \quad \begin{vmatrix} 1 & 3 \\ 4 & 6 \end{vmatrix} = -6, \quad \begin{vmatrix} 2 & 3 \\ 5 & 6 \end{vmatrix} = -3,$$

1, 2, 3, 4, 5, 6

By a *maximal minor* we mean a determinant of a sub-matrix of maximal size, i.e., the 2×2 matrices above.

We are now ready to state the main result:

Theorem 3.5 (Poles) *The pole polynomial for a system with transfer function matrix $G(s)$ is the least common denominator of all minors to $G(s)$. The poles of the system are the zeros of the pole polynomial.*

A proof of this result is given in Kailath (1980), Section 6.5.

Example 3.1:   A Third Order System

Consider the system (1.1)

$$G(s) = \begin{bmatrix} \frac{2}{s+1} & \frac{3}{s+2} \\ \frac{1}{s+1} & \frac{1}{s+1} \end{bmatrix}$$

The minors are

$$\frac{2}{s+1}, \quad \frac{3}{s+2}, \quad \frac{1}{s+1}, \quad \text{and}$$
$$\frac{2}{(s+1)^2} - \frac{3}{(s+1)(s+2)} = \frac{-s+1}{(s+1)^2(s+2)}$$

The least common denominator is

$$p(s) = (s+1)^2(s+2)$$

which gives the poles -2 and -1 (double pole).

The system is thus of third order and we can now easily find a state representation, e.g., with the states

$$x_1(t) = \frac{1}{p+1}u_1(t), \quad x_2 = \frac{1}{p+1}u_2(t), \quad x_3(t) = \frac{1}{p+2}u_2(t)$$

which gives

$$\dot{x} = \begin{bmatrix} -1 & 0 & 0 \\ 0 & -1 & 0 \\ 0 & 0 & -2 \end{bmatrix} x + \begin{bmatrix} 1 & 0 \\ 0 & 1 \\ 0 & 1 \end{bmatrix} u$$

$$y = \begin{bmatrix} 2 & 0 & 3 \\ 1 & 1 & 0 \end{bmatrix} x$$

(3.14)

Note that a pole of the multivariable system has to be a pole also of one of the SISO transfer functions, being the elements of the transfer function matrix. In order to find the poles as such it is then enough to find the least common denominator for the matrix elements. It is to decide the *multiplicity* of the poles that the minors have to be formed. The total number of poles (counted with multiplicity) is equal to the order of a minimum state representation, which is a useful number to know.

Zeros

For a SISO system, a zero is a value s that makes the transfer function zero, that is the zeros of $G(s)$ are the poles of $1/G(s)$. Zeros, thus in some sense, describe the inverse system dynamics. For a multivariable system with a square transfer function matrix (as many inputs as outputs) the zeros of $G(s)$ can be defined as the poles of $G^{-1}(s)$. In order to handle non-square transfer function matrices, however, we need a more general definition.

Definition 3.5 (Zeros) *For the system (3.1), the* **(transmission) zeros** *are defined as those values of s, for which the matrix*

$$M(s) = \begin{bmatrix} sI - A & B \\ -C & D \end{bmatrix}$$

(3.15)

does not have full rank. The polynomial that has these values as zeros is the **zero polynomial.**

For a system with the same number of inputs as outputs, the zero polynomial consequently is $\det M(s)$. The following result, shows how to compute the zeros directly from the transfer function matrix.

Theorem 3.6 *The zero polynomial of $G(s)$ is the greatest common divisor for the numerators of the maximal minors of $G(s)$, normalized to have the pole polynomial as denominator. The* **zeros** *of the system are the zeros of the zero polynomial.*

For a proof, see Kailath (1980), Section 6.5.

Note that the greatest common divisor also is called the *greatest common factor*.

Example 3.2: Zeros

Consider again the system (1.1)

$$G(s) = \begin{bmatrix} \frac{2}{s+1} & \frac{3}{s+2} \\ \frac{1}{s+1} & \frac{1}{s+1} \end{bmatrix}$$

The maximal minor is

$$\det G(s) = \frac{-s+1}{(s+1)^2(s+2)}$$

From Example 3.1 we know that the pole polynomial is

$$p(s) = (s+1)^2(s+2)$$

The determinant is thus already normalized to have the pole polynomial as denominator The zero polynomial thus is $-s+1$ and the system has a zero in 1.

In order to see the relationship with the inverse system we take

$$G^{-1}(s) = \begin{bmatrix} \frac{2}{s+1} & \frac{3}{s+2} \\ \frac{1}{s+1} & \frac{1}{s+1} \end{bmatrix}^{-1} = \begin{bmatrix} \frac{(s+1)(s+2)}{-s+1} & -\frac{3(s+1)^2}{-s+1} \\ -\frac{(s+1)(s+2)}{-s+1} & \frac{2(s+1)(s+2)}{-s+1} \end{bmatrix}$$

It can now be verified that this inverse system has the pole polynomial $-s+1$. Thus the relationship (zeros) $\Leftrightarrow$ (poles for inverse systems) holds.

Example 3.3: The Gripen Aircraft

Consider the aircraft model in Example 2.3. Choose the roll angle (x_4) and course angle (x_5) as outputs, so that

$$C = \begin{bmatrix} 0 & 0 & 0 & 1 & 0 & 0 & 0 \\ 0 & 0 & 0 & 0 & 1 & 0 & 0 \end{bmatrix}$$

The transfer function matrix is now

$$G(s) = \frac{1}{n(s)} \begin{bmatrix} t_{11}(s) & t_{12}(s) \\ t_{21}(s)/s & t_{22}(s)/s \end{bmatrix}$$

where

$$\begin{aligned} n(s) &= s^6 + 43.3s^5 + 543s^4 + 1778s^3 + 5203s^2 + 9213s + 177 \\ &= (s+20)^2(s+2.5876)(s+0.0194) \\ &\quad \times (s+0.3556 - 2.9458i)(s+0.3556 + 2.9458i) \end{aligned}$$

and $t_{ij}(s)$ are polynomials, which we do not write in detail here. Furthermore

$$\det G(s) = \frac{t_{11} \cdot t_{22} - t_{12} \cdot t_{21}}{s \cdot n \cdot n}$$

holds. Here

$$
\begin{aligned}
t_{11} \cdot t_{22} - t_{12} \cdot t_{21} = & - 46.9s^9 + 4287s^8 + 40896s^7 - 5.67 \cdot 10^6 s^6 \\
& - 1.036 \cdot 10^8 s^5 - 3.603 \cdot 10^8 s^4 - 1.099 \cdot 10^9 s^3 \\
& - 2.138 \cdot 10^9 s^2 - 4.417 \cdot 10^8 s - 7.7678 \cdot 10^6 \\
= & - 46.9(s - 77.47)(s - 57.4)(s + 0.2165)n(s)
\end{aligned}
$$

so that

$$\det G(s) = -46.9 \frac{(s - 77.47)(s - 57.4)(s + 0.2165)}{s \cdot n(s)}$$

From this follows that the pole polynomial is $s \cdot n(s)$ and the zero polynomial is $-46.9(s - 77.47)(s - 57.4)(s + 0.2165)$. The system thus has the poles

$$
\begin{aligned}
& 0 \\
& -2.5876 \\
& -0.0194 \\
& -0.3556 \pm 2.9458i \\
& -20 \quad \text{double, corresponding to rudder-servo dynamics}
\end{aligned}
$$

and the zeros

$$77.47, \quad 57.4, \quad -0.2165$$

The system has all poles in the left half plane (however, Gripen is unstable in the pitch channel at low speed) but has zeros far into the right half plane. The poles correspond to modes (see (3.3)) which are associated with certain motion of the aircraft. The pole in -2.5876 is the *roll mode*, which describes the rolling of the aircraft. This has a time constant of approximately 0.4 sec in Gripen. The pole in -0.0194 is the *spiral mode* which mainly describes the response to the turning angle. The complex conjugated pole-pair corresponds to the so called *Dutch-roll mode* which is a combined turning and rolling motion. The pole in the origin is the integration to the course angle from the underlying angular velocities.

3.4 Stability

Stability is a key concept in control. In Section 1.6 we have already treated input–output stability and stability of solutions in the state space. Here we will relate these definitions to linear, time invariant systems. For this, we first need a technical result on norms for powers of matrices.

Powers of Matrices

The following results show how the eigenvalues determine the size of matrix powers and matrix exponentials:

Theorem 3.7 *If the matrix A has the eigenvalues $\lambda_1, ..., \lambda_n$ (possibly multiple) the matrix $F = e^{At}$ has the eigenvalues*

$$e^{\lambda_1 t}, ..., e^{\lambda_n t}.$$

Proof: If A is diagonalizable, the proof is trivial. If A cannot be diagonalized, a matrix R exists so that

$$A = RUR^{-1}$$

where U is upper triangular with the eigenvalues along the diagonal (with zeros below). Then U^k is also upper triangular and has λ_i^k along the diagonal. From this we have that

$$e^{Ut} = \sum_{k=0}^{\infty} \frac{(Ut)^k}{k!}$$

is upper triangular with $e^{\lambda_i t}$ along the diagonal. These values are then also equal to the eigenvalues of the matrix. Then

$$e^{At} = Re^{Ut}R^{-1},$$

and therefore e^{At} and e^{Ut} have the same eigenvalues, which concludes the proof. $\qquad\qquad\square$

Theorem 3.8 *Let λ_i, $i = 1, ..., n$, be the eigenvalues of the matrix F. Assume that*

$$|\lambda_i| < \lambda, \ i = 1, ..., n$$

Then, for some positive constants C and c_i we have

$$|F^t| \le C\lambda^t \tag{3.16}$$

and

$$|F^t| \ge c_i|\lambda_i^t|, \ \ c_i > 0, \ \ i = 1, ..., n. \tag{3.17}$$

Here $|\cdot|$ is the matrix norm (operator norm) (1.24).

The proof is given in Appendix 3A.

Corollary 3.1 *Let* $\lambda_i, i = 1, \ldots, n$, *be the eigenvalues of the matrix* A. *Assume that*

$$\text{Re } \lambda_i < \sigma, i = 1, \ldots, n$$

For some constants C *and* c_i *we then have*

$$|e^{At}| \leq Ce^{\sigma t}$$

and

$$|e^{At}| \geq c_i|e^{\lambda_i t}| = c_i e^{\text{Re } \lambda_i t}, \quad c_i > 0, \quad i = 1, \ldots, n$$

Proof: We note that

$$e^{At} = (e^A)^t$$

and know from Theorem 3.7 that the eigenvalues of e^A are given by $\sigma_i = e^{\lambda_i}$ where λ_i are the eigenvalues of A. □

Input–Output Stability

From the corollary we see that the real parts of the eigenvalues of the system matrix play an important role for how $|e^{At}|$ behaves: if all the eigenvalues lie strictly in the left half plane, the norm decreases exponentially. If some eigenvalue is in the right half plane, the norm will instead grow exponentially. Therefore, we introduce the following definition:

Definition 3.6 (Stability Region) *For a continuous time system the stability region is equal to the left half plane, not including the imaginary axis.*

Input–output stability was defined in Section 1.6. For the linear system

$$y(t) = \int g(\tau)u(t - \tau)d\tau$$

we let the integration interval be the whole real axis and take $g(\tau) = 0$ for $\tau < 0$. We then have the triangle inequality

$$|y(t)| \leq \int |g(\tau)||u(t - \tau)|d\tau$$

and thereby

$$\|y\|^2 = \int |y(t)|^2 dt \le \int \int \int |g(\tau)| |g(\rho)| |u(t-\tau)| |u(t-\rho)| d\tau d\rho dt$$

$$\le \int \int |g(\tau)| |g(\rho)| d\tau d\rho \|u\|^2 = \left(\int |g(\tau)| d\tau \right)^2 \|u\|^2$$

The first equality is the definition of $\|y\|$, the next difference is the triangle inequality above, and in the third step we use Cauchy-Schwarz inequality applied to the integral over t. Input–output stability thus follows if the impulse response is absolutely integrable,

$$\int_0^\infty |g(\tau)| d\tau < \infty$$

From the Corollary 3.1 and (3.6) we know that if all eigenvalues of A are inside the stability region, the impulse response decreases exponentially. This means that the integral above is convergent and that the system thus is stable.

We know that the eigenvalues of the matrix A of a minimal state space representation are equal to the poles of the system. If any pole $p^\dagger$ lies outside the stability region, there is a controllable and observable state $x^\dagger$ such that $Ax^\dagger = p^\dagger x^\dagger$. With finite input energy we can always arrive at this state. If the input thereafter is zero the output will be of the character $Ce^{p^\dagger t}x^\dagger$ which gives an infinite output norm. (If $p^\dagger$ is complex valued we combine it with its complex conjugate and study pairs of real state vectors.) The following result has thereby been shown:

Theorem 3.9 (Stability of Linear Systems) *A linear, time invariant system is input–output stable if and only if its poles are inside the stability region.*

Remark. If the A-matrix has an eigenvalue outside the stability region, corresponding to a uncontrollable mode (and thereby not a pole), the matrix cannot be excited by the input. Thus the system is still input–output stable. If the unstable mode is included in the initial value, the output will however tend to infinity.

Stability of Solutions

Stability of solutions in the state space is defined in Definition 1.3. It requires that a small change in the initial state, $x(t_0)$, has a small influence on the continued behavior of the state vector.

From the general solution (3.2) we see that the difference between two solutions with different initial values, but the same input, is given by

$$x^*(t) - x(t) = e^{A(t-t_0)}(x^*(t_0) - x(t_0)) \qquad (3.18)$$

From this expression we see that the stability is determined only by the properties of A. It does not at all depend on the input u and also not on which initial value $x^*(t_0)$ we study. The stability is thus a *system property* for linear systems. We can therefore talk about *asymptotically stable systems* rather than asymptotically stable solutions.

The expression (3.18) gives together with Theorem 3.8 directly a criterion for stability of linear systems:

Theorem 3.10 (Stability of Solutions) *A linear system in state space form (2.9) is asymptotically stable if and only if all eigenvalues of the matrix A are inside the stability region. If the system is stable, all the eigenvalues are inside the stability region or on its boundary.*

If the system matrix A has an eigenvalue on the imaginary axis (the boundary of the stability region) the system can be either stable or unstable (but never asymptotically stable). It can be shown that the stability then depends on the number of linearly independent eigenvectors to the corresponding eigenvalue. If this number is equal to the multiplicity of the eigenvalue, the system is stable, otherwise not.

Test of Stability, Root Locus

To test the stability of a system means that the roots of the polynomial are calculated (or the eigenvalues of the system matrix A). For a given system this is most easily done numerically in a direct way (in MATLAB roots, eig).

There are also methods that can be used to investigate the stability of families of systems – most often closed loop systems obtained by proportional feedback in a given system. The *root locus method* and the *Nyquist criterion*, described in basic control courses are two well known such methods. They work also for discrete time systems. See Section 3.7

Non Minimum Phase System

A non minimum phase system is defined as follows.

Definition 3.7 (Non Minimum Phase System) *A system that has a zero outside the stability region is said to be* **non minimum phase.**

A non minimum phase system thus has, in some sense, an unstable inverse. This makes it harder to control, as we will see in Section 7.5. (Note that the system (1.1) that we found to be difficult to control in Example 1.1 indeed is non-minimum phase, according to Example 3.2.) Systems with unstable inverses are called "non-minimum phase" because of phase properties in their frequency function. We will return to this in Section 7.3.

3.5   Frequency Response and Frequency Functions

Consider a stable linear, multivariable transfer function matrix:

$$G(s) = \begin{bmatrix} G_{11}(s) & G_{12}(s) & \cdots & G_{1m}(s) \\ \vdots & \vdots & & \vdots \\ G_{p1}(s) & G_{p2}(s) & \cdots & G_{pm}(s) \end{bmatrix} \tag{3.19}$$

If this system is driven by an input

$$u_k(t) = \cos(\omega t)$$

(and the other inputs are zero), the output j will be in steady state

$$y_j(t) = A\cos(\omega t + \varphi)$$

where

$$A = |G_{jk}(i\omega)|, \qquad \varphi = \arg G_{jk}(i\omega) \tag{3.20}$$

This follows exactly as in the SISO case.

The matrix function $G(i\omega)$ thus gives information on how the system responds to pure sine och cosine inputs ("frequencies"). It is therefore called the *frequency response* or *frequency function*.

Gain for Multivariable Systems

From basic control theory we know that the *Bode plot* of a SISO system is important to understand its properties. The *amplitude curve* or the *gain*, i.e., the absolute value $|G(iw)|$ plotted as a function of frequency (typically in a log-log diagram) is particularly informative.

But what happens if G is a matrix? To study the different input-output combinations one at the time lies close at hand. This would involve making a Bode plot for each of the elements in $G(iw)$. This is of course informative, but connections between different sub-systems will not be displayed. Like in Example 1.1, important properties of the system can be missed. Another idea would be to study the eigenvalues of $G(s)$, but this would work only for a square G (as many inputs as outputs). Moreover, the eigenvalues are not necessarily a good measure of the gain of the matrix.

The solution is instead to study the *singular values* of the frequency function. We shall now define this concept.

Complex Matrices and Singular Values

Let us first consider a finite dimensional linear mapping

$$y = Ax \tag{3.21}$$

A is here a $p \times m$ matrix, y a p-dimensional vector and x an m-dimensional vector. Furthermore, we allow both y, x and A to have complex valued elements. For a complex valued matrix or vector we have the following definition:

Definition 3.8 *Let A be a given, complex valued matrix. The* **adjoint** *matrix A^* is obtained by taking the transpose of A and then complex conjugating its elements.*

For example, we have

$$A = \begin{bmatrix} 1+i \\ 2-3i \end{bmatrix}, \quad A^* = \begin{bmatrix} 1-i & 2+3i \end{bmatrix}$$

The rules for taking the adjoint are the same as for taking the transpose, thus $(AB)^* = B^*A^*$ etc.

Definition 3.9 *A complex valued matrix A is* **self-adjoint** *(or* **Hermitian***) if $A^* = A$.*

Hermitian matrices thus correspond to symmetric matrices and always have real eigenvalues. They can also always be diagonalized.

Let us now return to (3.21). How "large" is y compared to x? The 2-norm for x is

$$|x| = (\sum_{i=1}^{m} |x_i|^2)^{1/2} = \sqrt{x^* x}$$

Then

$$|y|^2 = |Ax|^2 = x^* A^* A x$$

The matrix $A^* A$ has the dimension $m \times m$ and is self-adjoint and positive semi-definite. It thus has real and non-negative eigenvalues. Call these eigenvalues $\lambda_1, \lambda_2, ..., \lambda_m$. Let λ_1 be the largest and λ_m the smallest eigenvalue. Then

$$\lambda_m |x|^2 \leq x^* A^* A x \leq \lambda_1 |x|^2 \tag{3.22}$$

This leads to the following definition:

Definition 3.10 (Singular values) *Given a matrix A. Its* **singular values** σ_i *are defined as* $\sigma_i = \sqrt{\lambda_i}$ *where* λ_i *are the eigenvalues of* $A^* A$. *The largest singular value of A is denoted by* $\bar{\sigma}(A)$ *and the smallest by* $\underline{\sigma}(A)$.

From this follows that if $y = Ax$ then

$$\underline{\sigma}(A) \leq \frac{|y|}{|x|} \leq \bar{\sigma}(A) \tag{3.23}$$

The *"gain"* of the matrix A thus lies between its largest and its smallest singular value. If we compare with the operator norm (1.24) we see that $|A| = \bar{\sigma}(A)$.

The singular values are not normally calculated numerically from the eigenvalues $A^* A$. Instead, A is represented as a factorization of matrices, so called *singular value factorization* (SVD)

$$A = U \Sigma V^* \tag{3.24}$$

If A is an $n \times m$ matrix then U is a unitary $n \times n$ matrix, , i.e., $UU^* = I$, Σ is an $n \times m$ matrix with the singular values of A along the diagonal and zeros everywhere else, while V is a unitary $m \times m$ matrix. If A is a complex valued matrix, U and V will also be complex valued, while the singular values are non negative real numbers.

Singular value decomposition is accomplished in MATLAB by

```
[U, Sigma, V] = svd(A)
```

Frequency Functions

For a stable system with transfer function $G(s)$ we have according to (2.3)

$$Y(i\omega) = G(i\omega)U(i\omega) \qquad (3.25)$$

where $Y(i\omega)$ and $U(i\omega)$ are the Fourier transforms of the output and input, respectively. How much is the input at frequency ω amplified by the system? As above we have

$$\underline{\sigma}(G(i\omega)) \leq \frac{|Y(i\omega)|}{|U(i\omega)|} \leq \bar{\sigma}(G(i\omega)) = |G(i\omega)| \qquad (3.26)$$

The *gain* of the system at ω thus lies between the largest and the smallest singular value of a frequency function $G(i\omega)$.

The counterpart to the amplitude Bode plot for SISO systems is to plot all the singular values of the frequency function as a function of frequency (typically in a log-log diagram).

The actual gain in a given instance depends on the "direction" of the vector $U(i\omega)$. If this vector is parallel to the eigenvector of $G^*(i\omega)G(i\omega)$ with the largest eigenvalue, the upper bound is valid for (3.26), etc. We will illustrate this in the following example.

Example 3.4:　The Gain of the Heat Exchanger

It is laborious to calculate the singular values of the frequency functions by hand, and computer help is needed. MATLAB has the command sigma for this. We illustrate its use on the heat exchanger in Example 2.2:

```
>> A=[-0.21 0.2;0.2 -0.21];
>> B=0.01*eye(2); C=eye(2);
>> D=zeros(2,2);
>> sigma(A,B,C,D)
```

This produces Figure 3.1. We see that for frequencies above approximately 1 rad/sec all inputs are damped equally much, and the gain decreases to zero as a typical first order system. For low frequencies, on the other hand, the difference is large (about 32 dB $\approx$ 40 times) for the gain, depending on whether the input vector is parallel to the eigenvector of the largest or smallest eigenvalue. Which these input directions are cannot been seen from the diagram – just that the gain will be between the given curves. Among other things we see that the *gain* of the system is 1.

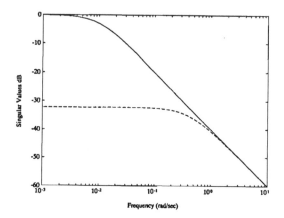

Figure 3.1:   The singular values of the frequency function for the heat exchanger

In order to study the directional dependency we have to calculate the eigenvalues of $G^*(i\omega)G(i\omega)$. We illustrate this for $\omega = 0$: According to (2.16)

$$G(0) = \frac{1}{0.41}\begin{bmatrix} 0.21 & 0.2 \\ 0.2 & 0.21 \end{bmatrix}$$

In MATLAB we then have

```
>> g0 = [0.21 0.2;0.2 0.21]/0.41;
>> [V,D]=eig(g0*g0')
V =
     0.7071    0.7071
    -0.7071    0.7071
D =
     0.0006         0
          0    1.0000
```

We see that the singular value 1 is associated with the eigenvector $\begin{bmatrix} 1 & 1 \end{bmatrix}^T$ while the singular value $\sqrt{0.0006} = 0.024$ is associated with the eigenvector $\begin{bmatrix} 1 & -1 \end{bmatrix}^T$. A stationary input of the kind

$$u_0 = \alpha \begin{bmatrix} 1 \\ -1 \end{bmatrix}$$

will thus have insignificant effect on the output in stationarity. Physically the input step above means that the temperature of the hot inflow is increased as much as the temperature of the cold one is decreased (or vice versa if α is negative). The stationary effect will be insignificant for the resulting outlet temperatures in the heat exchanger, which is natural.

An input step, parallel to $\begin{bmatrix} 1 & 1 \end{bmatrix}^T$, on the other hand means that the temperature of both the incoming hot and cold water increases by the same amount. It is then obvious that the output will increase as much – the temperature scale is just shifted. The gain in this direction thus has to be 1.

The Norm of the Frequency Function

As in (1.20) we introduce the infinity norm

$$\|G\|_\infty = \max_\omega |G(i\omega)| \tag{3.27}$$

for the largest singular value of the frequency function. For a SISO system $\|G\|_\infty$ is thus equal to the highest peak in the amplitude Bode plot. As in Example 1.8 we have

$$\|y\| \leq \|G\|_\infty \|u\| \tag{3.28}$$

with equality if the energy of the input is concentrated to the frequency where maximum in (3.27) is assumed, and if u is parallel to the eigenvector for the largest eigenvalue of $G^*(i\omega)G(i\omega)$. $\|G\|_\infty$ is thereby according to Definition 1.1 equal to the *gain of the system*.

3.6 Model Reduction

System models arrived at using physical modeling are often of high order. Some methods for control design also give controllers that are systems of high order. An important question is, if a system of lower order can be found, that well approximates a given system. This would simplify both analysis, control design and implementation.

A given system in state space form can be represented by a lower order model precisely when there are modes that are uncontrollable and/or unobservable. These modes (combination of state variables) can be eliminated without influencing the input–output relationship. Therefore, it appears to be a good starting point for model reduction to find modes that are "almost" uncontrollable or unobservable.

Balanced State Space Realization

A given controllable and observable system in state space form

$$\dot{x} = Ax + Bu$$
$$y = Cx + Du \tag{3.29}$$

can be represented in many different bases using the change of variables

$$\xi(t) = Tx(t) \tag{3.30}$$

We then have

$$\dot{\xi} = TAT^{-1}\xi + TBu$$
$$y = CT^{-1}\xi + Du \tag{3.31}$$

This freedom can be used in several different ways. We can for example choose T such that the system (3.31) will be in controller or observer canonical form, or in diagonal form, etc. We will now try to choose T such that (3.31) as clearly as possible shows state variables that do not influence the input–output relationship very much.

First let the input be an impulse in input i: $u_i(t) = \delta(t)$ and $x(0) = 0$. The state vector is then

$$x(t) = e^{At}B_i$$

where B_i is the ith column of B. See (3.3). Form the matrix of the state vectors for the impulse responses from each of the inputs: $X(t) = e^{At}B$. The "size" of the function $X(t)$ can be measured by forming the matrix

$$S_x = \int_0^\infty X(t)X^T(t)dt = \int_0^\infty e^{At}BB^Te^{A^Tt}dt$$

This matrix, the *controllability Gramian*, describes how much the different components of x are influenced by the input. It can be shown that the range of S_x is equal to the controllable subspace. Expressed in the variable ξ in (3.30) we have

$$S_\xi = TS_xT^T \tag{3.32}$$

Now, since S_x is symmetric, we can choose T such that S_ξ becomes diagonal:

$$S_\xi = \Sigma = \begin{bmatrix} \sigma_1 & 0 & \cdots & 0 \\ 0 & \sigma_2 & \cdots & 0 \\ \vdots & \vdots & \ddots & \vdots \\ 0 & 0 & \cdots & \sigma_n \end{bmatrix} \tag{3.33}$$

The diagonal element σ_k then measures how much the state variable ξ_k is influenced by the input. We can, so to speak, read the "relative controllability" in the state variables. We can, however, not dismiss variables corresponding to small values on σ_k. They might have a large influence on the output. Let us therefore study how the different state variables contribute to the energy of the output. If the input is zero, the output is

$$y(t) = Ce^{At}x_0$$

Its energy is measured using the scalar

$$\int_0^\infty y^T(t)y(t)dt = x_0^T \int_0^\infty e^{A^Tt}C^TCe^{At}dtx_0 = x_0^TO_xx_0 \qquad (3.34)$$

where the last step is a definition of the matrix O_x, called the *observability Gramian*. Expressed in the variable ξ we have

$$\int_0^\infty y^T(t)y(t)dt = \xi_0^TT^{-T}O_xT^{-1}\xi_0 = \xi_0^TO_\xi\xi_0 \qquad (3.35)$$

where we defined

$$O_\xi = T^{-T}O_xT^{-1} \qquad (3.36)$$

If T is selected so that O_ξ is a diagonal matrix Σ as in (3.33) we thus read in (3.35) exactly how the total energy of the output originates from the different components in the initial state ξ_0. We thus see the "relative observability" in the different state variables.

The point is now that *we can select T such that S_ξ and O_ξ are equal to the same diagonal matrix*. This is not obvious, but is proved in the following way:

Proof: Since O_x is positive semi-definite, it can be written

$$O_x = R^TR$$

for some matrix R. Now consider the matrix RS_xR^T. Since it is symmetric, it is diagonalized by a orthonormal matrix U ($UU^T = I$), and since it is positive semi-definite it has non-negative eigenvalues. It can thus be written as

$$RS_xR^T = U\Sigma^2U^T$$

with a diagonal matrix Σ. Now take $T = \Sigma^{-1/2}U^TR$ and insert into (3.32) and (3.36), which gives $S_\xi = O_\xi = \Sigma$. $\qquad \square$

Such a choice of T gives a representation (3.31) which is said to be a *balanced state space representation* of the system.

It should be added that units (scale factors) for input and output should be selected so that the different components of these vary approximately equally much. Otherwise it is not fair to use equally large impulses for the input, or to add together the different outputs without weighting. See also Section 7.1.

The advantage with a balanced state representation is that the relative significance of the different components of the state vector easily can be read. A state variable ξ_k corresponding to a small value of σ_k is only influenced insignificantly by the input and also influences the output insignificantly. Such a variable could thus easily be removed without essentially influencing the input–output properties.

Example 3.5:   The Heat Exchanger

In Example 2.1 we arrived at a simple model of a heat exchanger with the state variables x_1=temperature in the cold part and x_2=temperature in the hot part. This gave a state representation with

$$A = \begin{bmatrix} -0.21 & 0.2 \\ 0.2 & -0.21 \end{bmatrix}, \quad B = \begin{bmatrix} 0.01 & 0 \\ 0 & 0.01 \end{bmatrix}$$
$$C = I, \qquad\qquad D = 0$$

Calculations (in MATLAB) show that the change of variables

$$T = 7.07 \begin{bmatrix} -1 & -1 \\ 1 & -1 \end{bmatrix}$$

transforms the model to balanced form. This uses the state variables

$$\xi_1 = -7.07(x_1 + x_2), \qquad \xi_2 = 7.07(x_1 - x_2)$$

Clearly the difference in temperatures and average temperature, respectively, are more revealing state variables. With these variables the state representation becomes

$$\dot{\xi} = \begin{bmatrix} -0.01 & 0 \\ 0 & -0.41 \end{bmatrix} \xi + 0.0707 \begin{bmatrix} -1 & -1 \\ +1 & -1 \end{bmatrix} u$$
$$y = 0.0707 \begin{bmatrix} -1 & +1 \\ -1 & -1 \end{bmatrix} x$$

The common controllability and observability Gramians are

$$S_\xi = O_\xi = \begin{bmatrix} 0.5000 & 0 \\ 0 & 0.0122 \end{bmatrix}$$

The first state influences the input–output relationship 40 times more than the second state.

Elimination of States

Assume that we have a state with the state vector ξ. We want to keep the first r components of ξ and eliminate the $n - r$ last. The corresponding decomposition of the state matrices is

$$\begin{bmatrix} \dot{\xi_1} \\ \dot{\xi_2} \end{bmatrix} = \begin{bmatrix} A_{11} & A_{12} \\ A_{21} & A_{22} \end{bmatrix} \begin{bmatrix} \xi_1 \\ \xi_2 \end{bmatrix} + \begin{bmatrix} B_1 \\ B_2 \end{bmatrix} u$$

$$y = \begin{bmatrix} C_1 & C_2 \end{bmatrix} \begin{bmatrix} \xi_1 \\ \xi_2 \end{bmatrix} + Du \tag{3.37}$$

A simple attempt would be to just keep the "top part" of these equations, A_{11}, B_1, C_1 and D. With a bit more care we can keep the stationary properties of the system unchanged by replacing the dynamics in ξ_2 with the corresponding static relationship. If we put $\dot{\xi_2} = 0$ we have

$$0 = A_{21}\xi_1 + A_{22}\xi_2 + B_2 u$$

or

$$\xi_2 = -A_{22}^{-1}(A_{21}\xi_1 + B_2 u)$$

Now use this expression for ξ_2 in (3.37):

$$\dot{\xi_1} = A_r\xi_1 + B_r u \tag{3.38}$$
$$y = C_r\xi_1 + D_r u$$

where

$$A_r = A_{11} - A_{12}A_{22}^{-1}A_{21}, \qquad B_r = B_1 - A_{12}A_{22}^{-1}B_2$$
$$C_r = C_1 - C_2A_{22}^{-1}A_{21}, \qquad D_r = D - C_2A_{22}^{-1}B_2 \tag{3.39}$$

With these expressions it is now easy to achieve model reduction for any system. If the dynamics for ξ_2 is not too important for the input–output properties of the system, the reduced model is a good approximation to the original system. It is of course especially interesting to use the technique on a balanced state representation and eliminate states corresponding to relatively small values of σ. How well the approximation turns out can be verified by comparing step responses or Bode plots of the initial and the reduced model.

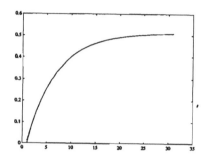

Figure 3.2: Solid line: The step response from input 1 to output 1 for the original second order model. Dashed curve: Corresponding curve for the reduced first order model. (They almost coincide.) The other step responses are similar.

Example 3.6:   Reduced Order Model for Heat Exchanger

In the model in Example 3.5 it is obviously interesting to get a first order approximation by elimination of the second state. Using the notation above we get

$$A_{11} = -0.01, \qquad A_{21} = A_{12} = 0, \qquad A_{22} = -0.41$$
$$B_1 = 0.0707 \begin{bmatrix} -1 & -1 \end{bmatrix}, \quad B_2 = 0.0707 \begin{bmatrix} 1 & -1 \end{bmatrix}$$
$$C_1 = 0.0707 \begin{bmatrix} -1 \\ -1 \end{bmatrix}, \qquad C_2 = 0.0707 \begin{bmatrix} 1 \\ -1 \end{bmatrix}, \qquad D = 0$$

which gives the model

$$\dot{\xi}_1 = -0.01\xi_1 - 0.0707 \begin{bmatrix} 1 & 1 \end{bmatrix} u$$
$$y = -0.0707 \begin{bmatrix} 1 \\ 1 \end{bmatrix} \xi_1 + 0.0122 \begin{bmatrix} 1 & -1 \\ -1 & 1 \end{bmatrix} u$$

Figure 3.2 shows that the first order approximation gives almost the same step responses.

3.7   Discrete Time Systems

In this section we will supply formulas and expressions for discrete time models. The main result is that this case is completely analogous to the continuous time model. The main difference is that s is replaced by $z = e^s$. It also means that the imaginary axis $i\omega$ is replaced by the unit circle $e^{i\omega}$ and the left half plane by the interior part of the unit circle.

Solution of the State Equations

A discrete time system in state space form

$$x(t + 1) = Ax(t) + Bu(t)$$
$$y(t) = Cx(t) + Du(t)$$
(3.40)

is its own solution algorithm. Computer simulations of (3.40) are straightforward. It is also easy to obtain an explicit expression for the solution by iteration:

$$
\begin{aligned}
x(t) &= Ax(t - 1) + Bu(t - 1) \\
&= A\left(Ax(t - 2) + Bu(t - 2)\right) + Bu(t - 1) \\
&= A^2 x(t - 2) + ABu(t - 2) + Bu(t - 1) = \ldots \\
&= A^{(t-t_0)} x(t_0) + \sum_{k=t_0}^{t-1} A^{(t-k-1)} Bu(k)
\end{aligned}
$$
(3.41)

Note the formal analogy to expression (3.2).

The definition of *modes* can of course be treated analogously to discrete time systems.

The impulse response is

$$
g(k) = \begin{cases} D & \text{for } k = 0 \\ CA^{k-1}B & \text{for } k > 0 \end{cases}
$$
(3.42)

Controllability and Observability

The definitions 3.1 and 3.2 apply without changes also to discrete time systems in state form. The criteria in Theorems 3.1 and 3.2 also apply without changes, which can be proved as follows:

Theorem 3.11 *The controllable states of the system (3.40) constitute a linear space, i.e., the range of the matrix (controllability matrix)*

$$S(A, B) = \begin{bmatrix} B & AB & A^2 B & \ldots & A^{n-1} B \end{bmatrix}$$
(3.43)

where n is the order of the system. The system is controllable if and only if S has full rank.

Proof: From (3.40) we have

$$x(1) = Bu(0)$$
$$x(2) = Ax(1) + Bu(1) = ABu(0) + Bu(1)$$
$$\vdots$$
$$x(N) = A^{N-1}Bu(0) + A^{N-2}Bu(1) + \ldots + Bu(N-1).$$

The last expression can be written

$$x(N) = \begin{bmatrix} B & AB & \ldots & A^{N-1}B \end{bmatrix} \begin{bmatrix} u(N-1) \\ u(N-2) \\ \vdots \\ u(1) \\ u(0) \end{bmatrix}$$

We see that by proper choice of $u(t)$, $x(N)$ may assume any value in the range space of the matrix

$$S_N = \begin{bmatrix} B & AB & \ldots & A^{N-1}B \end{bmatrix}.$$

The range space is equal to the vectors in the subspace spanned by the columns of the matrix. Cayley-Hamilton's theorem says that the matrix A satisfies its own characteristic equation. If $\det(\lambda I - A) = \lambda^n + a_1\lambda^{n-1} + \ldots + a_n$ then

$$A^n = -a_1 A^{n-1} - a_2 A^{n-2} \ldots - a_n I.$$

Higher powers of A than $n-1$ thus can be written as linear combinations of lower powers. The range S_N is thereby equal to the range for S_n for $N \geq n$, and the theorem follows. $\square$

Note that the theorem and the proof apply also if the system has several inputs ($u(t)$ is a vector). If u is a scalar, S is square and has full rank precisely when $\det S \neq 0$.

Theorem 3.12 *The unobservable states form a linear sub-space, i.e., the null space of the matrix (the observability matrix)*

$$\mathcal{O}(A, C) = \begin{bmatrix} C \\ CA \\ \vdots \\ CA^{n-1} \end{bmatrix} \qquad (3.44)$$

The system is thus observable if and only if $\mathcal{O}$ has full rank.

Proof: When the input is identically zero

$$y(0) = Cx(0)$$
$$y(1) = Cx(1) = CAx(0)$$
$$\vdots$$
$$y(N) = CA^N x(0)$$

which can be written

$$\begin{bmatrix} y(0) \\ y(1) \\ \vdots \\ y(N) \end{bmatrix} = \begin{bmatrix} C \\ CA \\ \vdots \\ CA^N \end{bmatrix} x(0) \triangleq \mathcal{O}_N x(0).$$

The output is thus identically zero for $0 \le t \le N$ exactly when $x(0)$ is in the null space of $\mathcal{O}_N$. As in the proof Theorem 3.1 it follows from Cayley-Hamilton's theorem that the null space for $\mathcal{O}_N$ is equal to the null space of $\mathcal{O}_{n-1}$ for all $N \ge n-1$. The theorem thus has been proved. $\square$

Poles and Zeros

The definitions of poles and zeros, as well as the calculation of them, is valid also for discrete time systems. The variable s just has to be exchanged for z in Section 3.3.

Stability

Comparing (3.41) with (3.2) we see that for discrete time systems the matrix power A^t plays the same role as the matrix exponential e^{At} for continuous time systems. According to Theorem 3.8, λ_i^t decides the qualitative behavior, where λ_i is its eigenvalue/poles. The crucial property is whether they have absolute values larger or smaller than 1. We therefore introduce the definition:

Definition 3.11 (Stability Region) *For a discrete time system the stability region is equal to the interior of the unit circle.*

Using this definition the main results, Theorems 3.9 and 3.10, are valid also for discrete time systems.

The Root Locus Method

The *root locus method* is a well known method to study how zeros of polynomials vary with some coefficient. See, for example, Franklin et al. (1994), Chapter 5. Investigating stability using this method works very well also for discrete time systems. For low order systems, it is actually easier to decide when some branch of the root locus passes through the unit circle than when it crosses the imaginary axis.

Example 3.7: Stability of a Discrete Time System, Obtained by Proportional Feedback

Consider the discrete time system with the transfer operator

$$G(q) = \frac{(q+1)}{(q-1)(q-0.5)}$$

The system is controlled using the feedback

$$u(t) = -K(y(t) - r(t))$$

which gives the closed loop system

$$\left(1 + K\frac{q+1}{(q-1)(q-0.5)}\right) y(t) = K\frac{q+1}{(q-1)(q-0.5)}r(t)$$

or

$$y(t) = \frac{K(q+1)}{(q-1)(q-0.5) + K(q+1)}r(t).$$

For which K is the closed loop system stable? The denominator polynomial is

$$(z-1)(z-0.5) + K(z+1).$$

The rules for drawing the root locus now state that it starts in the poles 0.5 and 1 and ends in the zeros. One branch will move to -1, and another to infinity. Points with an odd number of poles and zeros on the left hand side on the real axis belong to the root locus. This means that the root locus has to look like Figure 3.3.

For which value of K does it cross the unit circle? Since the constant term in the polynomial, i.e.,

$$K + 0.5$$

is equal to the product of the roots, this is equal to 1 when the roots cross the unit circle. (It is complex conjugated and the product is equal to the absolute value squared.) This happens for $K = 0.5$ and the system is stable for $0 < K < 0.5$.   □

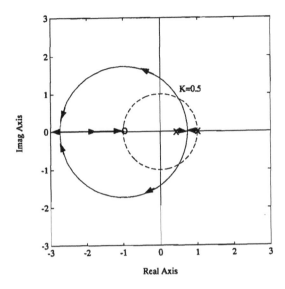

Figure 3.3: Root locus for the system in Example 3.7.

Another useful method is the *Nyquist criterion*, which is based on the argument principle for analytic functions. It is well known for continuous time systems, but can also be applied to discrete time systems.

The Discrete Time Nyquist Criterion

For a function, analytical except in a finite number of poles, the number of zeros and poles can be decided using the argument principle: If the function $f(z)$ has N zeros and P poles in an area Γ then $N - P =$ number of times the curve $f(z(\theta))$ encircles the origin when $z(\theta)$ follows the boundary of Γ in positive direction (counter clockwise)

Using this result, we can compute the number of zeros inside and outside the unit circle for an arbitrary polynomial.

The method is perhaps most useful when studying the stability of a closed loop system obtained by feedback with -1. Let the system

$$y(t) = G_0(q)u(t) \tag{3.45}$$

be a given, stable system. We assume as usual that the degree of the numerator in $G_0(q)$ is lower than the denominator. The system is closed by the feedback -1:

$$u(t) = r(t) - y(t).$$

The closed loop system thus is

$$y(t) = \frac{G_0(q)}{1 + G_0(q)} r(t) \tag{3.46}$$

This system is stable if it lacks poles outside the unit circle. Since G_0 is assumed to be stable, the question is if the function

$$R(z) = 1 + G_0(z)$$

has zeros outside the unit circle. The function $R(z)$ is rational with the same degree in numerator and denominator and all poles ($=$ the poles of G_0) are inside the unit circle. We want to test that also all zeros lay inside the unit circle, i.e., if $N - P = 0$. This is according to the argument principle the same as showing that the curve $R(e^{i\theta})$ does not encircle the origin when θ goes from 0 to 2π, since $z(\theta) = e^{i\theta}$ then encloses the unit circle. Alternatively we can test if the curve $G_0(e^{i\theta})$ encircles the point -1. This gives the following discrete time version of the Nyquist criterion:

> If the system $G_0(q)$ has no poles outside the unit circle, the closed loop system $\frac{G_0(q)}{1+G_0(q)}$ is stable if and only if, the curve $G_0(e^{i\theta})$, $0 \le \theta < 2\pi$ does not encircle the point -1.

Example 3.8:　The Nyquist Curve for a Discrete Time System

Consider the system

$$G(q) = \frac{K}{q - 0.5}$$

We have

$$G(e^{i\theta}) = \frac{K(\cos\theta - 0.5 - i\sin\theta)}{1.25 - \cos\theta}$$

This function is plotted for $0 \le \theta < 2\pi$ in Figure 3.4. We see that the point -1 is not encircled when $K < 3/2$. The closed loop system (3.46) is therefore stable for these values of K.

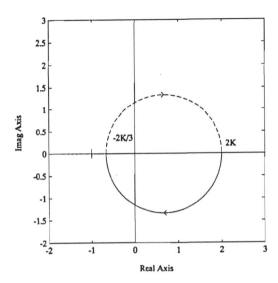

Figure 3.4: The Nyquist curve for the system in Example 3.8. The right crossing point with the real axis corresponds to $\theta = 0$, while the left one corresponds to $\theta = \pi$.

An interesting aspect of the Nyquist criterion is, that, as for the continuous time system, stability margins in terms of "how close the curves are to encircle" -1 can be defined. See Figure 3.5. The argument of the Nyquist curve, when it enters the unit circle (or rather its deviation from $-180°$) is such a measure. This angle, marked in Figure 3.5, is called the *phase margin* of the system. The interpretation is the same as for the continuous time systems. If $G_0(e^{i\omega})$ is plotted as a Bode plot (i.e., $\log|G_0(e^{i\omega})|$ and $\arg G_0(e^{i\omega})$ is plotted against $\log \omega$) the phase margin is thus

$$\pi + \arg G_0(e^{i\omega^*})$$

where ω^* (the crossover frequency) is defined by $|G_0(e^{i\omega^*})| = 1$.

The Frequency Function

Consider a discrete time stable transfer function matrix

$$G(z) = \begin{bmatrix} G_{11}(z) & G_{12}(z) & \cdots & G_{1m}(z) \\ \vdots & \vdots & \ddots & \vdots \\ G_{p1}(z) & G_{p2}(z) & \cdots & G_{pm}(z) \end{bmatrix} \tag{3.47}$$

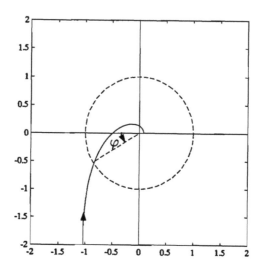

Figure 3.5: The phase margin is the angle φ.

Now let input number k be a discrete time cosine function:

$$u_k(\ell) = \cos(\ell\omega) = \text{Re } e^{i\omega\ell}, \qquad \ell = 0, \pm1, \pm2, \ldots$$

Let

$$G_{jk}(z) = \sum_{m=0}^{\infty} g_{jk}(m)z^{-m}$$

The output j is then

$$y_j(\ell) = \sum_{m=0}^{\infty} g_{jk}(m)u_k(\ell-m) = \sum_{m=0}^{\infty} g_{jk}(m)\text{Re } e^{i\omega(\ell-m)}$$

$$= \text{Re}\left\{\left(\sum_{m=0}^{\infty} g_{jk}(m)e^{-i\omega m}\right)e^{i\omega\ell}\right\} = \text{Re}\{G_{jk}(e^{i\omega})e^{i\omega\ell}\}$$

$$= |G_{jk}(e^{i\omega})|\cos(\ell\omega + \varphi)$$

where

$$\varphi = \arg G_{jk}(e^{i\omega})$$

The transfer function matrix evaluated at $e^{i\omega}$ thus gives the output response for a discrete time system when the input is a discrete time sine or cosine. Exactly as in the continuous time case, the absolute value of $G(e^{i\omega})$ describes the gain, while its argument gives the output phase lag compared to the input. For a discrete time system the *frequency response* or the *frequency function* thus is given by $G(e^{i\omega})$.

As for the continuous case, the size of multivariable frequency functions are measured by their singular values.

Model Reduction

In Section 3.6 we discussed how continuous time models are reduced. In discrete time the expressions are completely analogous. The S_x and O_x are defined as sums with powers of A in an obvious way. When eliminating the state ξ_2, without changing the static relationship, the relation $\xi_2(t+1) = \xi_2(t)$ is enforced. The corresponding expression for (3.39) needs only simple modifications.

3.8   Comments

The Main Points of the Chapter

Controllability, observability and frequency functions are concepts that can be generalized directly to multivariable systems. The most important aspect of multivariable frequency functions is that *the singular values of the frequency function matrices* play the same role as the amplitude curve for SISO systems. The gain of the system will lie between the largest and smallest singular value and it will depend on *the direction of the input vector* .

The calculation of the multiplicity of the poles and zeros is also more complicated in the multivariable case. A multivariable system can, for example, be non-minimum phase even if all its SISO partial systems are minimum phase.

Stability is decided by the position of the poles. For continuous time systems, the left half plane gives stability, as in the SISO case.

The meaning of all definitions for *discrete time systems* are the same as for the continuous time case. The main connection is in the relation

$$z = e^s$$

that states that the unit circle plays the same role in discrete time as the imaginary axis does in continuous time ($i\omega$ is replaced by $e^{i\omega}$).

Literature

The modern concepts concerning systems in state space form (controllability/observability, etc.) go back to Kalman's work around 1960, e.g, Kalman (1960c). The pole placement problem for multivariable systems (Theorem 3.4) was long unsolved. We followed Heymann (1968) in its proof. Fourier and frequency methods were developed for continuous time systems, into what today is known as "classic control theory" during the 1940s. These methods are summarized in, for example James, Nichols & Phillips (1947). Balanced realization and model reduction were developed in the early 80s, see, for example, Moore (1981).

Properties of linear systems are discussed in many textbooks. For a thorough review of structural properties in multivariable systems, see Kailath (1980). In Discrete time systems are covered in detail in, for example, Åström & Wittenmark (1997) and Franklin et al. (1990). Model reduction has become an important part of control construction and is discussed in textbooks on modern control design, as for example Maciejowski (1989).

Software

MATLAB contains the following commands for the methods we describe in this chapter:

ss2zp, zp2ss, tf2zp, zp Transformations between state representations (ss), transfer functions (tf) and poles and zero (zp). (In MATLAB 5 these transformations have been made transparent by object orientation.)

tzero Calculation of zeros according to Definition 3.5

lsim Simulation of continuous and discrete systems, respectively

eig, roots Eigenvalues and poles

bode Calculation of frequency functions

sigma Calculation of the singular values of frequency functions

obsv, ctrb Calculation of the matrices (3.8) and (3.7), respectively

gram Calculation of controllability and observability Gramians

balreal Calculation of balanced state form

modred Model reduction according to (3.38).

Appendix 3A: Proofs

Proof of Theorem 3.4

Proof: That the formulations in the theorem are equivalent follows by taking the transpose of the expression, identifying $A = A^T, C = B^T$ and using that the transpose of a matrix has the same eigenvalues as the matrix itself.

First we show the result for $m = 1$ in the first formulation, i.e., when B is a column vector. Let

$$\det(\sigma I - A) = a(\sigma) = \sigma^n + a_{n-1}\sigma^{n-1} + \ldots + a_0$$
$$\det(\sigma I - A + BL) = \beta(\sigma) = \sigma^n + \beta_{n-1}\sigma^{n-1} + \ldots + \beta_0$$

We have

$$\beta(\sigma) = \det\left((\sigma I - A)(I + (\sigma I - A)^{-1}BL)\right)$$
$$= \det(\sigma I - A)\det\left(I + (\sigma I - A)^{-1}BL\right) = a(\sigma)\left(1 + L(\sigma I - A)^{-1}B\right)$$

where we in the last column used that

$$\det(I + xy^T) = 1 + y^T x$$

if x and y are column vectors. This means that

$$\beta(\sigma) - a(\sigma) = a(\sigma)L(\sigma I - A)^{-1}B \qquad (3.48)$$

The following equality can be verified by multiplying both sides by $(\sigma I - A)$:

$$(\sigma I - A)^{-1} = \frac{1}{a(\sigma)}\left(\sigma^{n-1}\cdot I + \sigma^{n-2}(A + a_{n-1}\cdot I) + \ldots\right.$$
$$\left. + (A^{n-1} + a_{n-1}A^{n-2} + \ldots + a_1 \cdot I)\right)$$

Using this expression in (3.48) and after that identifying the coefficients for σ^k we obtain

$$\beta_{n-1} - a_{n-1} = LB$$
$$\beta_{n-2} - a_{n-2} = LAB + a_{n-1}LB$$
$$\vdots$$
$$\beta_0 - a_0 = LA^{n-1}B + a_{n-1}LA^{n-2}B + \ldots + a_1 LB$$

or

$$\left[\beta_{n-1} - a_{n-1} \;\vdots\; \beta_0 - a_0\right] = L\begin{bmatrix} B & AB & A^2B & \ldots & A^{n-1}B\end{bmatrix}$$
$$\times \begin{bmatrix} 1 & a_{n-1} & a_{n-2} & \ldots & a_1 \\ 0 & 1 & a_{n-1} & \ldots & a_2 \\ 0 & 0 & 1 & \ldots & a_3 \\ \vdots & \vdots & \vdots & \ddots & \vdots \\ 0 & 0 & 0 & \ldots & 1 \end{bmatrix}$$

The last matrix is invertible since it is triangular with ones along the diagonal. The other matrix is invertible according to the assumption of the theorem. Hence we can always find an L for every arbitrary left column, thus for every arbitrary characteristic polynomial $\beta(\sigma)$.

Let us now consider the case $m > 1$. We are going to show "Heyman's Lemma":

> If $S(A, B)$ has full rank
>
> there is an $m \times n$ matrix F for every column vector $b = Bv \neq 0$
>
> such that also $S(A + BF, b)$ has full rank.

According to what we just shown we can always select a row vector ℓ such that $A + BF - b\ell$ assumes arbitrary eigenvalues. By using $L = -F + v\ell$ the theorem will be proved.

Now it only remains to show Heynman's Lemma. We first show that we can always find a series of vectors $v_k, k = 1, \dots, n-1$ such that $x_1 = b; x_{k+1} = Ax_k + Bv_k$ gives a sequence of vectors x_k which spans the entire space. If this were not the case, it would mean that for some $k < n$

$$Ax_k + Bv_k \in \mathcal{L} \quad \forall v_k$$

where $\mathcal{L}$ is the space spanned by $x_1, \dots, x_k$. This means that $Ax_k \in \mathcal{L}$ (take $v_k = 0$) and therefore $Bv_k \in \mathcal{L}, \forall v_k$. In turn this would mean that $Ax_j = x_{j+1} - Bv_j \in \mathcal{L}; j = 1, \dots, k - 1$, thus $x \in \mathcal{L} \Rightarrow Ax \in \mathcal{L}$. The conclusion would be $\forall w, Bw, ABw, A^2 Bw, \dots, A^{n-1} Bw$ all would lie in $\mathcal{L}$ with rank $< n$. This contradicts the fact that $S(A, B)$ has full rank. Consequently, we can find a sequence v_k which gives a sequence x_k which spans the entire space. Now define F by $Fx_k = v_k$. (This can be done since the x_k-vectors span the entire space.) We can write $x_k = (A + BF)^{k-1}b$, which now proves that $S(A + BF, b)$ has full rank. The "if"-part of the proof is thus completed.

To show the "only if"-part, we use the PBH-test (3.10): If the system is not observable there is a value λ^*, such that the matrix in (3.10) does not have full rank. That is, there is a vector x such that

$$\begin{bmatrix} A - \lambda^* I \\ C \end{bmatrix} x = 0 \quad \text{i.e.,} \quad (A - \lambda^* I)x = 0 \quad \text{and} \quad Cx = 0$$

This means that $(A - KC)x = \lambda^* x$ for all values of K. The eigenvalue λ^* can thus not be changed, which proves the "only if"-part of the proof. $\qquad\square$

Proof of Theorem 3.8

Proof: First assume that F is diagonalizable. Then there is a matrix L such that

$$F = L \begin{bmatrix} p_1 & & 0 \\ & \ddots & \\ 0 & & p_n \end{bmatrix} L^{-1} \quad \text{och} \quad F^t = L \begin{bmatrix} p_1^t & & 0 \\ & \ddots & \\ 0 & & p_n^t \end{bmatrix} L^{-1}.$$

From this (3.16) and (3.17) follow directly. If F cannot be diagonalized, there is anyway a matrix L such that $F = LJL^{-1}$ with

$$J = \begin{bmatrix} p_1 & e_1 & 0 & 0 & \cdots & 0 \\ 0 & p_2 & e_2 & 0 & \cdots & 0 \\ 0 & 0 & p_3 & e_3 & \cdots & 0 \\ \vdots & & & & \ddots & e_{n-1} \\ 0 & 0 & 0 & 0 & \cdots & p_n \end{bmatrix}$$

where $e_i = 1$ or 0. e_i can also be 1 only if $p_i = p_{i+1}$. (This is on the condition that multiple eigenvalues are sorted together in J.) The form J is called the Jordan normal form. (For a diagonalizable matrix $e_i = 0$ for all i). By multiplication we see that J^t is a triangular matrix with zeros below the main diagonal. In the diagonal there are elements of the type p_i^t and in the super diagonal there are elements of the type $p(t)p_i^t$ where $p(t)$ is a polynomial in t of a degree not higher than $n_i - 1$ where n_i is the multiplicity of the eigenvalue p_i. Since

$$|p(t)p_i^t| \le C\lambda^t \text{ if } |p_i| < \lambda$$

t now follows that all elements in the matrix F^t are bounded by $C\lambda^t$. From this follows that also the norm of this matrix must satisfy the same bound (but perhaps with another C), so (3.16) follows. Also (3.17) follows from the fact that a matrix norm is at least as large as the largest eigenvalue of the matrix. □

Chapter 4

Sampled Data Systems

By a sampled data system we mean a discrete time system that is obtained by sampling a continuous one. It can be seen as an approximation, or as an alternative description of the system, where we are only interested in the values of the input and the output at a number of discrete time instances, often uniformly distributed in time

$$u(t), \quad t = \ell T; \quad \ell = \ldots, -2, -1, 0, 1, 2, \ldots$$

Here T is the sampling interval and $t = \ell T$ are the sampling instants. The frequency $\omega_s = 2\pi/T$ is called *sampling frequency*, while $\omega_N = \omega_s/2 = \pi/T$ is the *Nyquist frequency*. In this chapter we will discuss the relationships between a continuous time system, and the discrete time system describing the relationship between input and output at the sampling instants $t = \ell T$.

4.1 Approximating Continuous Time Systems

Today almost all control objects, that "physically" are continuous time systems, are controlled by sampled data (discrete time) controllers. We will discuss this further in Section 6.7. As we noted in Section 1.3, there are two main choices for obtaining such regulators ("Method I" and "Method II").

There are thus two typical situations for describing a continuous time system by means of a discrete time one:

I. We have calculated a controller as a continuous time system and want to implement it in a computer as a discrete time control algorithm.

II. We have a continuous time description of a control object, and want to use discrete time design theory to decide a sampled data controller. For this we need a sampled (discrete time) description of the control object.

The first situation is the question of digital implementation of controllers and is covered in detail in the literature on digital control. In short, the main idea is to replace the differentiation operator p in the continuous description by a suitable difference approximation, i.e., an expression in q. Euler's classic approximation corresponds to replacing p by $(q-1)/T$ i.e.,

$$\dot{x}(t) \approx \frac{(x(t+T) - x(t))}{T}$$

where T is the sampling interval. A better and often used approximation is the so called bilinear transformation, also called the trapezoidal rule or Tustin's formula

$$p = \frac{2}{T}\frac{q-1}{q+1} \tag{4.1}$$

In most cases, the resulting discrete time controller will give a closed loop system with properties close to what would be obtained using the original continuous time controller. One condition is that the sampling is "fast", typically about 10 times the bandwidth of the closed loop system.

In this chapter we will concentrate on the other situation, to describe a continuous control object using a discrete time system. If the input to the continuous time system can be reconstructed exactly from its values at the sampling times (for example if it is piecewise constant or piecewise linear between the sampling instances) it should be possible to obtain a discrete time representation exactly describing the sampled output.

4.2   System Sampling

To Go From a Continuous System to its Sampled Counterpart

Consider a continuous time system in state space form

$$\dot{x}(t) = Ax(t) + Bu(t) \tag{4.2a}$$
$$y(t) = Cx(t) \tag{4.2b}$$

We assume that the system is controlled by a sampled data controller with the sampling interval T. See also Section 6.7. The input $u(t)$ is then piecewise constant:

$$u(t) = u(kT), \qquad kT \le t < (k+1)T \tag{4.3}$$

According to (3.2) the solution of (4.2) can be written

$$x(t) = e^{A(t-t_0)}x(t_0) + \int_{t_0}^t e^{A(t-s)}B\,u(s)ds \tag{4.4}$$

If $kT \le t < (k+1)T$ we chose in (4.4) $t_0 = kT$ and get

$$x(t) = e^{A(t-kT)}x(kT) + \int_{kT}^t e^{A(t-s)}B\,u(kT)ds$$
$$= e^{A(t-kT)}x(kT) + \left(\int_{kT}^t e^{A(t-s)}ds\right)B\,u(kT) \tag{4.5}$$

In particular, at $t = (k+1)T$ we have

$$x((k+1)T) = e^{AT}x(kT) + \left(\int_{kT}^{(k+1)T} e^{A(kT+T-s)}ds\right)Bu(kT)$$
$$= Fx(kT) + Gu(kT) \tag{4.6}$$

with the matrices

$$F = e^{AT} \tag{4.7a}$$
$$G = \int_{kT}^{(k+1)T} e^{A(kT+T-s)}ds\,B$$
$$= \int_0^T e^{A(T-s)}ds\,B = \int_0^T e^{As}ds\,B = A^{-1}(e^{AT} - I)B \tag{4.7b}$$

The last equality holds if A is invertible. Since

$$y(kT) = C\,x(kT) \tag{4.8}$$

(4.6) and (4.8) give a description of the the sampled system . We have shown the following:

Theorem 4.1 (Sampling of systems) *If the system*

$$\dot{x} = Ax + Bu \tag{4.9a}$$
$$y = Cx \tag{4.9b}$$

is controlled by an input, piecewise constant over the sampling interval T, then the relationship among the values of the input, the state and the output at the sampling instants, is given by the **sampled data system**

$$x(kT + T) = Fx(kT) + Gu(kT) \tag{4.10a}$$
$$y(kT) = Hx(kT) \tag{4.10b}$$

Here

$$F = e^{AT}, \qquad G = \int_0^T e^{At}B\,dt, \qquad H = C \tag{4.10c}$$

The sampled system (4.10) thus only gives the output and state at the sampling instants. Their values in between these instants can be calculated using the general expression (4.5).

In the state space representation (4.10) there is always a delay because $u(kT)$ influences $y(kT + T)$ and not earlier values of y. This is natural considering the way the sampling normally is done: first the output $y(t)$ is read at $t = kT$, after that the input $u(kT)$ is chosen. There is no logical possibility for $u(kT)$ to influence $y(kT)$. On the other hand, $y(t)$ is normally influenced for $t > kT$, even if this is noticed first at the next sampling instance $t = kT + T$.

To find the sampled counterpart to a continuous system, it is only necessary to calculate the matrix e^{At}. There are several ways to do this. Here we only recall

$$e^{At} = \mathcal{L}^{-1}(sI - A)^{-1} \tag{4.11}$$

where $\mathcal{L}^{-1}$ is the inverse Laplace transformation. To compute F and G the following algorithm could be used:

$$S = \int_0^T e^{As} ds = IT + A\frac{T^2}{2} + A^2\frac{T^3}{3!} + \ldots + A^k\frac{T^{k+1}}{(k+1)!} + \ldots$$

$$F = I + AS$$

$$G = SB.$$

In the calculation of S the series expansion is truncated when the last term is smaller than a given ϵ. There are, however, several considerably better algorithms. See the literature references.

Example 4.1: The Double Integrator

Consider the system

$$y = \frac{1}{p^2}u$$

With the state variables

$$x_1 = y, \qquad x_2 = \dot{y}$$

the system can be written in state space form:

$$\begin{bmatrix} \dot{x}_1 \\ \dot{x}_2 \end{bmatrix} = \begin{bmatrix} 0 & 1 \\ 0 & 0 \end{bmatrix}\begin{bmatrix} x_1 \\ x_2 \end{bmatrix} + \begin{bmatrix} 0 \\ 1 \end{bmatrix}u$$

$$y = \begin{bmatrix} 1 & 0 \end{bmatrix}\begin{bmatrix} x_1 \\ x_2 \end{bmatrix}$$

(4.12)

We then have

$$A = \begin{bmatrix} 0 & 1 \\ 0 & 0 \end{bmatrix}, \quad B = \begin{bmatrix} 0 \\ 1 \end{bmatrix}, \quad C = \begin{bmatrix} 1 & 0 \end{bmatrix}$$

and

$$e^{At} = \mathcal{L}^{-1}\left(sI - \begin{bmatrix} 0 & 1 \\ 0 & 0 \end{bmatrix}\right)^{-1} = \mathcal{L}^{-1}\begin{bmatrix} s & -1 \\ 0 & s \end{bmatrix}^{-1}$$

$$= \mathcal{L}^{-1}\begin{bmatrix} 1/s & 1/s^2 \\ 0 & 1/s \end{bmatrix} = \begin{bmatrix} 1 & t \\ 0 & 1 \end{bmatrix}$$

This gives

$$F = \begin{bmatrix} 1 & T \\ 0 & 1 \end{bmatrix}$$

$$G = \int_0^T \begin{bmatrix} 1 & t \\ 0 & 1 \end{bmatrix}\begin{bmatrix} 0 \\ 1 \end{bmatrix} dt = \begin{bmatrix} T^2/2 \\ T \end{bmatrix}$$

The counterpart to (4.12), sampled by T is then given by

$$x(kT + T) = \begin{bmatrix} 1 & T \\ 0 & 1 \end{bmatrix} x(kT) + \begin{bmatrix} T^2/2 \\ T \end{bmatrix} u(kT)$$

$$y(kT) = \begin{bmatrix} 1 & 0 \end{bmatrix} x(kT).$$

(4.13)

The transfer operator is

$$G_T(q) = \begin{bmatrix} 1 & 0 \end{bmatrix} \left(qI - \begin{bmatrix} 1 & T \\ 0 & 1 \end{bmatrix} \right)^{-1} \begin{bmatrix} T^2/2 \\ T \end{bmatrix}$$

$$= \frac{1}{2} T^2 \frac{(q+1)}{q^2 - 2q + 1}$$

(4.14)

Example 4.2:   A Direct Current Motor

Consider a DC motor, with transfer operator given by

$$y = \frac{1}{p(p+1)} u$$

With the state variables $x_1 = y$, $x_2 = \dot{y}$ we have the state space representation

$$\begin{bmatrix} \dot{x}_1 \\ \dot{x}_2 \end{bmatrix} = \begin{bmatrix} 0 & 1 \\ 0 & -1 \end{bmatrix} \begin{bmatrix} x_1 \\ x_2 \end{bmatrix} + \begin{bmatrix} 0 \\ 1 \end{bmatrix} u$$

$$y = \begin{bmatrix} 1 & 0 \end{bmatrix} \begin{bmatrix} x_1 \\ x_2 \end{bmatrix}$$

(4.15)

This system is sampled using the sampling interval T. We then have

$$e^{At} = \mathcal{L}^{-1} \begin{bmatrix} s & -1 \\ 0 & s+1 \end{bmatrix}^{-1} = \begin{bmatrix} \mathcal{L}^{-1} \frac{1}{s} & \mathcal{L}^{-1} \frac{1}{s(s+1)} \\ \mathcal{L}^{-1} 0 & \mathcal{L}^{-1} \frac{1}{s+1} \end{bmatrix}$$

$$= \begin{bmatrix} 1 & 1 - e^{-t} \\ 0 & e^{-t} \end{bmatrix}$$

and

$$G = \int_0^T e^{At} B dt = \begin{bmatrix} \int_0^T (1 - e^{-t}) dt \\ \int_0^T e^{-t} dt \end{bmatrix} = \begin{bmatrix} (T - 1 + e^{-T}) \\ (1 - e^{-T}) \end{bmatrix}$$

The sampled data system is given by

$$x(T + kT) = \begin{bmatrix} 1 & 1 - e^{-T} \\ 0 & e^{-T} \end{bmatrix} x(kT) + \begin{bmatrix} (T - 1 + e^{-T}) \\ (1 - e^{-T}) \end{bmatrix} u(kT)$$

$$y(kT) = \begin{bmatrix} 1 & 0 \end{bmatrix} x(kT)$$

(4.16)

The transfer operator is

$$
\begin{aligned}
G_T(q) &= \begin{bmatrix} 1 & 0 \end{bmatrix} \left(qI - \begin{bmatrix} 1 & 1-e^{-T} \\ 0 & e^{-T} \end{bmatrix} \right)^{-1} \begin{bmatrix} T-1+e^{-T} \\ 1-e^{-T} \end{bmatrix} \\
&= \frac{(T-1+e^{-T})q + (1-e^{-T}(1+T))}{q^2 - (e^{-T}+1)q + e^{-T}}
\end{aligned}
\tag{4.17}
$$

To Go from a Sampled Data System to its Continuous Time Counterpart

We assume that a discrete time system is given

$$
x(t+T) = Fx(t) + Gu(t) \tag{4.18}
$$

We are looking for a continuous time system, that will give (4.18) when it is sampled with the sampling interval T. In principle, it is simple to find. We first solve the equation

$$
e^{AT} = F \tag{4.19}
$$

with respect to A, and then determine B as

$$
B = \left[\int_0^T e^{As} ds \right]^{-1} G \tag{4.20}
$$

This is done most easily if F is in diagonal form. There are two complications:

- (4.19) may have no solution

- (4.19) may have several solutions.

We illustrate this in two examples.

Example 4.3: A System without Continuous Counterpart

The scalar discrete time system

$$
\begin{aligned}
x(t+1) &= -x(t) + u(t) \\
y(t) &= x(t)
\end{aligned}
$$

does not have a continuous counterpart of order 1, since $e^{\alpha} = -1$ lacks a (real) solution α.

Example 4.4:   Harmonic Oscillator

Consider the sampled system (the harmonic oscillator)

$$x(t+T) = \begin{bmatrix} \cos \alpha T & \sin \alpha T \\ -\sin \alpha T & \cos \alpha T \end{bmatrix} x(t) + \begin{bmatrix} 1 - \cos \alpha T \\ \sin \alpha T \end{bmatrix} u(t).$$

It is easy to verify that

$$A = \begin{bmatrix} 0 & \beta \\ -\beta & 0 \end{bmatrix}, \quad B = \begin{bmatrix} 0 \\ \beta \end{bmatrix}$$

satisfies (4.19), (4.20) for this system if

$$\beta = \alpha + \frac{2\pi}{T} \cdot k, \quad k = 0, \pm 1, \ldots$$

There are thus infinitely many continuous systems that correspond to our sampled system. They have the eigenvalues (poles) in $\pm(\alpha + (2\pi/T) \cdot k)i$.

Generally speaking, complex conjugated poles in the continuous time system, cannot be calculated unambiguously from descriptions of the sampled system. Normally, when calculating the continuous time system, it is chosen to have complex poles with the smallest possible imaginary part. In the example above we then have the solution $\beta = \alpha$ if and only if

$$|\alpha T| < \pi. \tag{4.21}$$

This relationship gives the upper bound for the sampling interval to make reconstruction of the continuous time model possible. This is in agreement with the sampling theorem that states that we have to sample with a sampling interval $T < \pi/\omega^\dagger$ in order to be able to describe the frequency $\omega^\dagger$ correctly in a signal.

The relationship between the different continuous time and sampled representations for a system can now be summarized by the scheme in Figure 4.1.

4.3   Poles and Zeros

Poles

The poles of the continuous system are, if the system is controllable and observable, equal to the eigenvalues of the A matrix. When the system

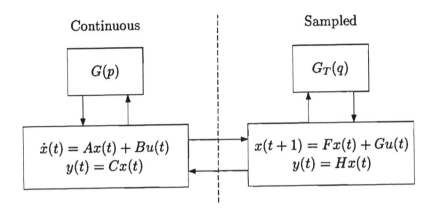

Figure 4.1: Relationship between different system models.

is sampled we form $F = e^{AT}$. Its eigenvalues are easily calculated from the A matrix according to Theorem 3.7. If also the sampled system is controllable and observable (see next section) the eigenvalues of the F matrix are equal to the poles of the sampled system. Under this condition we have shown

> *The poles of the sampled system are given by*
> $$e^{\lambda_i T}, \quad i = 1, \dots, n$$
> *where λ_i, $i = 1, \dots, n$, are the poles of the continuous system.*

Let us illustrate this relationship. We have

$$\lambda = \mu + i\omega \quad \Rightarrow \quad e^{\lambda T} = e^{\mu T}(\cos(\omega T) + i\ \sin(\omega T)). \tag{4.22}$$

From this expression we see first that when the sampling interval T is small, the poles of the sampled system will concentrate around 1. We also see that if the continuous system is asymptotically stable, $\mu < 0$, the sampled system will also be stable; $|e^{\lambda T}| = e^{\mu T} < 1$. The relationship (4.22) can also be illustrated as in Figure 4.2. Note that Figure 4.2c also can be interpreted as if λ is kept constant and T grows from 0 to infinity.

Zeros

It is considerably more difficult to describe the relationship between the zeros of the continuous time system and the discrete time system. Note in particular

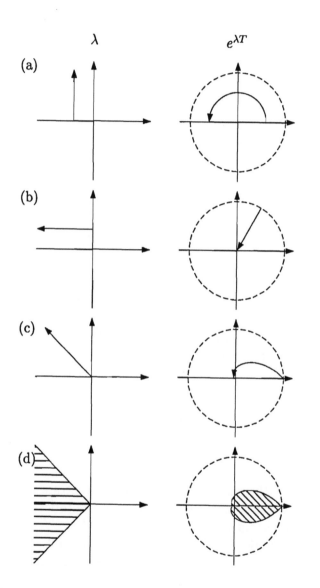

Figure 4.2: Relationship between the poles of the continuous and the sampled system.

- The sampled system can have both more and less (see Example 4.1) zeros than the continuous time system.

- The sampled system can be non-minimum phase even if the continuous is minimum phase and *vice versa*.

4.4   Controllability and Observability

Controllability means, for sampled as well as continuous systems, that the state zero can be controlled to an arbitrary state in finite time. When controlling the sampled system, we only have access to a subset of the input signals that can be used for the continuous system, i.e., piecewise constant signals. In order for a sampled system to be controllable the underlying continuous system has to be controllable. We also realize that if the choice of sampling interval is bad, it might happen that the sampled system cannot be controlled using the input signals that are available, even though the continuous system is controllable.

Observability means that only the state zero can produce an output sequence that is identically zero when the input is zero. For a continuous time system it is necessary that $y(t) \equiv 0$, while for the sampled system $y(kT) = 0$, $k = 0, 1, \dots$. It can happen that the output is zero at the sampling instants but not identically zero for all t. The sampled system can therefore be unobservable even if the continuous system is observable. With a badly selected sampling interval both controllability and observability can be lost.

If a continuous time system is unobservable there are concealed modes that do not show up in the output. For a sampled system unobservable modes do not have to be particularly concealed. They do not show in the output *at the sampling instants*, but in between they affect the output, if the underlying continuous system is observable. Unobservable modes can thus be highly visible during sampled data control. It is important to keep this in mind since observability can be lost by feedback. We give an example of this.

Example 4.5:   The Double Integrator

Consider control of the double integrator

$$G(p) = \frac{1}{p^2}$$

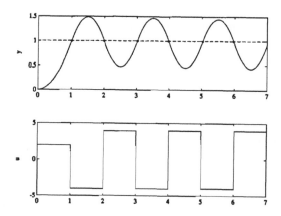

Figure 4.3: Output and input when (4.25) controls the double integrator $r(t) = 1$.

with a sampled data controller with sampling interval $T = 1$. The control algorithm sees the sampled system

$$y(t) - 2y(t-1) + y(t-2) = \frac{1}{2}(u(t-1) + u(t-2)) \qquad (4.23)$$

according to calculations in Example 4.1 (see equation (4.14)). If the output should follow a reference signal $r(t)$, it appears from (4.23) very suitable to select $u(t-1)$ such that

$$\frac{1}{2}(u(t-1) + u(t-2)) = -2y(t-1) + y(t-2) + r(t) \qquad (4.24)$$

since (4.24) inserted into (4.23) gives

$$y(t) = r(t),$$

i.e., the perfect servo properties. The controller (4.24) can be written

$$u(t) = -4y(t) + 2y(t-1) - u(t-1) + 2r(t+1) \qquad (4.25)$$

or, with operator notation

$$u(t) = \frac{-4 + 2q^{-1}}{1 + q^{-1}} y(t) + \frac{2}{1 + q^{-1}} r(t+1) \qquad (4.26)$$

Figure 4.3 shows how the control system behaves when the sampled controller (4.25) is applied to the double integrator for $r(t) = 1$.

From this figure we see two things:

1. The calculations have been correct: the output is identically equal to the reference signal at *the sampling instants*.

2. The controller is worthless if the control aim is to make the output follow the reference signal for all t, not only for $t = 1, 2, \ldots$.

The result implies that the feedback (4.26) has created unobservable modes. Let us analyze this. Inserting (4.26) into the system

$$y(t) = \frac{\frac{1}{2}(q^{-1} + q^{-2})}{1 - 2q^{-1} + q^{-2}} u(t)$$

we have

$$\left(1 - \frac{\frac{1}{2}(q^{-1} + q^{-2})}{1 - 2q^{-1} + q^{-2}} \cdot \frac{(-4 + 2q^{-1})}{1 + q^{-1}}\right) y(t)$$

$$= \frac{\frac{1}{2}(q^{-1} + q^{-2})}{1 - 2q^{-1} + q^{-2}} \cdot \frac{2}{1 + q^{-1}} r(t + 1)$$

i.e.,

$$y(t) = \frac{1 + q^{-1}}{(1 + q^{-1})(1 - 2q^{-1} + q^{-2} - q^{-1}(-2 + q^{-1}))} r(t)$$

$$= \frac{1 + q^{-1}}{1 + q^{-1}} r(t) = r(t)$$

In the closed loop system we can thus cancel the factor $1 + q^{-1}$. We know that this means that the closed loop system has a mode with the eigenvalue -1, that is uncontrollable and/or unobservable. The situation will be completely clarified by analyzing the feedback in state space form.

The double integrator can according to Example 4.1 be represented in state space form as

$$x(t + 1) = \begin{bmatrix} 1 & 1 \\ 0 & 1 \end{bmatrix} x(t) + \begin{bmatrix} 1/2 \\ 1 \end{bmatrix} u(t) \tag{4.27}$$

$$y(t) = \begin{bmatrix} 1 & 0 \end{bmatrix} x(t)$$

where

$$x(t) = \begin{bmatrix} y(t) \\ \dot{y}(t) \end{bmatrix} = \begin{bmatrix} x_1(t) \\ x_2(t) \end{bmatrix}$$

The first row in the state equation is

$$y(t + 1) = y(t) + x_2(t) + \frac{1}{2} u(t)$$

i.e.,

$$x_2(t) = y(t+1) - y(t) - \frac{1}{2}u(t).$$

From the second row we get

$$x_2(t+1) = x_2(t) + u(t) = y(t+1) - y(t) + \frac{1}{2}u(t).$$

The feedback law (4.25) can now be written

$$u(t) = -2y(t) - 2(y(t) - y(t-1) + \frac{1}{2}u(t-1)) + 2r(t+1)$$
$$= -2x_1(t) - 2x_2(t) + 2r(t+1) = -\begin{bmatrix} 2 & 2 \end{bmatrix} x(t) + 2r(t+1).$$

Inserted into (4.27) we get

$$x(t+1) = \begin{bmatrix} 1 & 1 \\ 0 & 1 \end{bmatrix} x(t) - \begin{bmatrix} 1 & 1 \\ 2 & 2 \end{bmatrix} x(t) + \begin{bmatrix} 1 \\ 2 \end{bmatrix} r(t+1)$$

i.e.,

$$x(t+1) = \begin{bmatrix} 0 & 0 \\ -2 & -1 \end{bmatrix} x(t) + \begin{bmatrix} 1 \\ 2 \end{bmatrix} r(t+1) = Fx(t) + Gr(t+1)$$
$$\tag{4.28a}$$
$$y(t) = \begin{bmatrix} 1 & 0 \end{bmatrix} x(t) = Hx(t). \tag{4.28b}$$

For (4.28) we have

$$\begin{bmatrix} H \\ HF \end{bmatrix} = \begin{bmatrix} 1 & 0 \\ 0 & 0 \end{bmatrix}$$

The system is thus unobservable. The unobservable subspace is spanned by $x^* = \begin{bmatrix} 0 \\ 1 \end{bmatrix}$, which according to (4.28) corresponds to a mode with the eigenvalue -1. This explains exactly our problem with the controller in (4.25).

4.5   Frequency Functions

Let the continuous system (4.9) have a transfer function matrix

$$G(s) = C(sI - A)^{-1}B + D$$

and let its sampled counterpart (4.10) have the transfer function matrix

$$G_T(z) = H(zI - F)^{-1}G + D$$

Let the system be driven by the piecewise constant cosine signal $u(\ell T) = \cos(\omega T \ell); \ell = 1, 2 \ldots$. The calculations following (3.47) show that the output at the sampling instants will be given by

$$y(\ell T) = |G_T(e^{i\omega T})| \cos(\omega T \ell + \arg G_T(e^{i\omega T}))$$

The *frequency function* for the sampled system is thus $G_T(e^{i\omega T})$. Note the factor T! It serves as normalization of the frequency: ω has the dimension "radians/time unit" while T has the dimension "time/sampling interval." The combination ωT – they always appear together – then has the dimension "radians/sampling interval" which can be understood as a normalized frequency.

Note that the piecewise constant cosine signal we have used is not a pure cosine. The lower the frequency ω the more similar will the original cosine signal be to its piecewise constant counterpart. This should mean that $G(i\omega)$ and $G_T(e^{i\omega T})$ will be approximately equal for small ωT. We have the following relationship:

Lemma 4.1

$$|G(i\omega) - G_T(e^{i\omega T})| \le \omega T \cdot \int_0^\infty |g(\tau)| d\tau \qquad (4.29)$$

where $g(\tau)$ is the impulse response of $G(s)$.

Proof: Let

$$G(i\omega) = \int_0^\infty g(\tau) e^{-i\tau\omega} d\tau$$

We have

$$G_T(e^{i\omega T}) = \sum_{k=1}^\infty \tilde{g}_k e^{-ik\omega T}$$

where $\tilde{g}_k$ is the impulse response of the sampled system. From

$$y(\ell T) = \int_0^\infty g(\tau) u(\ell T - \tau) d\tau = \sum_{k=1}^\infty \int_{(k-1)T}^{kT} g(\tau) d\tau \; u(\ell T - kT)$$

which holds if (4.3) is true, we realize that

$$\tilde{g}_k = \int_{(k-1)T}^{kT} g(\tau) d\tau \qquad (4.30)$$

Then

$$|G(i\omega) - G_T(e^{i\omega T})| = \left| \sum_{k=1}^{\infty} \int_{(k-1)T}^{kT} \left(g(\tau)e^{-i\tau\omega} - g(\tau)e^{-ik\omega T} \right) d\tau \right|$$

$$\leq \sum_{k=1}^{\infty} \max_{(k-1)T \leq \tau \leq kT} |e^{-i\omega\tau} - e^{-ik\omega T}| \cdot \int_{(k-1)T}^{kT} |g(\tau)| d\tau$$

$$\leq \omega T \cdot \sum_{k=1}^{\infty} \int_{(k-1)T}^{kT} |g(\tau)| d\tau = \omega T \int_{0}^{\infty} |g(\tau)| d\tau$$

applies, which gives (4.29).   □

The two frequency functions then agree at low frequencies. The rule of thumb is that up to one tenth of the sampling frequency ($\omega < 2\pi/(10T)$) the difference is small. For a system that is sampled fast compared to frequencies of interest, the sampled frequency function can be considered to give a good picture also of $G(i\omega)$.

Example 4.6:   The Sampled Frequency Function

Consider a continuous time system with the transfer function

$$G(s) = \frac{1}{s^2 + s + 1} \tag{4.31}$$

Its frequency function is shown in Figure (4.31). If we write (4.31) in state space form and sample according to Theorem 4.1 for some different sampling intervals T and then calculate the discrete time transfer function we get,

$$T = 0.1s: \quad G_T(q) = \frac{0.0048q^{-1} + 0.0047q^{-2}}{1 - 1.8953q^{-1} + 0.9048q^{-2}}$$

$$T = 0.5s: \quad G_T(q) = \frac{0.1044q^{-1} + 0.0883q^{-2}}{1 - 1.4138q^{-1} + 0.6065q^{-2}}$$

$$T = 1s: \quad G_T(q) = \frac{0.3403q^{-1} + 0.2417q^{-2}}{1 - 0.7859q^{-1} + 0.3679q^{-2}}$$

$$T = 2s: \quad G_T(q) = \frac{0.8494q^{-1} + 0.40409q^{-2}}{1 + 0.1181q^{-1} + 0.1353q^{-2}}$$

The frequency functions $G_T(e^{i\omega T})$ for these discrete time systems are also shown in Figure 4.4.   We see, as expected, that the shorter the sampling interval the better the correspondence between the sampled and the continuous time system.

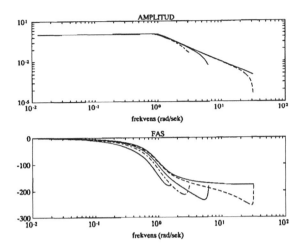

Figure 4.4: The continuous system's Bode diagram together with the sampled system. The curves are from top to bottom: the continuous system, $T = 0.1$s, $T = 0.5$s, $T = 1$s, $T = 2$s. Note that they end at the respective Nyquist frequencies.

4.6 Comments

The Main Points of the Chapter

A continuous time system can be approximated by a discrete time system in several ways. If the input of the continuous time system is constant over the sampling periods, the output at the sampling instances can be described exactly with a discrete time system by the sampling formula (4.10).

The relationship between a continuous time system and its sampled counterpart can be symbolized by

$$z = e^{sT}$$

where T is the sampling interval. This basic relationship is shown, for example, in the coupling between poles ($\lambda \rightarrow e^{\lambda T}$) and the argument of the frequency function ($i\omega \rightarrow e^{i\omega T}$).

Literature

The material in this chapter is covered in detail in standard textbooks on sampled data controllers, for example Åström & Wittenmark (1997),

Franklin et al. (1990). The relationship between the continuous and the sampled system's zeros is treated in Åström, Hagander & Sternby (1984). A good overview of numerical methods to calculate e^{AT} is given in Chapter 11 in Golub & Loan (1996). Sampling also has several practical aspects regarding alias, pre-filtering, etc. More thorough descriptions can be found in standard textbooks on digital signal processing, for example Oppenheim & Schafer (1989), as well as in the previously mentioned books on sampled data control.

Software

The transformations between continuous time and discrete time systems are obtained in MATLAB by the commands c2d, d2c.

In MATLAB 5 representation and properties of discrete time and continuous time systems are covered in parallel.

Chapter 5

Disturbance Models

5.1 Disturbances

We have mentioned earlier that the control objective is to make the system's output behave in a desirable way despite the influences of different disturbances. To achieve this, it is necessary to understand what disturbances are, and to describe their character in a suitable way. These types of questions will be discussed in this chapter.

We have described the relationship between control signal and output in transfer operator form as

$$y(t) = G(p)u(t) \tag{5.1}$$

In practice we are never so fortunate that the measured signal exactly follows (5.1). It will deviate from $G(p)u(t)$ in different ways. These deviations depend on *disturbances*, and the real output will be written

$$y(t) = G(p)u(t) + w(t) \tag{5.2}$$

where $w(t)$ is called the *disturbance signal*. From a system perspective, w is of course nothing but yet another input, to the system – it is the fact that it is not at our disposal to control the plant that leads to a distinction.

What is the origin of disturbances? In principle, the reason is that there are many factors in addition to our control input that influence the output. The sources of such influences, could be for example:

- *Load variations:* This is often the primary disturbance and the reason for control. It is a matter of signals beyond our control that influence the system, for example, wind gusts disturb aircraft control, the number of persons in a room directly influence the temperature, the load of an engine influences the speed, etc. In process industry a common disturbance of this character is variations in raw material.

- *Measurement errors:* The measurement of the output is influenced by noise, drifts in the measurement instruments, signal transmitting errors, inhomogeneity in the material being measured, etc.

- *Variations in plant and actuators:* Wear and changes in the plant as well as imperfections in the actuators, create disturbance sources.

- *Simplifications and model errors:* An important reason for the output not following (5.1) exactly, is of course that the model (5.1) contains errors and simplifications.

Realizing that the real output as in (5.2) can differ from that of the model is a way to make the controller less sensitive to model error. There is, however, reason to emphasize that such "disturbances" w are of different character than the external ones: they will depend on the input u, which means that it may be more difficult to predict their size and properties.

What does a typical disturbance look like? Let us look at some signals from actual control systems. Figure 5.1a shows temperature variations in a room, and Figure 5.1b the moisture content variations in manufactured paper in a paper machine.

How should the disturbance be described? In order to determine a suitable controller for the system (5.2) we need to describe the properties of the disturbance $w(t)$ in a suitable way. The classical approach is to use standardized disturbance types such as *impulses, steps, ramps* and *cosine signals*. In this chapter we will use a more general approach and discuss three aspects:

- To describe the "size" of the signal. (Section 5.2)

- To describe the frequency content of the signal. (Section 5.3)

- To describe the signal as an output from a linear system driven by "white noise" or an "impulse". (Section 5.4)

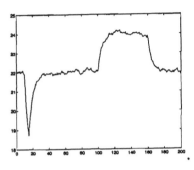
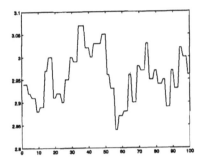

(a) Room temperature. Between time $t = 10$ and $t = 15$ the window has been opened. Between time $t = 100$ and $t = 160$ a number of persons are in the room.

(b) Moisture content in manufactured paper.

Figure 5.1: Output signals in some control systems. No control applied.

After that, in Section 5.6 we discuss how the disturbances can be incorporated in the system descriptions. How relevant information of the system can be estimated and reconstructed from the disturbed measured signals will be examined in Section 5.7.

5.2 Signal Size and Scaling

In Section 1.5 we defined the 2-norm for a signal $z(\cdot)$ as

$$\|z\|_2^2 = \int_{-\infty}^{\infty} |z(t)|^2 dt \tag{5.3}$$

Here $z(t)$ is a column vector and $|z(t)|^2 = z^T(t)z(t)$.

Scaling

The size measure (5.3) has weaknesses and it can be misleading if the different components are of completely different size. Assume that we have, for example,

$$|z(t)|^2 = z_1^2(t) + z_2^2(t)$$

with $z_1 = 100$ and $z_2 = 1$, thus $|z| = 100.005$. If z_2 is changed to 2, that is by 100%, $|z|$ is changed to 100.01 which is only 0.005%. The problem can be avoided by scaling the variables. If z^f represents the variables measured in the original physical units, scaled variables can be formed using

$$z_i = z_i^f / d_i, \quad i = 1, \ldots, n \tag{5.4}$$

where d_i is selected as the maximum of the absolute value of the component z_i^f. This guarantees that $|z_i| \leq 1$ for all i. We can also write the scaling as

$$z = D^{-1} z^f \tag{5.5}$$

where D is a diagonal matrix with d_i as ith diagonal element. To make the scaling effective, it is common to first deduct a mean value from z^f before the scaling (5.4).

The problem with weighting the components in z can also be handled using *weighted quadratic norms*

$$|z(t)|_P^2 = z^T(t) P z(t) \tag{5.6}$$

where P is a symmetric positive semi-definite matrix. $\|z\|_P$ then means the 2-norm according to (5.3) where $|z(t)|_P^2$ is used instead. The relationship between scalar variables according to (5.5) and weighted norms can obviously be written

$$\|z\| = \|z^f\|_{D^{-2}} \tag{5.7}$$

A Matrix Measure of Signal Size

For a vector signal $z(t)$ with m rows we can define the $m \times m$ matrix

$$R_z = \int_{-\infty}^{\infty} z(t) z^T(t) dt \tag{5.8}$$

as a variant of (5.3). We use the notation R_z to associate to the concept the "covariance matrix for $z(t)$", which is a good intuitive picture of (5.8). We will return to this interpretation in Section 5.3.

If R_z is known for a signal z, we can calculate the sizes of a number of related signals. If we, for example, form a scalar signal using a linear combination of the components of z, $s(t) = Lz(t)$, we obtain

$$\|s\|_2^2 = \int Lz(t) z^T(t) L^T dt = L \int z(t) z^T(t) dt L^T = L R_z L^T \tag{5.9}$$

Example 5.1:   Signal Size

Let the signal $u(t) = \begin{bmatrix} u_1(t) & u_2(t) \end{bmatrix}^T$ be of size

$$R_u = \begin{bmatrix} 15 & 3 \\ 3 & 1 \end{bmatrix}$$

The signal $z(t) = 3u_1(t) - u_2(t)$ than has the size

$$\begin{bmatrix} 3 & -1 \end{bmatrix} \begin{bmatrix} 15 & 3 \\ 3 & 1 \end{bmatrix} \begin{bmatrix} 3 \\ -1 \end{bmatrix} = 118$$

This means that the size of the kth component of z is given by the kth diagonal element in R_z. $\|z\|_2^2$ will then correspond to the *sum of the diagonal elements* in the matrix R_z. The sum of the diagonal elements of a square matrix A is called the *trace* of A and is denoted $\mathrm{tr}(A)$. For the trace we have the useful identity

$$\mathrm{tr}(AB) = \mathrm{tr}(BA) \tag{5.10}$$

The conclusion is that the usual norm (5.3) is related to the matrix measure R_z as

$$\|z\|_2^2 = \mathrm{tr}(R_z) \tag{5.11}$$

The different components can of course also be weighted as in (5.6) in the scalar measure. It is then easy to see that we have the general relation

$$\|z\|_P^2 = \mathrm{tr}(PR_z) \tag{5.12}$$

Power Measures

A drawback for all measures of signal size we have used so far, is that many signals do not have finite energy. This means that the integrals in (5.3) and corresponding expressions are not convergent. In such cases it is instead natural to consider the "mean power" in the signal, i.e., the limit (for simplicity we write it "one-sided" from 0 to ∞)

$$\|z\|_e^2 = \lim_{N \to \infty} \frac{1}{N} \int_0^N |z(t)|^2 dt \tag{5.13}$$

and correspondingly for the other measures. A drawback of (5.13) is that it has certain formal, mathematical limitations. The limit in question does not have to exist. Also $\| \cdot \|_e$ can be zero without the signal being zero. The measure is therefore not a formal norm. These situations are, however, exceptions. Normally $\| \cdot \|_e$ exists for all signals of interest. In the future, everything we say about the 2-norm also applies to $\| \cdot \|_e$.

5.3   Spectral Description of Disturbances

Spectral Properties

To describe the properties of a disturbance it is useful to consider its *spectrum* or *spectral density*. (The correct mathematical term is spectral density. In this book we will, however, follow engineering praxis and use the term "spectrum".)

The spectrum of a signal, in principle, indicates how its energy (or power) is distributed over different frequencies. The spectral concept is used and defined in several different contexts, and is, basically, the square of the absolute value of the Fourier transform of the signal (perhaps normed, averaged, etc.). Generally, this means that in the scalar case we assume the signal u to be formed by cosine signals according to

$$u(t) = \int_{-\infty}^{\infty} A(\omega) \cos(\omega t + \phi(\omega)) d\omega \tag{5.14}$$

The (energy)spectrum of u, denoted by $\Phi_u(\omega)$, is then (the expected value of) $A^2(\omega)$.

In this book we are not going to study definitions of spectra in depth, but refer to text books on signal theory. We only need three properties:

1. An m-dimensional signal $u(t)$ can be associated with a Hermitian (see Definition 3.9) $m \times m$-matrix function $\Phi_u(\omega)$, called its *spectrum*.

2. Let G be a linear, stable system. If u has spectrum $\Phi_u(\omega)$ and

$$y(t) = G(p)u(t) \tag{5.15a}$$

then y has spectrum

$$\Phi_y(\omega) = G(i\omega)\Phi_u(\omega)G^*(i\omega) \tag{5.15b}$$

3. The matrix (with dimension $m \times m$)

$$R_u = \frac{1}{2\pi} \int_{-\infty}^{\infty} \Phi_u(\omega) d\omega \qquad (5.15c)$$

is a measure of the size of the signal u.

In the sequel we shall only deal with signals that can be associated with a spectrum with the properties above. This is indeed a broad class of signals, as we shall exemplify shortly. The measure (5.15c) shows how the spectrum distributes the energy or power of the signal over the frequencies ω. That together with (5.15b) is all that is needed for a good theory for linear signal processing. The exact interpretation of R_u depends on the spectral definition.

Cross spectrum

Consider the spectrum of a signal partitioned into two parts:

$$z = \begin{bmatrix} y \\ u \end{bmatrix}, \qquad \Phi_z(\omega) = \begin{bmatrix} \Phi_{yy}(\omega) & \Phi_{yu}(\omega) \\ \Phi_{uy}(\omega) & \Phi_{uu}(\omega) \end{bmatrix} \qquad (5.16)$$

The diagonal blocks Φ_{yy} and Φ_{uu} are of course the spectra of y and u, respectively. The non-diagonal block Φ_{yu} is called the *cross spectrum* *between y and u*. Since Φ_z is Hermitian, we have

$$\Phi_{yu}(\omega) = \Phi_{uy}^*(\omega) \qquad (5.17)$$

Definition 5.1 *Two signals are said to be* **uncorrelated** *if their cross spectrum is identically zero.*

Basic Disturbance = White Noise

Disturbances with constant spectra are important as "basic disturbances". We therefore give them a name.

Definition 5.2 (White Noise) *A signal $e(t)$ is said to be* **white noise with intensity** R *if its spectrum is identically equal to a constant matrix R:*

$$\Phi_e(\omega) \equiv R \qquad (5.18)$$

White noise has its energy/power equally distributed over all frequencies. The exact meaning depends on which spectral definition we are working with. In any case, note that the measure (5.15c) shows that white noise in continuous time has infinite "size", so it may involve some idealization.

A noticeable feature in white noise, as we will see in Example 5.3, is that it cannot be predicted. This is also the best intuitive way to picture such a signal: *The past history of white noise does not contain any information about its future values.* Considering the state concept – to condense all relevant information about future system behavior to one state vector – it is obvious that white noise signals will play an important role for state representations of systems with disturbances.

Signals with Finite Energy

For a signal with finite energy, the spectrum is defined from its Fourier transform:

$$\Phi_u(\omega) = U(i\omega)U^*(i\omega), \quad U(i\omega) = \int_{-\infty}^{\infty} u(t)e^{-i\omega t}dt \tag{5.19}$$

For a scalar signal, the spectrum is thus the absolute square of the Fourier transform. It is well known, cf. (2.3), that (5.15b) holds for this definition, and (5.15c) gives by Parseval's identity

$$R_u = \frac{1}{2\pi} \int_{-\infty}^{\infty} \Phi_u(\omega)d\omega = \int_{-\infty}^{\infty} u(t)u^T(t)dt \tag{5.20}$$

This shows the connection to the size measures (5.8) and (1.21). For this family of signals, an *impulse*, i.e., the Dirac delta function $\delta(t)$ is a white noise signal, since it has a constant Fourier transform. (The idealization mentioned above shows up in the infinite amplitude/energy of the delta function.)

Signals with Well Defined Power

For signals with infinite energy, consider those where the following limit is well defined for all τ:

$$R_u(\tau) = \lim_{N\to\infty} \frac{1}{N} \int_0^N u(t)u^T(t-\tau)dt \tag{5.21}$$

The spectrum is then defined as

$$\Phi_u(\omega) = \int_{-\infty}^{\infty} R_u(\tau) e^{-i\omega\tau} d\tau \tag{5.22}$$

Again, it can be shown that (5.15b) holds, and Parseval's relation gives

$$R_u = \frac{1}{2\pi} \int_{-\infty}^{\infty} \Phi_u(\omega) d\omega = R_u(0) = \lim_{N\to\infty} \frac{1}{N} \int_0^N u(t) u^T(t) dt \tag{5.23}$$

showing the links to the mean power measure (5.13).

Stationary Stochastic Processes

In this treatment we do not rely upon any theory for stochastic processes. A reader who is familiar with this topic will of course recognize the connections. For a stationary stochastic process u the covariance function is defined as

$$R_u(\tau) = Eu(t)u^T(t-\tau) \tag{5.24}$$

where E denotes mathematical expectation. The spectrum is then defined by (5.22). The size measure (5.15c) is the covariance matrix of u:

$$R_u = \frac{1}{2\pi} \int_{-\infty}^{\infty} \Phi_u(\omega) d\omega = Eu(t)u^T(t) \tag{5.25}$$

Now, under quite mild conditions, stationary stochastic processes are also *ergodic*, which means that

$$Eu(t)u^T(t-\tau) = \lim_{N\to\infty} \frac{1}{N} \int_0^N u(t) u^T(t-\tau) dt \tag{5.26}$$

i.e., that the time averages converge towards mathematical averages (with probability 1). This is also known as the law of large numbers. All this shows the close connections between stationary stochastic processes and signals with well defined power, discussed above.

For stationary processes, we realize that white noise has a covariance function that is the delta function (times the intensity matrix R). This means, in the continuous time case, that it has infinite variance, and therefore is an idealization. To deal with this in a formally correct fashion requires quite a sophisticated mathematical machinery. It also means that the signal values at different time points are uncorrelated, which is often used as a definition of white noise. It ties nicely in with the property mentioned above, that *white noise is unpredictable*.

Although we will not be using the mathematical machinery of stochastic processes, but just the relations (5.15), it still is convenient and suggestive to use its language. We shall therefore also refer to R_u in (5.15c) as *the covariance matrix* of u, and use the alternative notation

$$Eu(t)u^T(t) = R_u \tag{5.27}$$

Filtering Signals

Using (5.15b), some well known equations can easily be derived:

Example 5.2:   Linear Filter Plus Noise

Assume that we have the relationship

$$y(t) = G(p)u(t) + H(p)e(t) \tag{5.28}$$

where u and e are uncorrelated scalar signals with spectra $\Phi_u(\omega)$ and $\Phi_e(\omega) \equiv \lambda$, respectively. With

$$\zeta(t) = \begin{bmatrix} u \\ e \end{bmatrix}$$

we have

$$z(t) = \begin{bmatrix} y(t) \\ u(t) \end{bmatrix} = \begin{bmatrix} G(p) & H(p) \\ 1 & 0 \end{bmatrix} \zeta(t), \qquad \Phi_\zeta = \begin{bmatrix} \Phi_u & 0 \\ 0 & \lambda \end{bmatrix}$$

and (5.15b) gives

$$\begin{aligned}
\Phi_z(\omega) &= \begin{bmatrix} \Phi_y(\omega) & \Phi_{yu}(\omega) \\ \Phi_{uy}(\omega) & \Phi_u(\omega) \end{bmatrix} \\
&= \begin{bmatrix} G(i\omega) & H(i\omega) \\ 1 & 0 \end{bmatrix} \begin{bmatrix} \Phi_u(\omega) & 0 \\ 0 & \lambda \end{bmatrix} \begin{bmatrix} G(-i\omega) & 1 \\ H(-i\omega) & 0 \end{bmatrix}
\end{aligned}$$

leading to

$$\Phi_y(\omega) = |G(i\omega)|^2 \Phi_u(\omega) + \lambda |H(i\omega)|^2 \tag{5.29}$$
$$\Phi_{yu}(\omega) = G(i\omega)\Phi_u(\omega) \tag{5.30}$$

Energy in the Impulse Response = Power in the White Noise Response

Since an impulse also is a white noise signal according to our definition, there is a close connection between the impulse response of a system and its response to other white noise signals. This is implicit in the expressions above, and to make it explicit, we do the following calculations:

Consider a SISO linear system $G(p)$ with impulse response $g(\tau)$. Let $z(t) = G(p)e(t)$ where e is white noise with intensity 1. Then the spectrum of z is $|G(i\omega)|^2$ according to (5.15b). We find that the energy in the impulse response is

$$\|g\|^2 = \int_0^\infty g^2(\tau)d\tau = \frac{1}{2\pi}\int_{-\infty}^\infty |G(i\omega)|^2 d\omega$$

$$= R_z = Ez^2(t) = \|z\|^2 = \lim_{N\to\infty}\frac{1}{N}\int_0^N z^2(t)dt$$

(5.31)

The second step follows from Parseval's equality, the third and fourth by the definition (5.15c) and the last from (5.26). The energy in the impulse response, $\|g\|^2$ is therefore equal to the power $\|z\|^2$ in the white noise response.

This has an important consequence: *When we construct linear filters and controllers for, for example, minimization of energy/power in a disturbance signal, it does not matter if we use impulses or white noise as the disturbance source.*

For a multivariable causal system G driven by white noise with intensity R, the power of the output is

$$R_g = \frac{1}{2\pi}\int_{-\infty}^\infty G(i\omega)RG^T(-i\omega)d\omega = \int_0^\infty g(t)Rg^T(t)dt \quad (5.32)$$

where $g(t)$ is the impulse response of the system (a $p \times m$-matrix) and the second step follows from Parseval's equality. How can the right hand side be interpreted "physically"? The response $y(t)$ to an impulse is a p-dimensional column vector. For an impulse in input k the response is $y(t) = g_k(t)$, where g_k is the kth column of g. The size of this signal is $\int g_k g_k^T dt$. The sum of the sizes of the impulse responses from u's different components is then

$$\sum_{k=1}^m \left[\int_0^\infty g_k(t)g_k^T(t)dt\right] = \int_0^\infty g(t)g^T(t)dt$$

With $R = I$ we can see (5.32) as the size of the output when it occurs one impulse at a time in each of the input signals, also assuming that the response to the impulse in one of the input signals has died out before the next occurs. If $R \neq I$ we write $u = R^{1/2}\tilde{u}$ and let the impulses occur in $\tilde{u}$'s components.

5.4 Description of Disturbances in the Time Domain

Disturbances as Output Signals of Systems Driven by White Noise

Consider a linear, stable system $G(p)$ with white noise input $e(t)$ with intensity R:

$$v(t) = G(p)e(t) \tag{5.33}$$

The spectrum of the output v is according to (5.15b)

$$\Phi_v(\omega) = G(i\omega)RG^*(i\omega) \tag{5.34}$$

In the scalar case we have

$$\Phi_v(\omega) = R \cdot |G(i\omega)|^2$$

A signal described by a spectrum $\Phi_v(\omega)$ in (5.34) thus can be seen as the output from the filter (5.33), subject to white noise. To *describe disturbances as outputs from linear stable systems, driven by white noise* is the basic disturbance model in the time domain.

The system G can be represented in several different forms, as described in Chapter 2. In this book we mostly use the state space form, and we will return to this in Section 5.6.

Spectral Factorization

Models describing the properties of disturbances can most often be constructed directly as (5.33), or as part of state space modeling. For example, if the disturbances are known to have their main energy below 5 Hz, G is selected as a low pass filter with that cut-off frequency, etc.

It might be interesting to start from a given spectrum Φ_v and try to find a stable filter G and a matrix R such that (5.34) holds. This problem is called *spectral factorization*. Here we will only give the result for the scalar case:

Theorem 5.1 (Spectral Factorization) *Assume that the real valued, scalar function $\Phi_v(\omega) \geq 0$ is a rational function of ω^2, finite for all ω. There is then a rational function $G(s)$, with real coefficients, and with all poles strictly in the left half plane, and all zeros in the left half plane or on the imaginary axis, such that*

$$\Phi_v(\omega) = |G(i\omega)|^2 = G(i\omega)G(-i\omega)$$

Proof: We can write

$$\Phi_v(\omega) = K\frac{\beta(\omega^2)}{\alpha(\omega^2)}$$

Factorize the polynomials α and β and cancel possible common factors. Seen as polynomials in ω^2 their zeros can be single and real, or double and real, or single and complex conjugated. Higher multiplicity is treated using repeated zeros. The factors of α and β are thus of the following kind:

$$(\omega^2 + a_k), \quad (\omega^2 - b_k)^2, \quad (\omega^2 + c_k)(\omega^2 + \overline{c_k})$$

where a_k and b_k are real and c_k is complex with $\overline{c_k}$ as its complex conjugate. Since $\Phi_v(\omega)$ is positive $\forall \omega$, each of the factors above must be positive, or sign changes would occur for some ω. This means that $K > 0$ and $a_k \geq 0$. b_k can be positive, but this cannot occur in the denominator polynomial, since Φ_v is finite for all ω. The four square roots of c_k and $\overline{c_k}$ are placed symmetrically on the four quadrants and none of them are on the imaginary axis. Let d_k and $\overline{d_k}$ be the two roots in the right half plane. We can now factorize each one of the three cases above as

$$(\omega^2 + a_k) = (i\omega + \sqrt{a_k})(-i\omega + \sqrt{a_k})$$
$$(\omega^2 - b_k)^2 = ((i\omega)^2 + b_k)((-i\omega)^2 + b_k)$$
$$(\omega^2 + c_k)(\omega^2 + \overline{c_k}) = (i\omega + d_k)(i\omega + \overline{d_k})(-i\omega + d_k)(-i\omega + \overline{d_k})$$
$$= ((i\omega)^2 + Re(d_k)(i\omega) + |d_k|^2)((-i\omega)^2 + Re(d_k)(-i\omega) + |d_k|^2)$$

Now define $B(s) = \sqrt{K}(s + \sqrt{a_k}) \ldots (s^2 + b_k) \ldots (s^2 + 2Re(d_k)s + |d_k|^2)$ and correspondingly for $A(s)$. These polynomials have definitely all zeros in the left half plane and the A polynomial has no zeros on the imaginary axis. The construction shows that

$$K\beta(\omega^2) = B(i\omega)B(-i\omega)$$
$$\alpha(\omega^2) = A(i\omega)A(-i\omega)$$

so, with $G(s) = B(s)/A(s)$ the proof is completed. $\qquad\square$

5.5   Estimation of one Signal from Another

Control design is about decision making to find a suitable control signal, based on observed measurements. This problem contains the sub-problem to estimate the controlled variable from the measured one. In Section 5.7 we will discuss how this is done for signals described in state space form. Here we will first solve a more general and more abstract problem.

We first recall two definitions: As system $G(p)$ is *stable and causal* if all poles are in the stability region. It can be written as

$$G(s) = \int_0^\infty g(\tau)e^{-\tau s}d\tau, \qquad \int_0^\infty |g(\tau)|d\tau < \infty \qquad (5.35)$$

A system $G(s)$ is *stable and anti-causal* if all its poles are outside the stability region. It can be written as

$$G(s) = \int_0^\infty g(-\tau)e^{\tau s}d\tau, \qquad \int_0^\infty |g(-\tau)|d\tau < \infty \qquad (5.36)$$

An arbitrary linear system lacking poles on the boundary of the stability region can always be written as the sum of an anticausal and a causal stable system

$$H(s) = L_c(s) + L_{ac}(s) \qquad (5.37)$$

To find this decomposition, H is expanded in partial fraction expansion and the terms with poles in the stability region are associated with the system L_c and the others with the system L_{ac}.

We have the following general result, which is the basis for all signal estimation theory.

Theorem 5.2 (The Wiener Filter) *Let $x(t)$ and $y(t)$ be two signals with cross spectrum $\Phi_{xy}(\omega)$. We observe $y(s), s \leq t$ and want to estimate $x(t)$ from these measurements with a linear, stable, causal, and strictly proper filter G:*

$$\hat{x}(t) = G(p)y(t) \qquad (5.38)$$

Let the spectrum of the signal $y(t)$ be factorized as

$$\Phi_y(\omega) = H(i\omega)H^*(i\omega) \qquad (5.39)$$

where the system H and its inverse H^{-1} are stable, causal, and proper. Consider the transfer function

$$L(s) = \Phi_{xy}(s)H^T(-s)^{-1} \tag{5.40}$$

where s in Φ_{xy} replaces $i\omega$. $\Phi_{xy}(s)$ is assumed to be strictly proper. Split L into causal and anticausal stable parts (both strictly proper):

$$L(s) = L_c(s) + L_{ac}(s) \tag{5.41}$$

The optimal choice of G, minimizing the error covariance matrix $R_{\tilde{x}} = E\tilde{x}(t)\tilde{x}^T(t)$ for the estimation error $\tilde{x}(t) = x(t) - \hat{x}(t)$ is then given by

$$G(s) = L_c(s)H^{-1}(s) \tag{5.42}$$

Remark. Minimization of the matrix $R_{\tilde{x}}$ means that every other choice of G gives an estimation error matrix larger than the minimal one (a matrix is said to be larger than another if their difference is positive semi-definite).

Proof: The estimation error $x(t) - \hat{x}(t)$ has according to (5.15b) the spectrum

$$\Phi_x - G\Phi_{xy}^* - \Phi_{xy}G^* + GHH^*G^*$$

which can be written as

$$(GH - \Phi_{xy}[H^*]^{-1})(GH - \Phi_{xy}[H^*]^{-1})^* + \Phi_x - \Phi_{yx}[H^*]^{-1}H^{-1}\Phi_{xy}^*$$

The two last terms do not depend on G and we rewrite the first one using (5.41)

$$(GH - L_c)(GH - L_c)^* - GHL_{ac}^* - L_{ac}H^*G^* + L_{ac}L_{ac}^*$$
$$+ L_{ac}L_c^* + L_cL_{ac}^*$$

Consider the second expression. We have

$$\int_{-R}^{R} G(i\omega)H(i\omega)L_{ac}^*(i\omega)d\omega = \frac{1}{i}\int_{\Gamma} G(s)H(s)L_{ac}^T(-s)ds$$
$$- \int_{\pi}^{-\pi} G(Re^{i\zeta})H(Re^{i\zeta})L_{ac}(Re^{-i\zeta})d\zeta \tag{5.43}$$

The complex integral in the right hand side is over a path Γ that includes the imaginary axis and a large semi circle (with the radius R) enclosing the right half plane. The function $G(s)H(s)L_{ac}^T(-s)$ has all poles inside the stability region and hence is analytic in the right half plane. The complex integral is therefore zero. The expression with the contribution along the large circle goes to zero when $R \to \infty$ since both G and L_{ac} are strictly proper, so that

$$\lim_{s \to \infty} |sG(s)H(s)L_{ac}^T(-s)| = 0$$

(Assumptions of properness are used to secure this limit; the assumptions in the theorem could thus be somewhat softened.) All this means that the integral in the left hand side of (5.43) is equal to zero for $R = \infty$. The contributions to the covariance matrix from the terms GHL_{ac}^* and $L_{ac}H^*G^*$ are therefore zero, and we can write

$$R_{\tilde{z}} = \frac{1}{2\pi} \int (GH - L_c)(GH - L_c)^* d\omega \quad + \quad (G - \text{independent terms})$$

which completes the proof. $\square$

Example 5.3:   White Noise Cannot be Predicted

Let $y(t)$ be a signal with constant spectrum R. Let $z(t) = y(t + \delta); \delta > 0$ and we want to estimate $z(t)$ from old values of y, i.e., predict y. We have the relationship

$$z(t) = e^{p\delta} y(t)$$

The cross spectrum between z and y is according to (5.30) $\Phi_{zy}(\omega) = e^{i\omega\delta} R$ and we have in Theorem 5.2

$$L(s) = e^{s\delta} R^{1/2}$$

The causal, stable part of L is thereby 0 and the best estimate (prediction) is $\hat{z}(t) = 0$ irrespective of the measurements y. These therefore have no information value for the prediction.

5.6   Measurement and System Disturbances

System Disturbances

In Section 1.4 we pointed to a number of external signals of different character which influence the control object:

- The control signal u

- Measurable system disturbances w_m

- Non-measurable system disturbances w_n

- Measurement disturbances n

We will use the notation w for all system disturbances, so that

$$w = \begin{bmatrix} w_m \\ w_n \end{bmatrix} \tag{5.44}$$

An important problem for control design, as we pointed out in Section 1.4, is to distinguish between system disturbances and measurement disturbances. Therefore, we have to discuss how the disturbances are best represented mathematically. We will use the state space form for this.

The controlled variable z is a linear combination of the state variables. These in turn are influenced by control signals and system disturbances:

$$\begin{aligned} \dot{x}_a &= A_a x_a + B_a u + N_a w \\ z &= M_a x_a + D_a u \end{aligned} \tag{5.45}$$

The character of the influence of the system disturbance w is often known on physical grounds and the model (5.45) can be built using this knowledge. For an aircraft, for example, the most important system disturbances are the forces from wind gusts, which influence the states (the position and speed of the aircraft) in a known way. If the physical influences are not known in detail, system disturbances can be added to the output, i.e.,

$$z(t) = G(p)u(t) + w(t)$$

which is then realized in state space form.

Example 5.4:   Aircraft Disturbances

Much work has been devoted to modeling the influences of the disturbances on the Gripen model in Example 2.3. The disturbance signals

w_1 = wind speed across the aircraft

w_2 = wind angular speed along the roll axis

w_3 = wind angular speed along the turning axis

w_4 = wind acceleration across the aircraft

are typically used. This gives the model $\dot{x} = Ax + Bu + N_a w$ with

$$N_a = \begin{bmatrix} 0.292 & 0.001 & 0.97 & 0.0032 \\ 0.152 & -2.54 & 0.552 & 0.000043 \\ -0.0364 & -0.0688 & -0.456 & -0.00012 \\ 0 & 0 & 0 & 0 \\ 0 & 0 & 0 & 0 \\ 0 & 0 & 0 & 0 \\ 0 & 0 & 0 & 0 \end{bmatrix} \tag{5.46}$$

and A, B given as in Example 2.3.

Example 5.5:   Room Temperature Control I

Consider the heating of a room using radiators fed by hot water. The control signal u (°C) is the temperature of the incoming water, while the flow α (m³/hour) is constant. The room temperature z (°C) is the controlled variable. It is influenced by the outside temperature w (°C), which is a disturbance signal, as well as by the control signal. We assume that we do not have access to an outdoor thermometer. This disturbance signal is therefore described as non-measurable.

Let the temperature in the radiator be x_1. Heat balance in the room then gives

$$\dot{z} = K_1(x_1 - z) - K_2(z - w)$$

where K_i are positive constants, dependent on the heat transfer coefficient between radiator and air, and through the outer wall, respectively. Heat balance in the radiator gives

$$\dot{x}_1 = K_3(u - x_1) - K_4(x_1 - z)$$

where the first term describes added heat from the water flowing in. With the state vector

$$x = \begin{bmatrix} x_1 \\ z \end{bmatrix}$$

we have

$$\dot{x} = \begin{bmatrix} -K_3 - K_4 & K_4 \\ K_1 & -K_1 - K_2 \end{bmatrix} x + \begin{bmatrix} K_3 \\ 0 \end{bmatrix} u + \begin{bmatrix} 0 \\ K_2 \end{bmatrix} w \tag{5.47a}$$

$$z = \begin{bmatrix} 0 & 1 \end{bmatrix} x \tag{5.47b}$$

which is a description of the type (5.45).

Is (5.45) given "in state space form"? A distinctive feature is that the state vector at time t, $x_a(t)$, should include "all that is worth knowing" about the future behavior of the system, if we have access to future values of u. If we solve (5.45) we have (see (3.2))

$$z(t+T) = M_a e^{A_a T} x_a(t)$$
$$+ \left(\int_t^{t+T} M_a e^{A_a(t+T-\tau)} B_a u(\tau) d\tau + D_a u(t+T) \right)$$
$$+ \int_t^{t+T} M_a e^{A_a(t+T-\tau)} N_a w(\tau) d\tau$$

The first term is known if $x_a(t)$ is known, and the second if future values of u are known. If the future values of w cannot be predicted from old information there is no relevant information about the third term in the expression above, available at time t. In that case $x_a(t)$ in (5.45) really is a state. If, on the other hand, w can be predicted from available observations up to time t, there is information of relevance for the future behavior of the system in addition to $x_a(t)$. To avoid this situation we could thus let (5.45) be driven by *unpredictable* signals w. According to the previous section this means that w should be *white noise*

If this is not the case, we represent w as in (5.33)

$$w(t) = H(p)v_w(t) \tag{5.48}$$

where v_w is white noise. In state space form we have

$$\dot{x}_w = A_w x_w + B_w v_w \tag{5.49}$$
$$w = C_w x_w + D_w v_w$$

We can now combine (5.45) and (5.49) into

$$\dot{x}_b = A_b x_b + B_b u + N_b v_w \tag{5.50a}$$
$$z = M_b x_b + D_b u \tag{5.50b}$$

with

$$x_b = \begin{bmatrix} x_a \\ x_w \end{bmatrix}, \qquad A_b = \begin{bmatrix} A_a & N_a C_w \\ 0 & A_w \end{bmatrix}, \qquad B_b = \begin{bmatrix} B_a \\ 0 \end{bmatrix},$$

$$N_b = \begin{bmatrix} N_a D_w \\ B_w \end{bmatrix}, \qquad M_b = \begin{bmatrix} M_a & 0 \end{bmatrix}, \qquad D_b = D_a$$

Example 5.6: Room Temperature Control II

In Example 5.5 it is not particularly natural to describe the disturbance w ($=$ the outdoor temperature) as white noise. Instead, we know that it typically has a rather periodic behaviour, with a period of 24 hours. The spectrum of the signal therefore has a peak at $\omega^* = \pi/12$ rad/hour. We can describe it as the output from a filter with its poles close to $\pm i\omega^*$:

$$w = \frac{1}{p^2 + 0.01p + 0.068} v_w$$

where v_w is white noise. This description can be realized in state space in, for example, the controller canonical from (2.17), which together with (5.47) gives the description

$$\dot{x}_b = \begin{bmatrix} -K_3 - K_4 & K_4 & 0 & 0 \\ K_1 & -K_1 - K_2 & 0 & K_2 \\ 0 & 0 & -0.01 & -0.068 \\ 0 & 0 & 1 & 0 \end{bmatrix} x_b + \begin{bmatrix} K_3 \\ 0 \\ 0 \\ 0 \end{bmatrix} u + \begin{bmatrix} 0 \\ 0 \\ 1 \\ 0 \end{bmatrix} v_w$$

$$z = \begin{bmatrix} 0 & 1 & 0 & 0 \end{bmatrix} x_b \tag{5.51}$$

The state variable x_4 now is to the outdoor temperature.

Measurement Noise

If the measurement noise is additive to the controlled variable, the measured output is

$$y(t) = z(t) + n(t) \tag{5.52}$$

Are (5.50) and (5.52) together in state space form? Is there anything reasonable to say about future values of $y(t + \tau)$ that is not based on $x(t)$ and future values of u? Again, this depends on whether $n(t)$ can be predicted from information other than $x(t)$. If n also is white noise, this is not the case, and we have a state space representation. If n does not have constant spectrum we write

$$n(t) = H_n(p)v_2(t) \tag{5.53}$$

for a white noise signal v_2, and represent H_n as

$$\begin{aligned} \dot{x}_n &= A_n x_n + B_n v_2 \\ n &= C_n x_n + D_n v_2 \end{aligned} \tag{5.54}$$

In the sequel we will assume that the white noise source v_2 has been normalized such that we can take $D_n = I$. We can now combine (5.54) with (5.50):

$$
\begin{aligned}
\dot{x} &= Ax + Bu + Nv_1 \\
z &= Mx + D_z u \\
y &= Cx + D_y u + v_2
\end{aligned}
\qquad (5.55)
$$

where

$$
x = \begin{bmatrix} x_b \\ x_n \end{bmatrix}, \qquad v_1 = \begin{bmatrix} v_w \\ v_2 \end{bmatrix}, \qquad A = \begin{bmatrix} A_b & 0 \\ 0 & A_n \end{bmatrix}, B = \begin{bmatrix} B_b \\ 0 \end{bmatrix}
$$

$$
N = \begin{bmatrix} N_b & 0 \\ 0 & B_n \end{bmatrix}, \quad M = \begin{bmatrix} M_b & 0 \end{bmatrix}, \quad C = \begin{bmatrix} M_b & C_n \end{bmatrix}
$$

With this we have concluded the long discussion of how the controlled and disturbed system is represented in state space form. The expression (5.55) with v_1 and v_2 as white noises, will be our standard representation of systems. Note especially that the controlled variable z and the measured output y often are different combinations of the state variables. This allows also relationships between y and z that are more complicated than (5.52).

Remark. That signal values prior to t are of no importance for the future $(s > t)$ behavior of $x(s)$ if the vector $x(t)$ is known, also means that x is a *Markov process*. The interest in Markov processes within mathematical statistics has the same reasons as our interest for the state vector: x is its own "memory".

Example 5.7: Room Temperature Control III

We return to Example 5.6. Assume that we measure the room temperature with two thermometers. One is a robust sensor that is not influenced very much by random disturbances, but has an unknown calibration error. We describe this measurement signal as

$$
y_1 = z + n_1
$$

where the calibration error n_1 is modeled as integrated white noise (or an integrated impulse) v_n:

$$
n_1 = \frac{1}{p} v_n
\qquad (5.56)
$$

The second thermometer has no calibration error, but is mounted such that it is influenced by wind gusts from the window. These measurements are described as $y_2 = z + v_2$, where v_2 is white noise.

Begin with the state representation (5.51). Add a fifth state variable $x_5 = n_1$ which satisfies $\dot{n}_1 = v_n$ according to (5.56). With this addition to (5.51) we have

$$\dot{x} = Ax + \begin{bmatrix} K_3 \\ 0 \\ 0 \\ 0 \\ 0 \end{bmatrix} u + \begin{bmatrix} 0 & 0 \\ 0 & 0 \\ 1 & 0 \\ 0 & 0 \\ 0 & 1 \end{bmatrix} \begin{bmatrix} v_w \\ v_n \end{bmatrix} \tag{5.57a}$$

$$z = \begin{bmatrix} 0 & 1 & 0 & 0 & 0 \end{bmatrix} x \tag{5.57b}$$

$$y = \begin{bmatrix} 0 & 1 & 0 & 0 & 1 \\ 0 & 1 & 0 & 0 & 0 \end{bmatrix} x + \begin{bmatrix} 0 \\ 1 \end{bmatrix} v_2 \tag{5.57c}$$

$$A = \begin{bmatrix} -K_3 - K_4 & K_4 & 0 & 0 & 0 \\ K_1 & -K_1 - K_2 & 0 & K_2 & 0 \\ 0 & 0 & -0.01 & -0.068 & 0 \\ 0 & 0 & 1 & 0 & 0 \\ 0 & 0 & 0 & 0 & 0 \end{bmatrix} \tag{5.57d}$$

If the outdoor temperature also is measured and denoted by y_3 the consequence is that (5.57c) is changed to

$$y = \begin{bmatrix} 0 & 1 & 0 & 0 & 1 \\ 0 & 1 & 0 & 0 & 0 \\ 0 & 0 & 0 & 1 & 0 \end{bmatrix} x + \begin{bmatrix} 0 \\ 1 \\ 0 \end{bmatrix} v_2 \tag{5.58}$$

Spectrum and Size for States

Consider the system

$$\dot{x} = Ax + Nv \tag{5.59}$$

where v is white noise with intensity R. We can then write

$$x = (pI - A)^{-1} Nv$$

which, if A is stable, via (5.15b) shows that the spectrum of x is

$$\Phi_x(\omega) = (i\omega I - A)^{-1} N R N^T (-i\omega I - A)^{-T} \tag{5.60}$$

where we have introduced the convention $A^{-T} = [A^{-1}]^T$. The size measure (the covariance matrix), for x is then

$$\Pi_x = R_x = \frac{1}{2\pi} \int_{-\infty}^{\infty} \Phi_x(\omega) d\omega \tag{5.61}$$

A simpler expression can be derived for Π_x:

Theorem 5.3 (State Size) *Let Π_x be defined by (5.59)-(5.61), where all eigenvalues of A are strictly inside the stability region. Then Π_x is the solution to the matrix equation*

$$A\Pi_x + \Pi_x A^T + NRN^T = 0 \tag{5.62}$$

Proof: We begin with (5.60)-(5.61) and will rewrite the integral using Parseval's equality. The Fourier transform of the function

$$f(t) = \begin{cases} e^{At} & t \geq 0 \\ 0 & t < 0 \end{cases}$$

is $(i\omega I - A)^{-1}$ if A is a stable matrix. We then have

$$\Pi_x = \frac{1}{2\pi} \int_{-\infty}^{\infty} (i\omega I - A)^{-1} NRN^T (-i\omega I - A)^{-T} d\omega$$

$$= \int_0^{\infty} e^{At} NRN^T e^{A^T t} dt \tag{5.63}$$

Multiply the expression above by A and perform partial integration:

$$A\Pi_x = \int_0^{\infty} [Ae^{At}] NRN^T e^{A^T t} dt = [e^{At} NRN^T e^{A^T t}]_0^{\infty}$$

$$- \int_0^{\infty} e^{At} NRN^T [e^{A^T t} A^T] dt = -NRN^T - \Pi_x A^T$$

which gives the desired result. To show that (5.62) has a unique solution, we note that this is a linear equation system in the elements of the matrix Π_x. We have just shown that this has a solution (5.63) for each right hand side NRN^T. The square coefficient matrix in the equation system then must be of full rank, which also means that the solution is unique.   □

Example 5.8:   A Second Order System

Consider the system

$$\dot{x} = \begin{bmatrix} -1 & 2 \\ -1 & 0 \end{bmatrix} x + \begin{bmatrix} 1 \\ 0 \end{bmatrix} v$$

where v is white noise with variance 1, i.e., it has a spectrum identically equal to 1. What is the covariance matrix of x? According to Theorem 5.3 Π_x satisfies

$$\begin{bmatrix} -1 & 2 \\ -1 & 0 \end{bmatrix} \begin{bmatrix} \Pi_{11} & \Pi_{12} \\ \Pi_{12} & \Pi_{22} \end{bmatrix} + \begin{bmatrix} \Pi_{11} & \Pi_{12} \\ \Pi_{12} & \Pi_{22} \end{bmatrix} \begin{bmatrix} -1 & -1 \\ 2 & 0 \end{bmatrix} + \begin{bmatrix} 1 \\ 0 \end{bmatrix} \begin{bmatrix} 1 & 0 \end{bmatrix} = \begin{bmatrix} 0 & 0 \\ 0 & 0 \end{bmatrix}$$

Here we use the fact that the covariance matrix is symmetric. The (1,1), (1,2) and (2,2) elements in the matrix equation give

$$2(-\Pi_{11} + 2\Pi_{12}) + 1 = 0, \quad -\Pi_{12} + 2\Pi_{22} - \Pi_{11} = 0, \quad -2\Pi_{12} = 0$$

We now have three equations and three unknowns. The solution is

$$\Pi_x = \begin{bmatrix} 1/2 & 0 \\ 0 & 1/4 \end{bmatrix}$$

In MATLAB the problem is solved using

$$\text{lyap}([-1, 2; -1, 0], [1; 0] * [1, 0])$$

The command name "lyap" stems from the fact that the equation (5.62) is called a *Lyapunov equation*.

5.7   Observers and Kalman Filters

State Reconstruction

As we have seen, the state of a system description is important. Typically, the state vector is not observed, but only the measurable output y. The problem is to estimate or reconstruct the state from observations of y and u. This is normally done using an *observer*. (See, for example, Section 7.5 in Franklin et al. (1994).) The idea of an observer can be explained as follows: Start by simulating the system, using the known input:

$$\dot{\hat{x}}(t) = A\hat{x}(t) + Bu(t)$$

The quality of the estimate $\hat{x}(t)$ can be assessed by the quantity $y(t) - C\hat{x}(t) - Du(t)$. If $\hat{x}(t)$ is equal to $x(t)$ and there is no measurement noise, this quantity would be equal to zero. By feeding back this measure of the quality of $\hat{x}$ to the simulation, we obtain the observer

$$\dot{\hat{x}}(t) = A\hat{x}(t) + Bu(t) + K(y(t) - C\hat{x}(t) - Du(t)) \tag{5.64}$$

Here K is an $n \times p$ matrix. How should it be selected? Since the purpose is that $\hat{x}(t)$ should be close to the true state we form the estimate error

$$\tilde{x}(t) = x(t) - \hat{x}(t) \qquad (5.65)$$

From (5.64) and (5.55) we now have after simple manipulations

$$\dot{\tilde{x}}(t) = (A - KC)\tilde{x}(t) + Nv_1(t) - Kv_2(t) \qquad (5.66)$$

Here we see that K affects the estimation error in two ways. On the one hand, K determines the matrix $A - KC$, which indicates how fast effects from old errors die out. The eigenvalues of $A - KC$ should be as "far into the stability area" as possible. On the other hand, K influences the error by multiplying the measurement error v_2 by K. A large K therefore gives significant influence from the noise v_2.

In practice the choice of K is a trade-off between how fast we want the state reconstruction to happen (the eigenvalues of $A - KC$) and how sensitive we can be to the measurement disturbances.

From Theorem 3.4 we also know that if the system (A, C) is observable, we can select K such that the eigenvalues of $A - KC$ are placed in arbitrary (complex conjugated) positions.

Observer with Direct Term

In the observer (5.64) y does not directly influence $\hat{x}$, but the influence comes via $\dot{\hat{x}}$. Put differently, the transfer functions from y to the components $\hat{x}_i$ are strictly proper, and thus of low pass character. This is normally reasonable, since high frequency noise in y then will not affect the estimate.

In some cases it can be justifiable to have direct influence from y on $\hat{x}$, and we will denote such an estimate $\hat{x}(t|t)$. This is easily achieved by modifying the observer

$$\dot{\hat{x}}(t) = A\hat{x}(t) + Bu(t) + K(y(t) - C\hat{x}(t) - Du(t))$$
$$\hat{x}(t|t) = \hat{x}(t) + \tilde{K}(y(t) - C\hat{x}(t) - Du(t)) \qquad (5.67)$$

The estimation error is then

$$x(t) - \hat{x}(t|t) = \tilde{x}(t|t) = (I - \tilde{K}C)\tilde{x}(t) - \tilde{K}v_2(t) \qquad (5.68)$$

where $\tilde{x}(t)$ is determined by (5.66).

The choice of K and $\tilde{K}$ is a balance between disturbance sensitivity and reconstruction speed. If, for example v_2 is white, continuous time noise, and therefore has infinite variance, it is always optimal to select $\tilde{K} = 0$. Otherwise a special possibility exists: The matrix $C\tilde{K}$ has dimension $p \times p$ and we can therefore normally choose $\tilde{K}$ such that $C\tilde{K} = I$ (if C has full rank and $p \leq n$). Multiplying (5.68) with C we have

$$C\tilde{x}(t|t) = (I - C\tilde{K})C\tilde{x}(t) - C\tilde{K}v_2(t) = -v_2(t)$$

i.e., $C\tilde{x}(t) + v_2(t) = 0$. This means that we have

$$C\hat{x}(t|t) + Du(t) = y(t) \tag{5.69}$$

even if there are measurement disturbances. This also means that p equations – corresponding to the estimate of Cx – in the original observer (5.67) can be removed. This leads to a so called *reduced order observer*. Both the observer with direct term and the reduced order observer is of particular interest for discrete time systems, and we will return to these issues with examples in Section 5.8.

The Kalman Filter

The choice of K in the observer is a balance between sensitivity to the measurement disturbances ("small K") and adaptability to the influence of the system disturbances ($(A - KC)$ "very stable"). If we know the properties of the disturbances, this balance can of course be formalized. Using (5.66) and Theorem 5.3 we can calculate how large the estimation error is. If

$$\begin{bmatrix} v_1 \\ v_2 \end{bmatrix} \text{ is white noise with intensity } \begin{bmatrix} R_1 & R_{12} \\ R_{12}^T & R_2 \end{bmatrix} \tag{5.70}$$

$Nv_1 - Kv_2$ will also be white noise (use (5.15b)), but with intensity

$$R = NR_1N^T + KR_2K^T - NR_{12}K^T - KR_{12}^TN^T \tag{5.71}$$

According to (5.62) and (5.66) the size of the estimation error $\tilde{x}$ (its covariance matrix) is equal to the solution P of the equation

$$(A - KC)P + P(A - KC)^T + NR_1N^T + KR_2K^T$$
$$-NR_{12}K^T - KR_{12}^TN^T = 0 \tag{5.72}$$

if $A - KC$ is stable. This expression defines the estimation error P as a function of K, which should be selected so that P is minimized. We first give the result as a formal lemma about matrices:

Lemma 5.1 Let A, C, R_1, R_2, R_{12} be given matrices, such that R_2 is symmetric och positive definite and $\tilde{R}_1 = R_1 - R_{12}R_2^{-1}R_{12}^T$ is positive semidefinite. Assume that (A, C) is detectable (Definition 3.3). Assume also that $(A - R_{12}R_2^{-1}C, \tilde{R}_1)$ is stabilizable. Then the matrix equation

$$AP + PA^T + R_1 - (PC^T + R_{12})R_2^{-1}(PC^T + R_{12})^T = 0 \quad (5.73)$$

has a unique positive semidefinite, symmetric solution $P = P^\diamond$. Let

$$K^\diamond = (P^\diamond C^T + R_{12})R_2^{-1} \quad (5.74)$$

Let the symmetric matrix $P(K)$ be defined as a function of the matrix K by

$$(A - KC)P(K) + P(K)(A - KC)^T + R_1 + KR_2K^T \\ -R_{12}K^T - KR_{12}^T = 0 \quad (5.75)$$

$A - K^\diamond C$ is then stable and

$$P(K) \geq P(K^\diamond) \quad \text{for all } K \text{ such that } A - KC \text{ is stable} \quad (5.76)$$

Remark 1. By (5.76) is meant that $P(K) - P(K^\diamond)$ is a positive semidefinite matrix.

Remark 2. If we omit the condition that $(A - R_{12}R_2^{-1}C, \tilde{R}_1)$ be stabilizable, the equation (5.73) may have several positive semidefinite solutions. There is however only one such solution that makes the corresponding $K^\diamond$ in (5.74) stable. This is the solution that should be used in the Kalman filter below, in case stabilizability does not hold.

The proof is given in Appendix 5A.

The lemma can now be applied directly to find the optimal K value in the observer. Such an optimal observer is called a *Kalman filter*.

Theorem 5.4 (The Kalman Filter: Continuous Time)
Consider the system

$$\dot{x} = Ax + Bu + Nv_1$$
$$y = Cx + Du + v_2 \quad (5.77)$$

where v_1 and v_2 are white noises with intensity R_1 and R_2, respectively. The cross spectrum between v_1 and v_2 is constant and equal to R_{12}. For A, C, R_1, R_2 and R_{12} the same assumptions as in Lemma 5.1 apply. The observer that minimizes the estimation error $\tilde{x}(t) = x(t) - \hat{x}(t)$ is given by

$$\dot{\hat{x}}(t) = A\hat{x}(t) + Bu(t) + K(y(t) - C\hat{x}(t) - Du(t)) \qquad (5.78)$$

where K is given by

$$K = (PC^T + NR_{12})R_2^{-1} \qquad (5.79)$$

where P is the symmetric positive semidefinite solution of the matrix equation

$$\begin{aligned} AP + PA^T - (PC^T + NR_{12})R_2^{-1}(PC^T + NR_{12})^T \\ +NR_1N^T = 0 \end{aligned} \qquad (5.80)$$

The minimal estimation error covariance matrix is given by

$$E\tilde{x}(t)\tilde{x}^T(t) = P \qquad (5.81)$$

The observer (5.78) with this choice of K is called the *Kalman filter*. Equation (5.80), which is the basis for the calculation of K is a (stationary, or algebraic) *Riccati equation*. In Section 9.2 we will discuss how to best solve it numerically.

Use of Kalman Filters

The Kalman filter is used in many different contexts. It forms the foundation for *signal fusion*. Note that the different measurement signals in y can be completely different physical quantities and (5.78) shows how these signals can be merged to estimate a related, interesting, but not directly measurable signal x. There is extensive literature around all aspects of the Kalman filter – see Section 5.10 – but in this book we primarily view it as an optimal state observer.

The Innovations

The prediction error we feed back in the observer/the Kalman filter (5.64), i.e., $\nu(t) = y(t) - C\hat{x}(t) - Du(t)$ can be said to the "the new information" in the measurement $y(t)$, that is the information about

$y(t)$ that is unavailable in the old measurements. This signal $\nu(t)$ is therefore called the *innovations* of the measurement signal. It turns out that the innovation signal is white noise with intensity R_2. To show this, we first provide a formal result on the solution to the Riccati equation:

Lemma 5.2 *Assume that K is determined as*

$$K = PC^T R_2^{-1}$$
$$AP + PA^T + NR_1 N^T - PC^T R_2^{-1} PC = 0 \tag{5.82}$$

Then

$$(I + C(i\omega I - A)^{-1}K)R_2(I + C(-i\omega I - A)^{-1}K)^T$$
$$= R_2 + C(i\omega I - A)^{-1}NR_1 N^T(-i\omega I - A^T)^{-1}C^T \tag{5.83}$$

Proof: If the difference between the two members of (5.83) is formed, the indicated multiplications of expressions within parentheses are expanded, and the definition of P is used, all the terms cancel each other.

□

Using this result we can show the following theorem:

Theorem 5.5 *Assume that the conditions in Theorem (5.4) are satisfied. The innovation from the Kalman filter, $\nu(t) = y(t) - C\hat{x}(t) - Du(t)$ is then white noise with intensity R_2.*

Proof: We only show the theorem in the special case $R_{12} = 0$. The general case is completely analogous. From (5.84) we have

$$y(t) = (C(pI - A)^{-1}K + I)\nu(t) + (C(pI - A)^{-1}B + D)u(t)$$

and from (5.77)

$$y(t) = C(pI - A)^{-1}Nv_1(t) + v_2(t) + (C(pI - A)^{-1}B + D)u(t)$$

Let $y_\nu = (C(pI - A)^{-1}K + I)\nu = C(pI - A)^{-1}Nv_1 + v_2$. According to the second expression and (5.15b) the spectrum of y_ν is

$$\Phi_{y_\nu}(\omega) = C(i\omega I - A)^{-1}NR_1 N^T(-i\omega I - A^T)^{-1}C^T + R_2$$

Also $\nu(t) = (C(pI - A)^{-1}K + I)^{-1}y_\nu(t)$ and therefore, still according to (5.15b), the spectrum of ν is

$$\Phi_\nu(\omega) = (C(i\omega I - A)^{-1}K + I)^{-1}\Phi_{y_\nu}(\omega)(C(-i\omega I - A)^{-1}K + I)^{-T}$$
$$= R_2$$

where the last equation follows from (5.83).

□

State Representation in Innovations Form

In terms of $\nu(t)$, (5.64) can be written as

$$\dot{\hat{x}}(t) = A\hat{x}(t) + Bu(t) + K\nu(t) \tag{5.84a}$$
$$y(t) = C\hat{x}(t) + Du(t) + \nu(t) \tag{5.84b}$$

Such a model is said to be *in innovations form*. It is, of course, also its own observer and Kalman filter: If $\nu(t)$ is white noise, Theorem 5.4 can be applied to (5.84) to calculate its Kalman filter. This gives, as expected, a K-matrix equal to K in (5.84a), provided that $A - KC$ is a stable matrix.

The Kalman Filter is the Optimal Linear Filter

We first show a useful property of the Kalman filter estimate:

Lemma 5.3 *Let $x, \hat{x}$, and $\tilde{x}$ be defined as in Theorem 5.4. Then the corresponding covariance matrices obey*

$$Ex(t)x^T(t) = E\hat{x}\hat{x}^T + E\tilde{x}\tilde{x}^T \tag{5.85}$$

Proof: The covariance matrix of $\hat{x}$, $\Pi_{\hat{x}}$, can be computed from (5.84) and (5.62) (for $u=0$):

$$A\Pi_{\hat{x}} + \Pi_{\hat{x}}A^T + KR_2K^T = 0$$

Similarly, the covariance matrix for the state, Π_x is obtained from (5.77) as

$$A\Pi_x + \Pi_x A^T + NR_1N^T = 0$$

Forming the difference between these equations and defining $\Delta = \Pi_x - \Pi_{\hat{x}}$ gives

$$0 = A\Delta + \Delta A^T + NR_1N^T - KR_2K^T = A\Delta + \Delta A^T - AP - PA^T$$

Here, P is the solution to (5.80). In the last step we inserted the expression (5.79) for K and then used the equation (5.80). The resulting equation has the solution $\Delta = P$, and the lemma is proved. $\square$

If we adopt a stochastic viewpoint, the lemma implies that the two random vectors $\hat{x}(t)$ and $\tilde{x}(t)$ are *uncorrelated*. Equation (5.66) then shows that also $\hat{x}(t)$ and $\tilde{x}(t + \tau), \tau \geq 0$ must be uncorrelated (since future values of v_1 and v_2 are uncorrelated with $\hat{x}$). Hence $\tilde{x}(t)$ is uncorrelated will all past estimates $\hat{x}(s), s \leq t$. This, in turn implies that $\tilde{x}(t)$ is uncorrelated with past observations, so $\hat{x}(t)$ must be the orthogonal projection of $x(t)$ onto the space of past observations. From least squares estimation theory, this means that the Kalman filter is not only the optimal observer, but also the causal system among all linear filters, that minimizes $E\tilde{x}(t)\tilde{x}^T(t)$.

If the signals v_1 and v_2 are *normally (Gaussian) distributed* this also means that $\hat{x}$ and other quantities are normally distributed, since a linear combination of Gaussian variables is also Gaussian. Now, uncorrelated Gaussian variables are also always *independent*. The relationships above then mean that the corresponding quantities are independent, and not just uncorrelated. The Kalman filter error $\tilde{x}$ is consequently *independent* of past observations. This means that in the Gaussian case, the Kalman filter estimate $\hat{x}(t)$ equals the conditional expectation of $x(t)$ given past observations and is therefore the optimal estimate also if nonlinear filters are allowed.

5.8 Discrete Time Systems

The theory for discrete time disturbance signals is entirely parallel to that for continuous time signals. In this section we will point to differences in formulas and expressions.

Signal Size and Scaling

With obvious reinterpretations, such as

$$\|z\|_2^2 = \sum_{t=1}^{\infty} |z(t)|^2, \qquad \|z\|_e^2 = \lim_{N\to\infty} \frac{1}{N} \sum_{t=1}^{N} |z(t)|^2$$

everything in Section 5.2 is valid also for discrete time signals.

Spectral Description

The spectrum of a discrete time signal is also defined as the absolute square of its Fourier transform (possibly normed and/or averaged). The spectrum of a signal sampled using the sampling interval T is only defined up to the Nyquist frequency π/T (and can be defined using periodic continuation for other frequencies). If u has the spectrum

$\Phi_u(\omega)$ then the following expressions

$$y(t) = G(q)u(t) \tag{5.86a}$$

$$\Phi_y(\omega) = G(e^{i\omega T})\Phi_u(\omega)G^*(e^{i\omega T}) \tag{5.86b}$$

$$R_u = \frac{1}{2\pi T}\int_{-\pi/T}^{\pi/T} \Phi_u(\omega)d\omega \tag{5.86c}$$

hold instead of (5.15). *White noise* is as in the continuous time case a signal with constant spectrum. A stochastic white noise signal is a sequence of independent stochastic variables:

$$e(t), \ t = 1, 2, \ldots, \quad Ee(t)e^T(t) = R, \quad Ee(t)e^T(s) = 0 \quad \text{if } s \neq t$$

In contrast to the continuous time case, discrete time white noise thus does not involve technical difficulties of mathematical nature with infinite covariances.

Spectral factorization

Theorem 5.1 is valid also for continuous time signals, if the function Φ_v is rational in $\cos\omega$. The system $G(q)$ then has all its poles strictly within the unit circle and all zeros on or inside the unit circle.

Wiener Filters

Theorem 5.2 and its proof is valid also in the discrete time case if

- s is replaced by z

- $i\omega$ is replaced by $e^{i\omega T}$

- The decomposition (5.41) is done so that $L_{ac}(z)$ contains the factor z.

Example 5.9:   One Step Prediction of Discrete Time Signals

We have observed $y(k)$, $k = \ldots, t-1, t$ and from these measurements we want to predict $y(t+1)$. In terms of Theorem 5.2 we have $x(t) = y(t+1) = qy(t)$. If the spectrum of y is $\Phi_y(\omega) = |H(e^{i\omega})|^2$ then the cross spectrum is $\Phi_{xy}(\omega) = e^{i\omega}|H(e^{i\omega})|^2$. We assume that

$$H(z) = \sum_{k=0}^{\infty} h(k)z^{-k}$$

with $h(0) = 1$. With the notations of the theorem we have

$$L(z) = \Phi_{xy}(z)/H^*(z) = zH(z)H^*(z)/H^*(z) = zH(z)$$
$$= z(H(z) - 1) + z$$

The last step corresponds to (5.41), since $z(H(z) - 1)$ is a causal, stable filter and z (q) is an anticausal, stable filter. The optimal predictor is thus

$$G(q) = q(H(q) - 1)/H(q) = q - q/H(q)$$

If, for example, y is given by

$$y(t) + ay(t - 1) = e(t)$$

where e is white noise, we have $H(q) = q/(q + a)$, which gives the predictor

$$G(q) = q - (q + a) = -a \quad \text{i.e.,} \quad \hat{y}(t + 1) = -ay(t)$$

State Space Models

The treatment of measurement and system disturbances that led to the basic description (5.55) is completely identical in the discrete time case. We illustrate this with a simple example.

Example 5.10: Inventory Control

Consider the inventory control problem in Example 2.6. The order $u_1(t) = u(t)$ is the control signal of the system, while the sales $u_2(t) = w(t)$ is the disturbance signal. How should it be described? One possibility is to say that $w(t)$ has constant spectrum, i.e., it is white noise. This corresponds to the assumption that sales cannot be predicted. From (2.30) we can get a representation of the kind (5.50) directly:

$$x(t + 1) = \begin{bmatrix} 1 & 1 \\ 0 & 0 \end{bmatrix} x(t) + \begin{bmatrix} 0 \\ 1 \end{bmatrix} u(t) + \begin{bmatrix} -1 \\ 0 \end{bmatrix} v(t) \tag{5.87a}$$

$$y(t) = \begin{bmatrix} 1 & 0 \end{bmatrix} \tag{5.87b}$$

We here have $x(t) = \begin{bmatrix} y(t) & u(t - 1) \end{bmatrix}^T$ and we have taken $u_2(t) = w(t) = v(t)$.

If, on the other hand, we have noticed that the sales on consecutive days are positively correlated, we could describe them as

$$u_2(t) = w(t) = v(t) + 0.9v(t - 1) \tag{5.88}$$

Here $v(t)$ is a sequence of independent stochastic variables with zero mean and variance 1, i.e., v is white noise with intensity 1. This in turn means that the sales on consecutive days are positively correlated with the correlation coefficient $Ew(t)w(t+1)/Ew^2(t) = 0.9/1.81$. The inventory model (2.28) can then be described as

$$y(t) = y(t-1) + u(t-2) - v(t-1) - 0.9v(t-2). \tag{5.89}$$

What does (5.89) look like in state space form? We introduce the states

$$x_1(t) = y(t)$$
$$x_2(t) = u(t-1) - 0.9v(t-1)$$

We then have

$$x_1(t+1) = y(t+1) = y(t) + u(t-1) - v(t) - 0.9v(t-1)$$
$$= x_1(t) + x_2(t) - v(t)$$
$$x_2(t+1) = u(t) - 0.9v(t)$$

which in standard form is

$$x(t+1) = \begin{bmatrix} 1 & 1 \\ 0 & 0 \end{bmatrix} x(t) + \begin{bmatrix} 0 \\ 1 \end{bmatrix} u(t) + \begin{bmatrix} -1 \\ -0.9 \end{bmatrix} v(t) \tag{5.90a}$$

$$y(t) = \begin{bmatrix} 1 & 0 \end{bmatrix} x(t) \tag{5.90b}$$

For the state space model

$$x(t+1) = Ax(t) + Nv(t) \tag{5.91}$$

where v is white noise with intensity R, Theorem 5.3 applies but with the expression

$$\Pi_x = A\Pi_x A^T + NRN^T \tag{5.92}$$

for Π_x.

Observers

The whole discussion in (5.64)–(5.69) is valid without changes in discrete time, simply by replacing p by q ($\dot{x}(t)$ by $x(t+1)$). The observer is then

$$\hat{x}(t+1) = A\hat{x}(t) + Bu(t) + K(y(t) - C\hat{x}(t) - Du(t)) \tag{5.93}$$

Example 5.11:   Dead-beat Reconstruction

In the discrete time case an interesting choice of K is to place all the eigenvalues of $A - KC$ in the origin. Cayley-Hamilton's theorem (see the proof of Theorem 3.11) then tells us that $(A - KC)^n = 0$ (the null matrix) if n is the dimension of the matrix A. The consequence is that all effects of the initial estimate $\hat{x}$ have disappeared after n samples. If there are no disturbances influencing the system, the estimation error then will be zero. This is the fastest possible reconstruction in discrete time, but in some sense also the most disturbance sensitive. This selection of eigenvalues for the observer is called *dead-beat reconstruction*.

Consider the system

$$x(t+1) = \begin{bmatrix} 1 & 2 \\ 1 & 0 \end{bmatrix} x(t) + \begin{bmatrix} 1 \\ 0 \end{bmatrix} u(t)$$

$$y(t) = \begin{bmatrix} 3 & 4 \end{bmatrix} x(t)$$

We select an observer gain $K = \begin{bmatrix} k_1 & k_2 \end{bmatrix}^T$. The eigenvalues of $A - KC$ are given by the solution of the characteristic equation

$$\det(\lambda I - A + KC) = \det \begin{bmatrix} \lambda - 1 + 3k_1 & -2 + 4k_1 \\ -1 + 3k_2 & \lambda + 4k_2 \end{bmatrix}$$

$$= \lambda^2 + (4k_2 - 1 + 3k_2)\lambda + 4k_1 + 2k_2 - 2$$

If both eigenvalues must be 0, this means that

$$4k_2 + 3k_1 = 1$$
$$4k_1 + 2k_2 = 2$$

i.e., $k_1 = 0.6$ and $k_2 = -0.2$.

Measurement and Time Updates

We see that there is a time delay from y to $\hat{x}$ in the basic observer (5.93). We can emphasize this with the notation $\hat{x}(t) = \hat{x}(t|t-1)$. To base the state estimate on the latest available data, the observer with direct term (5.67) has to be used. This gives

$$\hat{x}(t+1|t) = A\hat{x}(t|t-1) + Bu(t)$$
$$+ K(y(t) - C\hat{x}(t|t-1) - Du(t)) \tag{5.94a}$$
$$\hat{x}(t|t) = \hat{x}(t|t-1) + \tilde{K}(y(t) - C\hat{x}(t|t-1) - Du(t)) \tag{5.94b}$$

The first equation can be written as

$$\hat{x}(t+1|t) = A\hat{x}(t|t) + Bu(t)$$
$$+ (K - A\tilde{K})(y(t) - C\hat{x}(t|t-1) - Du(t)) \qquad (5.94c)$$

The calculation of $\hat{x}$ can thus be divided into two steps, the *measurement update* (5.94b) from $\hat{x}(t|t-1)$ to $\hat{x}(t|t)$, and the *time update* (5.94c) from $\hat{x}(t|t)$ to $\hat{x}(t+1|t)$.

We see that a natural, but not necessary, choice is $K = A\tilde{K}$, which gives a straightforward time updating and the following expression for $\hat{x}(t|t)$:

$$\hat{x}(t+1|t+1) = A\hat{x}(t|t) + Bu(t)$$
$$+ \tilde{K}(y(t+1) - CA\hat{x}(t|t) - (CB + D)u(t)) \qquad (5.95)$$

In this case the error $\tilde{x}(t|t)$ in (5.68) satisfies

$$\tilde{x}(t|t) = (A - \tilde{K}CA)\tilde{x}(t-1|t-1) - \tilde{K}v_2(t) + Nv_1(t) \qquad (5.96)$$

Reduced Order Observer

If $\tilde{K}$ is selected as $C\tilde{K} = I$ such that (5.69) is valid, the states that always coincide with the output can be taken away, and we end up with fewer updating equations. We illustrate the use of such a reduced order observer with an example.

Example 5.12: Sampled Double Integrator

According to Example 4.1 the sampled double integrator is described by

$$x(t+T) = \begin{bmatrix} 1 & T \\ 0 & 1 \end{bmatrix} x(t) + \begin{bmatrix} T^2/2 \\ T \end{bmatrix} u(t)$$
$$y(t) = \begin{bmatrix} 1 & 0 \end{bmatrix} x(t).$$

We are going to estimate the state vector with the observer (5.95). The observer dynamics is determined by the matrix

$$A - \tilde{K}CA = \begin{bmatrix} 1 & T \\ 0 & 1 \end{bmatrix} - \begin{bmatrix} \tilde{k}_1 \\ \tilde{k}_2 \end{bmatrix} \begin{bmatrix} 1 & 0 \end{bmatrix} \begin{bmatrix} 1 & T \\ 0 & 1 \end{bmatrix}$$

Since $x_1(t) = y(t)$ we want to make sure that $\hat{x}_1(t) = y(t)$. This means that $\tilde{K}$ should be selected so that

$$1 = C\tilde{K} = \begin{bmatrix} 1 & 0 \end{bmatrix} \begin{bmatrix} \tilde{k}_1 \\ \tilde{k}_2 \end{bmatrix} = \tilde{k}_1.$$

With $\tilde{k}_1 = 1$

$$A - \tilde{K}CA = \begin{bmatrix} 0 & 0 \\ -\tilde{k}_2 & -\tilde{k}_2 T + 1 \end{bmatrix}$$

If we want to have dead-beat reconstruction we use $\tilde{k}_2 = 1/T$, since the matrix above will have both eigenvalues in the origin. This gives the observer

$$\hat{x}(t+1|t+1) = \begin{bmatrix} 1 & T \\ 0 & 1 \end{bmatrix} \hat{x}(t|t) + \begin{bmatrix} T^2/2 \\ T \end{bmatrix} u(t)$$

$$+ \begin{bmatrix} 1 \\ 1/T \end{bmatrix} \left(y(t+1) - \begin{bmatrix} 1 & 0 \end{bmatrix} \begin{bmatrix} 1 & T \\ 0 & 1 \end{bmatrix} \hat{x}(t|t) - \frac{T^2}{2} u(t) \right)$$

The first row in this expression after simplification is

$$\hat{x}_1(t+1|t+1) = y(t+1).$$

This is also how we have chosen $\tilde{K}$. The second row is the actual observer and is, after reduction and introduction of $\hat{x}_1(t|t) = y(t)$:

$$\hat{x}_2(t+1|t+1) = \frac{1}{T}[y(t+1) - y(t)] + \frac{T}{2}u(t) \tag{5.97}$$

This is thus the only equation necessary for reconstruction.

Kalman Filters

Using the same technique we can carry out the corresponding calculation which led to the continuous time Kalman filter. The result is given by the following theorem.

Theorem 5.6 (Kalman filter: Discrete Time) *Consider the system*

$$x(t+1) = Ax(t) + Bu(t) + Nv_1(t)$$
$$y(t) = Cx(t) + Du(t) + v_2(t)$$

where v_1 and v_2 are white noises with intensities R_1 and R_2, respectively. The cross spectrum between v_1 and v_2 is constant and equal to R_{12}. The observer minimizing the estimation error $x(t) - \hat{x}(t)$ is given by

$$\hat{x}(t+1|t) = A\hat{x}(t|t-1) + Bu(t)$$
$$+ K(y(t) - C\hat{x}(t|t-1) - Du(t)) \tag{5.98}$$

where K is given by

$$K = (APC^T + NR_{12})(CPC^T + R_2)^{-1} \tag{5.99}$$

and P is the symmetric, positive, semidefinite solution to the matrix equation

$$P = APA^T + NR_1N^T \tag{5.100}$$
$$- (APC^T + NR_{12})(CPC^T + R_2)^{-1}(APC^T + NR_{12})^T$$

This P is also equal to the covariance matrix of the optimal estimation error $x(t) - \hat{x}(t|t-1)$.

Here (5.100) is the discrete time Riccati equation and corresponds to (5.80).

The estimate according to (5.98) is based on all old measurements $y(s), s \le t - 1$, and is therefore denoted by $\hat{x}(t|t-1)$. If we want to include the last measurement $y(t)$ in the estimation, we use (5.94b). The error is then according to (5.68)

$$E\tilde{x}(t|t)\tilde{x}^T(t|t) = (I - \tilde{K}C)P(I - \tilde{K}C)^T + \tilde{K}R_2\tilde{K}^T$$

since $v_2(t)$ and $\tilde{x}(t|t-1)$ are independent. If the right hand side is minimized with respect to $\tilde{K}$ we get

$$\tilde{K} = PC^T(CPC^T + R_2)^{-1} \tag{5.101}$$

Note that $K - A\tilde{K} = NR_{12}(CPC^T + R_2)^{-1}$, which implies that the measurement and time update (5.94bc) can be written

$$\hat{x}(t|t) = \hat{x}(t|t-1) + \tilde{K}\nu(t) \tag{5.102a}$$
$$\hat{x}(t+1|t) = A\hat{x}(t|t) + Bu(t) + \hat{v}_1(t|t) \tag{5.102b}$$
$$\nu(t) = y(t) - C\hat{x}(t|t-1) - Du(t) \tag{5.102c}$$
$$\hat{v}_1(t|t) = NR_{12}(CPC^T + R_2)^{-1}\nu(t) \tag{5.102d}$$
$$\tilde{K} = PC^T(CPC^T + R_2)^{-1} \tag{5.102e}$$

where we introduced $\hat{v}_1(t|t)$ as the best estimate of the system disturbance $v_1(t)$, and $\nu(t)$ as the innovation. Note that $\hat{v}_1(t|t) = 0$ if there is no correlation between system noise and measurement noise.

5.9   Practical Aspects of Signal Sampling

There are many important practical aspects in connection with sampling of signals and systems. In this section we will briefly point to some essentials. See the literature list for more detailed descriptions.

If the sampling interval is T, $\omega_s = 2\pi/T$ is the *sampling (angular) frequency*. Half the sampling frequency, $\omega_n = \pi/T$, is called the *Nyquist frequency*. The main property is that a sampled version of a continuous time sinusoid with frequency above the Nyquist frequency cannot be distinguished from a signal with frequency below this. This is the *alias phenomenon*.

When sampling a continuous signal some information is normally lost. The continuous signal can typically not be recreated from the sampled one. In principle, information about frequencies higher than the Nyquist frequency is lost. Therefore we should select a sampling interval T which makes the frequency content above the Nyquist frequency uninteresting – just "noise".

If there is high frequency noise in the signal, this will influence the sampled value. In order to reduce this influence and give a more correct picture of the signal, it has to be pre-filtered. This is done using a low pass filter (pre-sampling filter) with pass band up to the Nyquist frequency. This means that some kind of average is formed over the sampling interval.

A formal description of the influence of high frequency noise is given by *Poisson's summation formula*: If a continuous time stochastic process u with spectrum $\Phi_u(\omega)$ is sampled using the sampling interval T, the spectrum of the sampled signal is

$$\Phi_u^{(T)}(\omega) = \Phi_u(\omega) + \sum_{r \neq 0} \Phi_u(\omega + 2r\pi/T), \quad -\pi/T \leq \omega \leq \pi/T$$

$$(5.103)$$

Here we see that contributions from the sum, i.e., frequencies above the Nyquist frequency, give a sampled spectrum with a faulty picture of the spectrum of the continuous signal. This is a result of the alias phenomenon. The reason for using the pre-sampling filter can also be described as eliminating these higher frequency contributions.

5.10   Comments

The Main Points of the Chapter

- A disturbance signal w is naturally characterized by its spectrum $\Phi_w(\omega)$.

- If $u(t) = G(p)w(t)$ then $\Phi_u(\omega) = G(i\omega)\Phi_w(\omega)G^T(-i\omega)$.

- The standard representation in the time domain is $w(t) = G(p)v(t)$, where v is white noise with intensity R and $\Phi_w(\omega) = G(i\omega)RG^T(-i\omega)$

- A control object with disturbances is represented in state space form by

$$\dot{x} = Ax + Bu + Nv_1$$
$$z = Mx + D_z u$$
$$y = Cx + D_y u + v_2$$

  where v_1 and v_2 are white noises with spectra R_1, R_2 and cross spectrum R_{12}.

- The state vector is estimated using an observer

$$\dot{\hat{x}} = A\hat{x} + Bu + K(y - C\hat{x} - Du)$$

- The choice of K which minimizes the error in the state estimation is given by the Kalman filter in Theorem 5.4.

- Discrete time/sampled signals are treated in the same way. Then $i\omega$ is exchanged for $e^{i\omega T}$ in the filter expressions and $\dot{x}(t)$ for $x(t + T)$. Theorem 5.6 gives the discrete time Kalman filter.

Literature

The description of disturbances as stochastic processes developed during the 1930s, mainly in Wiener's works. They are summarized in the classic book Wiener (1949). The treatment of disturbances in state space form and the Kalman filter originate from Kalman (1960b) and Kalman & Bucy (1961).

There are many good books on signal theory, signal models and signal estimation. See, for example, Papoulis (1977) for general signal theory and Anderson & Moore (1979) for Kalman filtering. Oppenheim & Schafer (1989) is a standard textbook on signal processing and sampling of signals is discussed in detail in Chapter 3. Among a number of books on the use of Kalman filters Minkler & Minkler (1993) can be suggested. Ljung & Glad (1994) also contains the basics on spectra and signal models. Some formalities around signals subject to (5.21) are dealt with in Ljung (1999).

The mathematical theory necessary for correct treatment of continuous time white noise is described in, for example, Wong (1983). Spectral factorization in the multivariable case is covered in Anderson (1967).

Observers and discrete time Kalman filters are covered in detail in Åström & Wittenmark (1997).

Software

The MATLAB routines corresponding to the methodology in this chapter are

lyap Solution of equation (5.62)
dlyap Solution of equation (5.92)
place Calculation of K for placement of the eigenvalues of $A - KC$.
kalman Forming a Kalman filter according to Theorems 5.4 or 5.6.
lqe, dlqe Calculation of K according to the theorems above.

Appendix 5A: Proofs

Proof of Lemma 5.1

Proof: We carry out the proof for $R_{12} = 0$. We then show how this special case can be extended to the general case. We also assume that a positive, semidefinite solution P° to (5.73) exists, and we return later to how this is proved.

We will start out by showing that $A - K^\circ C$ is a stable matrix: equations (5.73)-(5.74) can be arranged as

$$(A - K^\circ C)P^\circ + P^\circ(A - K^\circ C)^T + K^\circ R_2 (K^\circ)^T + R_1 = 0$$

It is easy to show that if (A, R_1) is stabilizable (which we assume in the theorem) then $(A - KC, R_1 + KZK^T)$ is stabilizable for all K, C and all positive definite Z. (See below.) We can then use Lemma 12.1 to draw the conclusion that $A - K^\circ C$ is stable.

Now let K be an arbitrary stabilizing matrix and denote $P(K) = P$ and $P(K^\diamond) = P^\diamond$. We have

$$(A - KC)P + P(A - KC)^T + R_1 + KR_2K^T = 0$$
$$(A - K^\diamond C)P^\diamond + P^\diamond(A - K^\diamond C)^T + R_1 + K^\diamond R_2 K^{\diamond T} = 0$$

Form the difference and expand $(A - KC)P - (A - K^\diamond C)P^\diamond = (A - KC)P - (A - KC)P^\diamond + (A - KC)P^\diamond - (A - K^\diamond C)P^\diamond$. This results in

$$0 = (A - KC)(P - P^\diamond) + (P - P^\diamond)(A - KC)^T + KR_2K^T - K^\diamond R_2 K^{\diamond T}$$
$$+ K^\diamond CP^\diamond - KCP^\diamond + P^\diamond C^T K^{\diamond T} - P^\diamond C^T K^T = (A - KC)(P - P^\diamond)$$
$$+ (P - P^\diamond)(A - KC)^T + (K - K^\diamond)R_2(K - K^\diamond)^T$$

In the last equality the definition $K^\diamond = P^\diamond C^T R_2^{-1}$ is used. Moreover a number of terms eliminate each other. The last matrix is positive semidefinite (since R_2 is positive definite). Since $A - KC$ is stable, Theorem 5.3 can be applied, which shows that $(P - P^\diamond)$ is a covariance matrix, and therefore positive semidefinite.

To complete the proof only the three promised details remain:

1. (A, R_1) detectable $\Rightarrow (A - BL, R_1 + L^T ZL)$ detectable if Z is positive definite. (We choose to show the result in terms of detectability, rather than of stabilizability). If the latter system would not be detectable there would exist an unstable mode with

$$\dot{x} = (A - BL)x$$
$$y = (R_1 + L^T ZL)x \equiv 0$$

Since R_1 and $L^T ZL$ are positive semidefinite, each term of y has to be identically zero: $R_1 x \equiv 0$ and $L^T ZLx \equiv 0$. Since Z is positive definite, it means that $Lx \equiv 0$, and that x fulfills $\dot{x} = Ax; R_1 x \equiv 0$, which contradicts that (A, R_1) is detectable.

2. The existence of a positive semidefinite solution of (5.73): Select K_1 so that $A - K_1 C$ is stable. This is always possible, since (A, C) is assumed to be detectable. Define recursively P_k and K_k as

$$(A - K_k C)P_k + P_k(A - K_k C)^T + R_1 + K_k R_2 K_k^T = 0$$
$$K_{k+1} = P_k C^T R_2^{-1}$$

The first equation is linear in P_k for a given K_k and has an unique positive semidefinite solution (the covariance matrix, according to Theorem 5.3) if $A - K_k C$ is stable. In the same way as above we can now prove that $P_{k+1} \leq P_k$ and that $A - K_k C$ is stable $\forall k$. A decreasing sequence of positive semidefinite matrices must converge and we realize that $\lim_{k \to \infty} P_k = P^\diamond$ has to satisfy (5.73). That the solution is unique follows from (5.76).

3. R_{12}-term: Introduce $\tilde{A} = A - R_{12}R_2^{-1}C$, $\tilde{K} = K - R_{12}R_2^{-1}$ and $\tilde{R}_1 = R_1 - R_{12}R_2^{-1}R_{12}^T$. Then (5.75) can be written

$$(\tilde{A} - \tilde{K}C)(P(\tilde{K}) + P(\tilde{K})(\tilde{A} - \tilde{K}C)^T + \tilde{R}_1 + \tilde{K}R_2\tilde{K}^T = 0$$

This is exactly the problem we have examined in the proof and we can therefore compute the optimal $\tilde{K}$. Going back to the initial matrices A and K we get (5.73) and (5.74).

$\square$

PART II: LINEAR CONTROL THEORY

In this part we shall study properties and design of linear controllers. The discussion contains three steps:

- *What do we want to achieve?*

- *What is possible to achieve?*

- *How is it done?*

We will discuss specifications (Chapter 6), theoretical limitations (Chapter 7), and design methods (Chapters 8 to 10). The design methods which will be covered are

- *Configuration of multivariable controllers.*

- *Methods with internal model (Internal Model Control, IMC).*

- *Minimization of quadratic criteria (Linear Quadratic Gaussian Control, LQG).*

- *Formal methods for loop shaping (so called $\mathcal{H}_\infty$-methods)*

In Part III we will discuss additional design methods for nonlinear systems.

Chapter 6

The Closed Loop System

As we discussed in Chapter 1, the controller is a system driven by the measured signal y and by the reference signal r. It produces the control signal u to be fed into the control object. A *linear controller* can be represented as

$$u(t) = F_r(p)r(t) - F_y(p)y(t) \tag{6.1}$$

with the feedback controller F_y and prefilter F_r. Traditional, analog, controllers are continuous time systems as above, while sampled controllers (see Section 6.7) are discrete time systems (replace p with q). In this chapter we will study the *closed loop system*, i.e., the system obtained when (6.1) is fed back to the control object. Many of the expressions obtained look like the SISO, continuous time result known from basic control theory.

6.1 The Transfer Functions of the Closed Loop System

Consider the system

$$\begin{aligned}
z(t) &= G(p)u(t) + w(t) \\
y(t) &= z(t) + n(t)
\end{aligned} \tag{6.2}$$

controlled by

$$\bar{u}(t) = F_r(p)r(t) - F_y(p)y(t) \tag{6.3}$$

147

and with a disturbance on the input signal

$$u(t) = \bar{u}(t) + w_u \tag{6.4}$$

Note that many load disturbances typically occur at the input side of the system. See Figure 6.1. As we discussed in Chapter 5 it is often

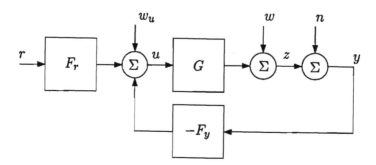

Figure 6.1: The closed loop system.

natural to model the disturbance as the output of a linear system.

$$w(t) = G_d(p)d(t) \tag{6.5}$$

Here d can be a signal with physical interpretation and G_d a model for how it is transferred in the system to the exit. For example, if d is a disturbance at the input side we have $G_d = G$. The transfer function may also be chosen to reflect the frequency distribution of the disturbance. Then G_d may have been constructed by spectral factorization (Section 5.4), with d as a white noise signal.

By inserting (6.3) into (6.2) we obtain the closed loop system

$$z = GF_r r - GF_y z - GF_y n + w + Gw_u$$

i.e.,

$$z = (I + GF_y)^{-1}GF_r r + (I + GF_y)^{-1}w - (I + GF_y)^{-1}GF_y n$$
$$+ (I + GF_y)^{-1}Gw_u \tag{6.6}$$

The control error is as before $e = r - z$. The *loop gain* is the gain obtained when following a signal around the loop, i.e., GF_y or

F_yG, depending on the starting point. We will also use the following notations

$$z = G_c r + Sw - Tn + GS_u w_u \tag{6.7a}$$
$$e = (I - G_c)r - Sw + Tn - GS_u w_u \tag{6.7b}$$

Here G_c is the *closed loop system*

$$G_c = (I + GF_y)^{-1}GF_r \tag{6.8}$$

S is the *sensitivity function*

$$S = (I + GF_y)^{-1} \tag{6.9}$$

T the *complementary sensitivity function*

$$T = (I + GF_y)^{-1}GF_y \tag{6.10}$$

and S_u the *input sensitivity function*

$$S_u = (I + F_yG)^{-1} \tag{6.11}$$

In (6.7) and (6.11) the identity $(I + AB)^{-1}A = A(I + BA)^{-1}$, is used. It is easily verified, for example, by multiplying $(I + AB)$ from the left and $(I + BA)$ from the right.

Note that

$$S + T = I \text{ (the identity matrix)} \tag{6.12}$$

In the common case $F_y = F_r$

$$T = G_c \tag{6.13}$$

It is interesting to study the input u to the system (including the disturbance w_u)

$$u = (I + F_yG)^{-1}F_r r - (I + F_yG)^{-1}F_y(w + n) + (I + F_yG)^{-1}w_u$$
$$= S_u F_r r - S_u F_y(w + n) + S_u w_u = G_{ru}r + G_{wu}(w + n) + S_u w_u \tag{6.14}$$

where

$$G_{ru} = (I + F_yG)^{-1}F_r \tag{6.15}$$
$$G_{wu} = -(I + F_yG)^{-1}F_y \tag{6.16}$$

The closed loop system can be seen as a system fed by the external signals r, w, w_u and n and producing the interesting signals z (or the control error $e = r - z$) and u as outputs. The transfer function for the closed loop system is thus given by (6.7) and (6.14).

6.2 Stability of the Closed System

Of course, the controller has to be selected so that the closed system becomes stable. Since cancellations can happen in the rational expressions for G_c, we must consider what is to be meant by "stable". We will start with a couple of examples.

Example 6.1: Cancellation of a Non-observable Mode

Consider the system $G(s) = \frac{s-1}{s+1}$. It is controlled by $F_y(s) = F_r(s) = \frac{1}{s-1}$. The closed system is

$$G_c = \frac{\frac{s-1}{s+1} \cdot \frac{1}{s-1}}{1 + \frac{s-1}{s+1} \cdot \frac{1}{s-1}} = \frac{1}{s+2}$$

which is stable. There may be some concern about the factor $\frac{1}{s-1}$ that "disappeared" The transfer function from r to u is

$$G_{ru} = \frac{\frac{1}{s-1}}{1 + \frac{s-1}{s+1}\frac{1}{s-1}} = \frac{s+1}{(s-1)(s+2)}$$

which is unstable. The input u thus grows towards infinity, while the output remains limited. There then has to be some combination of states driven by u, which does not show up in the output. Evidently the mode corresponding to $s - 1$ is not observable in the closed system G_c.

Example 6.2: Cancellation of a Non-controllable Mode

Consider the system $G(s) = \frac{1}{s-1}$. It is controlled by $F_y(s) = F_r(s) = \frac{s-1}{s+1}$ The closed system then becomes

$$G_c = \frac{\frac{1}{s-1}\frac{s-1}{s+1}}{1 + \frac{1}{s-1}\frac{s-1}{s+1}} = \frac{1}{s+2}$$

The transfer function from r (and w) to u is

$$G_{ru} = \frac{\frac{s-1}{s+1}}{1 + \frac{1}{s-1}\frac{s-1}{s+1}} = \frac{s-1}{s+2}$$

and the sensitivity function is

$$S = \frac{1}{1 + \frac{1}{s-1}\frac{s-1}{s+1}} = \frac{s+1}{s+2}$$

All these transfer functions are stable. The cancelled mode $s - 1$ should all the same cause some problem in some way. That this is the case can be seen as follows. The transfer function from the disturbance at the input w_u to y (see Figure 6.1) is according to (6.11) and (6.14)

$$G_{w_u y}(s) = (I + F_y G)^{-1} G \qquad (6.17)$$

which in our case is

$$G_{w_u y}(s) = \frac{\frac{1}{s-1}}{1 + \frac{1}{s-1}\frac{s-1}{s+1}} = \frac{s+1}{(s-1)(s+2)}$$

This transfer function is unstable. A small disturbance at the input results in the output tending to infinity. The reference signal r as such does not excite this unstable mode. Apparently it corresponds to a *non-controllable* mode in the closed loop system G_c.

Example 6.3: Harmless Pole/Zero Cancellation

Again, consider the system $G(s) = \frac{1}{s-1}$, and use the feedback controller $F_r = F_y = 3$. This results in

$$G_c = \frac{\frac{3}{s-1}}{1 + \frac{3}{s-1}} = \frac{3}{s+2}$$

We also get

$$G_{ru} = \frac{3}{1 + \frac{3}{s-1}} = \frac{3(s-1)}{s+2}$$

$$S = \frac{1}{1 + \frac{3}{s-1}} = \frac{s-1}{s+2}$$

$$G_{w_u y} = \frac{\frac{1}{s-1}}{1 + \frac{3}{s-1}} = \frac{1}{s+2}$$

All these transfer functions are stable, even if unstable factors (perhaps) have been cancelled when they were formed. This means that both u and y have stable response to all incoming signals in Figure 6.1.

Based on these examples we now introduce the following definition.

Definition 6.1 (Internal Stability of a Closed Loop System)
A closed loop system according to Figure 6.1 is said to be **internally stable** *if the following transfer functions all are stable (after all possible cancellations).*

$$\begin{aligned}
w_u \mapsto u, && S_u = G_{w_u u} = (I + F_y G)^{-1} && (6.18) \\
w_u \mapsto y, && G_{w_u y} = (I + G F_y)^{-1} G && (6.19) \\
w \mapsto u, && G_{wu} = -(I + F_y G)^{-1} F_y && (6.20) \\
w \mapsto y, && S = G_{wy} = (I + G F_y)^{-1} && (6.21)
\end{aligned}$$

and if F_r also is stable.

If F_y and G are unstable, all four have to be checked. If one of of them is stable, less work has to be done. If, for example, the controller $F_y(s)$ is stable it is sufficient to check that

$$w_u \mapsto y, \qquad G_{w_u y} = (I + GF_y)^{-1}G \tag{6.22}$$

and F_r are stable.

6.3   Sensitivity and Robustness

Sensitivity

In practice, the system is not known in detail. Various model errors result in a different closed loop system than expected. Assume that the true system is given by G_0 and that is differs from the nominal model G in the following way

$$G_0 = (I + \Delta_G)G \tag{6.23}$$

(We do not enter the argument: the calculations are valid without change for both continuous time (p) and discrete time (q)).

Remark In the scalar case Δ_G is simply the relative model error. For multivariable systems there is a difference in letting the model error be a multiplicative factor at the exit (Δ_G is a $p \times p$ matrix as above) and letting it influence the input according to

$$G_0 = G(I + \Delta_G), \quad \Delta_G \text{ is } m \times m \tag{6.24}$$

For the case (6.24) the input-sensitivity function S_u (6.11) should be used instead of S in the expressions to follow below.

If we for the moment disregard disturbances, the actual output will be

$$z_0 = (I + G_0F_y)^{-1}G_0F_r r \tag{6.25}$$

instead of

$$z = (I + GF_y)^{-1}GF_r r \tag{6.26}$$

that the model G predicts.

The following now applies for the difference between z and z_0

$$z_0 = (I + \Delta_z)z \qquad\qquad (6.27a)$$
$$\Delta_z = S_0 \Delta_G \qquad\qquad (6.27b)$$
$$S_0 = (I + G_0 F_y)^{-1} \qquad\qquad (6.27c)$$

The sensitivity function corresponding to the true system (see (6.9)) therefore describes how the relative model error Δ_G is transformed to a relative output error. The closed loop system is thus not so sensitive to model errors in the frequency band where the sensitivity function is small.

To show (6.27a)-(6.27c) we see that (6.25) and (6.26) can be written as

$$z_0 + G_0 F_y z_0 = G_0 F_r r$$
$$z + G_0 F_y z - \Delta_G G F_y z = G F_r r$$

Subtract the equations:

$$z_0 - z + G_0 F_y (z_0 - z) = \Delta_G (G F_r r - G F_y z)$$

But $G F_r r = z + G F_y z$, so we have $(z_0 - z) = (I + G_0 F_y)^{-1} \Delta_G z$, or $\Delta_z z = S_0 \Delta_G z$ which gives (6.27c).

Robustness

Another important question is *robustness*, i.e., what model errors Δ_G can be allowed without endangering the stability in the closed loop system. To study this, we assume the relationship $G_0 = (I + \Delta_G)G$ and that Δ_G is a *stable* transfer function. This means that G and G_0 have the same stability properties (the same amount of unstable poles, etc.). We also assume that the feedback F_y has been chosen so that the nominal closed system is internally stable.

The actual closed loop system $(I + G_0 F_y)^{-1} G_0$, can then be illustrated as in Figure 6.2. We will concentrate on the loop gain and take $r \equiv 0$.

We can perceive this system as feedback between $\Delta_G G$ and the dashed block with x as input and ξ as output. What is the transfer function of this block? We see that

$$\xi = -F_y x - F_y G \xi \quad \text{so that} \quad \xi = -(I + F_y G)^{-1} F_y x$$

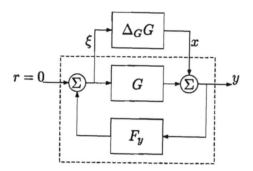

Figure 6.2: The true closed loop system.

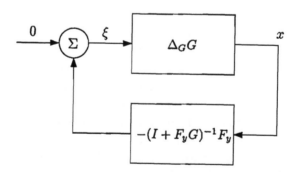

Figure 6.3: The true closed loop system after changes in the block diagram.

Figure 6.2 now can be redrawn as Figure 6.3. This is feedback between the linear systems $\Delta_G G$ and $-(I + F_y G)^{-1} F_y$. The corollary to the Small Gain Theorem 1.1 tells us that stability is assured if these two systems are stable and if the gain of $\Delta_G G (I + F_y G)^{-1} F_y$ is less than one.

Note that

$$G(I + F_y G)^{-1} F_y = GF_y(I + GF_y)^{-1} = (I + GF_y)^{-1} GF_y = T$$

i.e., the complementary sensitivity function (6.10). (In the first equality we use $(I + AB)^{-1}A = A(I + BA)^{-1}$). Since both Δ_G and T are stable transfer functions (the nominal system was assumed to be internally stable), we draw the conclusion that the true closed loop system is stable if

$$\boxed{\|\Delta_G T\|_\infty < 1}$$ (6.28)

This conditions is implied by

$$|T(i\omega)| < \frac{1}{|\Delta_G(i\omega)|}, \quad \forall \omega$$ (6.29)

The argument is valid for discrete as well as continuous time.

Note that the Small Gain Theorem is also valid for non-linear blocks. Robustness results of the type (6.28) can therefore be developed also for non-linear model errors.

Remark. To show internal stability for the true closed loop system all transfer functions according to Definition 6.1 should be tested. The stability of the remaining transfer function is shown using the stability in the main loop as above and the internal stability of the nominal closed system.

6.4   Specifications for the Closed Loop System

The control problem can be described as follows. "Choose a controller such that the controlled variable tracks the reference signal as close as possible, despite disturbances, measurement errors and model errors, while reasonably large control signals for the system are used". In this and the following sections we will formalize these words.

From (6.7), (6.14), (6.27c), and (6.28) we have the basic important relations (take $w_u = 0$; its influence can always be included in w):

$$e = (I - G_c)r - Sw + Tn \tag{6.30a}$$

$$u = G_{ru}r + G_{wu}(w + n) \tag{6.30b}$$

$$\Delta_z = S_0\Delta_G \tag{6.30c}$$

$$\|\Delta_G T\|_\infty < 1 \tag{6.30d}$$

The basic demands are easy to list:

1. The closed loop system has to be close to I (i.e., $|I - G_c|$ small) for the controlled variable to follow the reference signal.

2. The sensitivity function S has to be small so that system disturbances and model errors have little influence on the output

3. The complementary sensitivity function T has to be small so that measurement disturbances have little influence on the output, and so that model errors do not endanger the system stability.

4. The transfer functions G_{ru} and G_{wu} should not be too large.

In Chapter 7 we will discuss limitations and conflicts in these specifications in more detail, but already the simple relations

$$S + T = I \tag{6.31}$$

$$G_c = GG_{ru} \tag{6.32}$$

show the conflicts inherent in the demands above.

We will now discuss the balance between the different demands in the time domain on the one hand, and the frequency domain on the other.

6.5   Specifications in the Time Domain

Step Response

One way of characterizing the distance between two transfer operators is to compare the output from the two systems for a given input. A step is often selected as input. To capture $G_c \approx I$ we thus measure how close the step response is to the reference step itself.

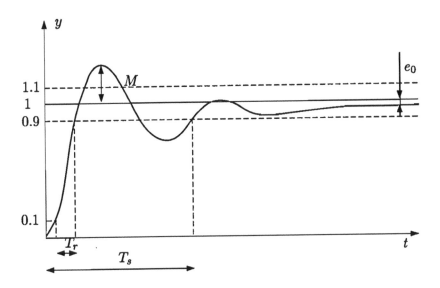

Figure 6.4: Step response with specifications.

Figure 6.4 shows the step response of G_c. Its deviation from the unit step is traditionally measured using the numbers M, e_0, T_r, T_s, which are called overshoot, static error, rise time and settling time. All these numbers should be small. A typical design criterion is to state upper bounds for them:

Design Criterion 6.1 *Choose the controller so that M, e_0, T_r and T_s for the step response of G_c is smaller than given values.*

The step response of the sensitivity function S can be examined in the same way.

For multivariable systems all step responses from different combinations of input and output should be considered. Note that the nondiagonal elements in G_c ideally should be close to zero. This means that each reference signal should influence only one output.

Static Errors and Error Coefficients

The static error for a unit step is equal to

$$e_0 = \lim_{t \to \infty} e(t) = I - G_c$$

evaluated at the frequency zero (thus $I - G_c(0)$ for continuous time systems). This quantity is called the static *error coefficient*. In the common case $F_y = F_r$ we have $I - G_c = S$ so then $e_0 = S(0)$. The controller is usually designed so that $e_0 = 0$ for the nominal model G. To avoid static errors even if the true system G_0 differs from the model, it is necessary according to (6.30c) that the sensitivity function S_0 is zero at zero frequency. Since

$$S_0 = (I + G_0 F_y)^{-1}$$

this implies that $G_0 F_y$ assumes the value infinity for zero frequency, i.e., either the system or the controller contain integrations (the factor $1/s$). If the error coefficient $e_0 = 0$, we can study the remaining error when the reference signal is a ramp. According to the final value theorem (2.4), this error is given by

$$e_1 = \lim_{t \to \infty} e(t) = \lim_{s \to 0} \frac{1}{s}(I - G_c(s)) \tag{6.33}$$

This gives the error coefficient for ramp reference. Higher error coefficients can be defined analogously.

Control errors after disturbances can be handled analogously by considering the sensitivity function. See (6.30a). A step disturbance and a ramp disturbance, respectively, result in the steady state errors

$$s_0 = -S(0) \tag{6.34a}$$
$$s_1 = -\lim_{s \to 0} S(s)/s \tag{6.34b}$$

Translation to Poles and Zeros

A systematic way to handle specifications is to state explicit suitable transfer functions G_c and S that provide acceptable time responses. We then try to calculate a controller F_y and F_r which gives these functions G_c and S.

Design Criterion 6.2 *Select the controller such that G_c and S are equal to given transfer functions.*

Since there are clear connection between the poles of the system and its time response, it is often enough to specify the poles:

Design Criterion 6.3 *Select the controller so that G_c and S have their poles placed within given areas.*

To be able to use such a criterion in a reasonable way we must of course know the relationship between the step response and the location of the poles. Such relationships are discussed for *continuous time* systems in, for example, Franklin et al. (1994), Section 3.4. Complex conjugated poles along the lines bisecting the left hand quadrants, $p = -\lambda \pm i\lambda$, $\lambda > 0$ give "nice looking" step responses.

Quantitative Expressions of Control Error and Input

Recall the definitions of signal size in Section 5.2:

$$\|z\|_P^2 = \lim_{N \to \infty} \frac{1}{N} \int_0^N z^T(t) P z(t) dt = E z^T(t) P z(t) = \text{tr}(R_z P)$$

$$R_z = \frac{1}{2\pi} \int \Phi_z(\omega) d\omega \tag{6.35}$$

where Φ_z is the spectrum of the signal. See (5.6), (5.12), (5.13) and (5.15c). The matrix P weighs together the different components of the vector z.

To trade off the size of the control error (e) against the input energy, the following design criterion is natural:

Design Criterion 6.4 *Select the controller such that*

$$\|e\|_{Q_1}^2 + \|u\|_{Q_2}^2$$

is minimized.

If the scaling of the control signals and control variables has been done according to the recommendations of Section 7.1, Q_1 and Q_2 should be approximately unit matrices.

6.6 Specifications in the Frequency Domain

How well the closed loop system G_c approximates I in the frequency domain is captured by the concepts of *bandwidth* ω_B and *resonance peak* M_p. The *resonance frequency* is the frequency where the peak occurs. See Figure 6.5. The definitions can of course be used for any system, for example, the complementary sensitivity function T and the sensitivity function S. The latter has, however, typically the reversed behavior: it grows from a low value to a resonance peak at a certain resonance frequency and then levels out at level 1. The

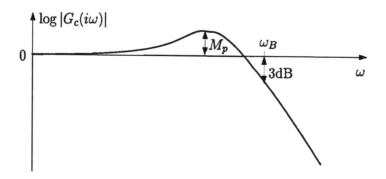

Figure 6.5: The bandwidth ω_B of the closed system and the resonance peak M_p.

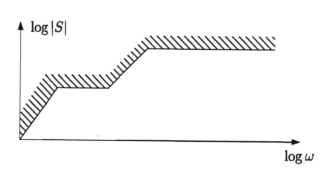

Figure 6.6: Typical upper limit specifications for the sensitivity function.

bandwidth is then the frequency interval where disturbance rejection takes place, i.e., $|S(i\omega)| < 1$. For multivariable systems, the singular values of the transfer function have to be considered, which may lead to several peaks and bandwidths. Then it is customary to consider worst case performance, i.e., the highest resonance peak and the smallest bandwidth.

We will see more precisely from (6.43) and (6.44) below how the specifications on the frequency functions should be matched to the frequency contents in reference and disturbance signals. This leads to requirements on the sensitivity function of the type

$$|S(i\omega)| \leq |W_S^{-1}(i\omega)| \quad \forall \omega \tag{6.36}$$

where the function $|W_S^{-1}(i\omega)|$ can be specified graphically as in Figure 6.6. Most often we want to suppress the sensitivity at low frequencies. If the system is multivariable, (6.36) means that all singular values for S have to be smaller than $|W_S^{-1}(i\omega)|$.

In the multivariable case, it is reasonable to have different demands for different output signals and use a matrix valued function W_S (often diagonal) so that conditions of the type (6.36) can be expressed as

$$||W_S S||_\infty \leq 1 \tag{6.37}$$

$(||\cdot||_\infty$ is defined in (3.27))

Example 6.4: Limits for the Sensitivity Function

We want the sensitivity to satisfy

$$\begin{aligned} |S(i\omega)| &\leq 0.1\omega, & \omega < 1 \\ |S(i\omega)| &\leq 2, & \forall\omega \end{aligned} \tag{6.38}$$

Take, for example

$$W_S(s) = \frac{s+20}{2s}$$

Note that this actually sharpens the conditions in the frequency interval $1 < \omega < 20$. At $\omega = 1$ our upper bound (6.38) takes a step from $|S| = 0.1$ to $|S| = 2$, and this cannot be captured very well using a function W_S of low order.

We also have demands on the complementary sensitivity T not to be too large, so that measurement errors are rejected, according to (6.43). Another reason is that (6.30d) should be valid, so that stability is guaranteed also under model errors. This relation holds if

$$|T(i\omega)| < \frac{1}{|\Delta_G(i\omega)|}, \quad \forall\omega$$

These demands are usually stricter at high frequencies, since measurement errors as well as relative model errors often are of high frequency character. To assure this, an upper bound for $|T|$, like in Figure 6.7, can be introduced. As in (6.37) such a specification can be written

$$||W_T T||_\infty < 1 \tag{6.39}$$

for some suitably chosen function W_T.

In the same way the conditions that bound the size of the input can be formulated as

$$||W_u G_{ru}||_\infty < 1 \tag{6.40}$$

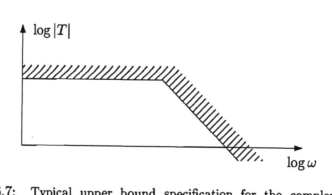

Figure 6.7: Typical upper bound specification for the complementary sensitivity function.

for some function W_u. Note that in the common case $F_y = F_r$ we have

$$I - G_c = S, \quad G_{ru} = -G_{wu}, \quad GG_{ru} = T = I - S \qquad (6.41)$$

The condition (6.36) or (6.37) then also is a condition on $G_c \approx I$, and (6.40) describes the influence on the input both from reference signal and noise. The three conditions then form a complete specification of the closed loop system:

Design Criterion 6.5 *Choose the controller so that*

$$\|W_S S\|_\infty < 1, \quad \|W_T T\|_\infty < 1, \quad \|W_u G_{ru}\|_\infty < 1$$

This is, in principle, the design conditions used in classic lead-lag compensation, even if the demands normally are not formalized in this way.

It may happen that the conditions in this design criterion are impossible to meet. A compromise could then be made, for example, by weighting the three criteria together and finding the controller that minimizes the sum:

Design Criterion 6.6 *Choose the controller so that*

$$\int_{-\infty}^{\infty} |W_S(i\omega)S(i\omega)|_2^2 + |W_T(i\omega)T(i\omega)|_2^2 + |W_u(i\omega)G_{ru}(i\omega)|_2^2 d\omega$$

$$= \|W_S S\|_2^2 + \|W_T T\|_2^2 + \|W_u G_{ru}\|_2^2$$

is minimized.

Here

$$|A|_2^2 = \text{trace}(A^*A) \qquad (6.42)$$

is the 2-norm of the matrix A. Note that the criterion 6.6 in practice coincides with the criterion 6.4, which is evident from the following example:

Example 6.5: Conditions on the Loop Functions

Consider a design problem where the reference signal r is identically zero, and $w_u = 0$ (we can always let w include disturbances at the input). The control error $e = r - z$ is then given by

$$e = -Sw + Tn$$

If the spectra of w and n are $\Phi_w(\omega)$ and $\Phi_n(\omega)$, respectively, and these signals are uncorrelated, equation (5.15b) gives the spectrum of the control error

$$\Phi_e(\omega) = |S(i\omega)|^2 \Phi_w(\omega) + |T(i\omega)|^2 \Phi_n(\omega) \qquad (6.43)$$

in the SISO case. Analogously, the spectrum of u is

$$\Phi_u(\omega) = |G_{wu}(i\omega)|^2 (\Phi_w(\omega) + \Phi_n(\omega)) \qquad (6.44)$$

See expression (6.30b) for u. We want to keep the output as close to zero as possible, and use the design criterion 6.4

$$\min(\|u\|^2 + \|e\|^2) = \min \int (\Phi_u(\omega) + \Phi_e(\omega)) d\omega$$
$$= \min \int \big(|G_{ru}(i\omega)|^2 (\Phi_w(\omega) + \Phi_n(\omega)) $$
$$+ |S(i\omega)|^2 \Phi_w(\omega) + |T(i\omega)|^2 \Phi_n(\omega) \big) d\omega$$

Here we have used (6.43) and (6.44) and (6.41). This is, however, identical to design criterion 6.6 if we take

$$|W_S(i\omega)|^2 = \Phi_w(\omega), \qquad |W_T(i\omega)|^2 = \Phi_n(\omega),$$
$$|W_u(i\omega)|^2 = \Phi_w(\omega) + \Phi_n(\omega)$$

The "upper bound" given for S, T and G_{ru} in the criteria 6.5 and 6.6 are thus coupled directly to the assumed distrurbance spectra. The coupling is also intuitively obvious: to assume that we have large system disturbances in a certain frequency band Ω ($\Phi_w(\omega)$ large when $\omega \in \Omega$) means that we demand that the sensitivity function be small in this band ($|W_S^{-1}(i\omega)|$ small for $\omega \in \Omega$) etc.

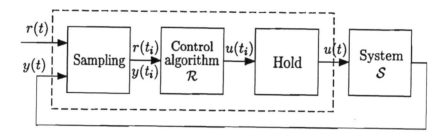

Figure 6.8: A sampled data controller.

6.7   Sampled Data Controllers

A controller is a system that creates the control signal u from the external signals r and y. This system can be either a continuous time or a discrete time system.

There is a multitude of methods for implementing the control relationship represented by the controller. Traditional techniques have used mechanical, pneumatical, hydraulical, and electrical devices for realization of the differential equations (P, I and D functions, lead-lag compensators and so on) that define the controller.

Today digitally implemented controllers dominate completely. It is simpler and cheaper to realize signal relationships in a computer, micro processor, or with digital VLSI techniques.

Signal Flow in a Sampled Data Controller

Digital implementation of a controller has an important consequence. A computer works sequentially with a finite amount of measurement data. It reads the values of $y(t)$ and $r(t)$ at discrete time instants $t_i, i = 1, 2, \dots$ and calculates the control signal for the process based on these values. A controller working in this fashion is called a *sampled data controller* since its control signal is based on samples of the output of the process and the reference signal.

The basic signal flow of such a controller is shown in Figures 6.8 and 6.9. The block "Sampling" receives the continuous time signals $y(t)$ and $r(t)$, samples them and produces a sequence of numbers, $y(t_i)$ and $r(t_i)$, $i = 1, 2, \dots$ which are the values of the signals at the sampling instances. The block "Control algorithm" works in discrete time and calculates from the sequences $r(t_i)$ and $y(t_i)$ another sequence

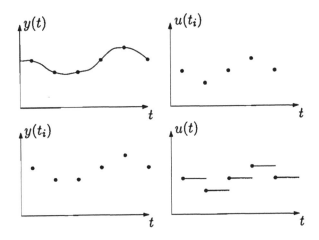

Figure 6.9: The signals in Figure 6.8.

of numbers, $u(t_i)$; $i = 1, 2, \ldots$, according to a rule we, for the time being, symbolize by $\mathcal{R}$. The block "Hold" receives the number sequence $u(t_i)$, $i = 1, \ldots$ and produces a continuous time signal $u(t)$ with a value equal to $u(t_i)$ for $t_i \leq t < t_{i+1}$. This signal $u(t)$ is input to the continuous time system that delivers the output $y(t)$, etc.

Note that the block "Control algorithm" in Figure 6.8 is a *discrete time system* according to our definition in Section 1.4. The whole sampled controller – the dashed block in Figure 6.8 – is, on the other hand, a continuous time system.

In addition to computer control there are several other examples of sampled data controllers. In economic systems the control decisions, like interest rates, are based on data sampled, for example, monthly.

Another case of sampled control is when the output of the process only *can* be measured at discrete time instants. In, for example, paper pulp processing, important measurement values can only be obtained by laboratory analysis of samples from the process. The control signals (chemical ingredients, temperature, etc.) are then based on the analysis results.

Advantages and Disadvantages of Sampled Control

In principle, it is harder to control a system using a sampled data controller than using a continuous time controller. A sampled data controller can use only a part of the control signals a continuous time controller can use, i.e., piecewise constant signals and cannot provide better control.

The sampled data controller does not have complete information of the behavior of the system. If the sampling interval (i.e., the time between sampling instants) is too long, important things can occur between sampling instants which the controller does not see and therefore cannot react to.

Instability in the closed loop system stems from "trusting old information too much." Sampled data controllers use information where the latest information can be up to one sampling interval old. The risk for instability in systems controlled by sampled data controllers is thus greater, especially if the sampling interval is long.

The most important advantage of a sampled data controller is that it is easier and cheaper to use computer implementation. Also, nonlinearities and switches of different kinds can easily be introduced. Another advantage is that systems containing time delays are handled more easily using sampled data control. If the effect of an input signal does not show up in the output until after a certain amount of time, it is important to remember which input was used in the system during this period to achieve good control. With sampled data control, the input is piecewise constant, and it can be kept in memory by storing a finite amount of numbers.

Remark. More general holding circuits can also be considered, as long as they supply an input over the sampling interval, that is unambiguously determined by $u(t_i)$.

Discretization of Continuous Controllers

Technical and physical systems are most often time continuous in nature. It is therefore usually most natural to analyze them as such and also calculate continuous time controllers for them. The controllers then typically become differential equations (linear or nonlinear) fed by y and r. These cannot be realized exactly in a sampled data controller, so the differential equations are discretized some way or another, to obtain a discrete time control algorithm. See Section 4.1.

Continuous Systems and Sampled Data Controllers

Suppose a continuous time system is controlled by a sampled data controller. The resulting closed loop system is not time invariant. The

properties of the system at a certain time depend on how it is phased
in comparison to the periodically recurring sampling. It is therefore
somewhat complicated to describe the closed loop system formally.
There are two main possibilities:

1. Regard the sampled data controller approximately as a continuous time controller and use the expressions of Section 6.1. This
   often works well if the sampling is fast compared to the time
   constants of the system.

2. Describe the control object in sampled data form as a relationship between the input sequence $\{u(kT)\}$ and output sequence
   $\{y(kT)\}$ at the sampling instants. The expressions for this were
   given in Section 4.2:

$$z(kT) = G_T(q)u(kT) + w(kT)$$
$$y(kT) = z(kT) + n(kT)$$
(6.45)

The input at the sampling instant is given analogously to (6.3)-(6.4) as

$$u(kT) = F_r(q)r(kT) - F_y(q)y(kT) + w_u(kT) \tag{6.46}$$

If we close the system (6.45)-(6.46) the expressions for the closed
system will be completely analogous to the continuous time case (6.7)–
(6.16). The transfer operator algebra does not depend on whether the
argument is p or q.

These expressions uniquely describe the input, since it is piecewise
constant between the sampling instants. It is, however, important to
remember that the output only is described at the sampling instants.
No direct information is obtained from (6.45)-(6.46) of what happens
between these.   For this we have to go back to the continuous
description and solve these equations over the sampling interval. Keep
Figure 4.3 in mind!

Example 6.6:   Sampled Data Control of a DC motor

Consider the DC motor in Example 4.2

$$G(s) = \frac{1}{s(s+1)} \tag{6.47}$$

It is controlled by a sampled data P-controller

$$u(t) = K(r(\ell T) - y(\ell T)), \quad \ell T \le t \le (\ell+1)T \tag{6.48}$$

i.e., $F_r(q) = F_y(q) = K$

What does the closed loop system look like? By sampling the system (6.47) we have

$$y(\ell T) = G_T(q)u(\ell T)$$

$$G_T(q) = \frac{(T - 1 + e^{-T})q + (1 - e^{-T}(1 + T))}{q^2 - (e^{-T} + 1)q + e^{-T}}$$

(see (4.17)). The behavior of the closed system at the sampling instants is given by

$$y(\ell T) = G_s(q)r(\ell T) \tag{6.49}$$

$$G_s(q) = (1 + G_T(q) \cdot K)^{-1} \cdot G_T(q) \cdot K = \frac{T(q)}{N(q)}$$

$$T(q) = K(T - 1 + e^{-T})q + K(1 - e^{-T}(1 + T))$$

$$N(q) = q^2 + (KT - K - 1 + e^{-T}(K - 1))q$$
$$+ (K - e^{-T}(K + KT - 1))$$

These expressions describe exactly how the output for the closed system behaves *at the sampling instants*. To find out what happens in between we have to solve the time varying linear differential equation that results from (6.47) being combined with (6.48).

Poles of Sampled Control Systems

For a sampled system, a reference step is a piecewise constant input to the closed loop system. This means that we get exactly the same output as for its continuous counterpart. The relationships between poles and step responses for continuous time systems can be directly translated to sampled systems using Theorem 3.7 and Figure 4.2. For the design criterion 6.3 we mentioned that poles along the line $-\lambda \pm i\lambda$ give nice step responses for continuous time systems. Figure 6.10 shows this line translated to sampled systems.

According to Theorem 3.8 the control error decreases as ϱ^k where ϱ is the largest absolute value of the poles (eigenvalues) and k is the number of sample steps. We introduce the time t and the sampling interval T, $k = t/T$. We then have

$$e(t) \sim \varrho^{t/T}, \quad t : \text{time}, \quad t/T : \text{number of samples.} \tag{6.50}$$

This means that the distance to the origin of a pole must be put in relation to the length of the sampling interval. It is $\varrho^{1/T}$ that determines the decay rate, not ϱ itself.

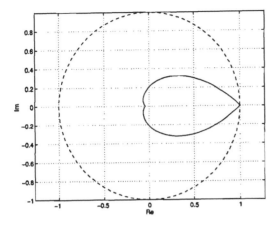

Figure 6.10: Suitable pole placement for sampled systems: the area inside the solid line gives good damping.

6.8 Comments

The Main Points of the Chapter

- The expression for the closed system in Figure 6.1:

$$z = G_c r + Sw - Tn + GS_u w_u$$
$$u = S_u F_r r - S_u F_y(w + n) + S_u w_u$$

- Influence of model error:

$$G_0 = (I + \Delta_G)G, \qquad z = G_c r$$
$$z_0 = (I + S_0 \Delta_G)z, \qquad S_0 = (I + G_0 F_y)^{-1}$$
$$\text{Stability if } \|\Delta_G T\|_\infty < 1$$

- We have discussed several different ways to formalize the condition that the control error should be small and the input reasonably large, despite disturbances and model errors. The basic idea in the time domain is to quantify deviations from the ideal response. This includes descriptions of the upper bounds for (the greatest singular values of) the basic loop functions in the frequency domain, and/or to quantify how to make compromises between such conditions.

Chapter 7

Basic Limitations in Control Design

In this chapter we will discuss limitations and conflicts between different goals for the design. We will mostly consider the SISO case since the fundamental trade-offs are most clear there. We will however indicate how the basic results can be carried over to the MIMO situation.

7.1 Scaling of Variables

When performing rule-of-thumb calculations or analyzing what is reasonable to achieve, it is often advisable to scale the variables so that they all vary over approximately the same interval. It is then much easier to interpret what is meant by a transfer function being "big" or "small".

We saw in Chapter 5 that it could be useful to scale variables according to (5.5). Suppose the original model is given in the form (6.2), (6.5)

$$z^f(t) = G^f(p)u^f(t) + G_d^f(p)d^f(t) \tag{7.1a}$$

$$y^f(t) = z^f(t) + n^f(t) \tag{7.1b}$$

$$e^f(t) = r^f(t) - z^f(t) \tag{7.1c}$$

where index f denotes a variable measured in physical units. We can then introduce scaled variables according to

$$D_u u = u^f, \quad D_d d = d^f \tag{7.2}$$

171

where D_u and D_d are diagonal matrices if u and d are vectors, and scalars otherwise. As we said in the discussion of (5.5), it is often natural to choose the scaling so that each component of $u(t)$ or $d(t)$ is in the interval between -1 and 1. (If d is white noise with intensity I as in section 5.4, then this is often a natural scaling and further scaling is unnecessary.) The signals z^f, y^f, n^f, r^f and e^f are related through addition and subtraction. To preserve these relations they should have the same scaling.

$$Dz = z^f, \quad Dy = y^f, \quad Dn = n^f, \quad Dr = r^f, \quad De = e^f \qquad (7.3)$$

If we are primarily interested in the control error, then one natural choice of D is to make the tolerated error approximately equal to one.

In the scaled variables we have the relationships

$$z(t) = G(p)u(t) + G_d(p)d(t) \qquad (7.4a)$$
$$y(t) = z(t) + n(t) \qquad (7.4b)$$
$$e(t) = r(t) - z(t) \qquad (7.4c)$$

with G and G_d defined by

$$G = D^{-1}G^f D_u, \quad G_d = D^{-1}G_d^f D_d \qquad (7.5)$$

An advantage with the scaling is that it is now possible to discuss whether a gain of G or G_d is large or small. We will also see that it becomes simple to make rule-of-thumb calculations to see if control is possible at all.

7.2 Intuitive Analysis

We begin by considering some limitations of system performance which can be deduced intuitively. Below we will relate performance to the *gain crossover* ω_c. Recall that it is defined as the frequency where the magnitude of the loop gain is 1. When making approximate calculations one often assumes that the gain crossover is approximately equal to the bandwidth of the sensitivity function or the closed loop transfer function.

Unstable Systems

Let us assume that the system $G(s)$ has a pole in $s = p_1$ ($p_1 > 0$). The open loop response then contains a term of the form $e^{p_1 t}$. The size of the output signal is then approximately tripled in a time interval of length $1/p_1$. Intuitively it is clear that the controller in order to stop the instability from growing must react in a time scale faster than $1/p_1$. Since time and frequency domain quantities are approximately each other's inverses, this reasoning leads to the following gain crossover requirement

$$\omega_c > \text{approximately } p_1 \qquad (7.6)$$

Below we will formulate more precise specifications. However, we conclude that an unstable system can put severe restrictions on the choice of regulator, even if there are no performance specifications except stabilization.

Systems with Time Delay

Let the controlled system have a time delay T_d. This means that a control action at time t has no effect until $t + T_d$. Intuitively it is then clear that it is not possible to counteract disturbances or reference signal variations in a time scale shorter than T_d. In the frequency domain this corresponds to angular frequencies higher than $1/T_d$. It is thus unrealistic to achieve a bandwidth or a crossover frequency higher than that, so we have the performance limitation

$$\omega_c < \text{approximately } 1/T_d \qquad (7.7)$$

Non-minimum Phase Systems

A non-minimum phase system is characterized by zero(s) in the right half plane and in the time domain by unusual step responses. For systems with precisely one zero in the right half plane it is easy to see (by using the initial value and final value theorems) that the step response initially is in the opposite direction of the final value. In Figure 7.1 we show the step response of

$$\frac{-s + 1}{(s + 1)(0.5s + 1)(0.1s + 1)}$$

An intuitive interpretation of this step response is that the gain for rapid inputs has opposite sign to the static gain. This indicates that a controller that handles slow signals well might have to refrain from

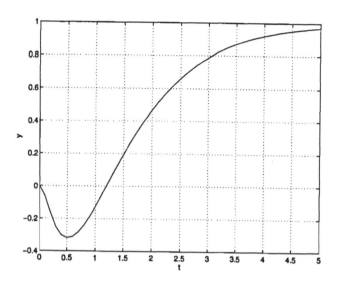

Figure 7.1: Step response of a non-minimum phase system.

trying to influence the fast behavior. There will then be an upper limit to the gain crossover.

$$\omega_c < \text{limit determined by right half plane zeros} \qquad (7.8)$$

Below we will try to be more precise about this limit.

Bounded Control Signal

In practise the control signal amplitude is bounded by various constraints. Therefore it is impossible to compensate for disturbances above a certain level. There is also a limit to how fast the system can react. It may be easy to forget this when working with linear models, because by definition they have the same qualitative response regardless of signal magnitude. One way of being aware of the importance of signal magnitude is to scale the variables as in Section 7.1.

The Role of the System Inverse

Let us assume that we want to control a system given by (6.2), that is

$$z(t) = G(p)u(t) + w(t)$$
$$y(t) = z(t) + n(t)$$

under the following ideal conditions.

1. The control signal is bounded

$$|u(t)| \leq u_{max}$$

   but we can choose u_{max} at will. (We can use as powerful motors, pumps, valves, etc., as we want.)

2. We can use sensors of arbitrary precision. The disturbance n can then be neglected and we have $y = z$.

3. We can describe $G(p)$ with arbitrary accuracy.

It is an interesting fact that we cannot always, even under these ideal and unrealistic circumstances, control the system exactly as we want. Consider the controller

$$u = F_r r - F_y y$$

where, as usual, r is the reference signal. From (6.14) we get the control signal

$$u = \frac{F_r}{1 + F_y G} r - \frac{F_y}{1 + F_y G} w = G^{-1}(G_c r - (1 - S)w) \qquad (7.9)$$

Perfect control is achieved for $G_c = 1$ and $S = 0$, which gives

$$u = G^{-1}(r - w) \qquad (7.10)$$

Note that if we had assumed w to be measured, we had immediately obtained (7.10) as a feed-forward solution. It is now clear that whether we think in terms of feed-forward or feedback solutions, we require the system inverse to get perfect control. This gives a different explanation of why time delays and non-minimum phase systems are difficult to control. A time-delay has an inverse which normally cannot be realized since it requires knowledge of future signals. A system with right half plane zeros has an unstable inverse, so perfect control might require a control signal to grow without bound.

7.3   Loop Gain Limitations

Sensitivity versus Complementary Sensitivity

In Section 6.1 the sensitivity function S, (6.9), and the complementary sensitivity function T, (6.10) were introduced. To prevent disturbances and minor modeling errors from affecting the output too much, the sensitivity function S should be small. To minimize the influence of measurement noise and to prevent modeling errors from destroying stability, the complementary sensitivity T should be small. We discussed these aspects in Section 6.3. There is obviously a severe conflict here since

$$S + T = 1 \qquad\qquad\qquad (7.11)$$

We conclude that S and T cannot both be small at the same frequency. (They can however both have a large magnitude, since complex numbers are added.) In most cases the conflict is resolved by requiring

- S to be small at low frequencies (and possibly some other frequency where disturbances are concentrated)

- T to be small at high frequencies

These requirements reflect typical control specifications and typical frequency properties of disturbances, measurement noise and model uncertainty as we discussed in Section 6.6.

S and T are uniquely determined by the *loop gain* GF_y. If ε is a small number, we have approximately

$$|S| < \varepsilon \Longleftrightarrow |GF_y| > \frac{1}{\varepsilon} \qquad\qquad (7.12)$$

$$|T| < \varepsilon \Longleftrightarrow |GF_y| < \varepsilon \qquad\qquad (7.13)$$

In the multivariable case, the interpretation is that all singular values should satisfy the requirements.

A typical loop gain requirement is given in Figure 7.2. Often we want the transition from high to low gain to take place over a narrow frequency interval. "If the sensitivity cannot be small, it might just as well be close to 1 so that T can be small". We will see, however, that there is a theoretical limit for how narrow the zone between ω_0 and ω_1 in Figure 7.2 can be.

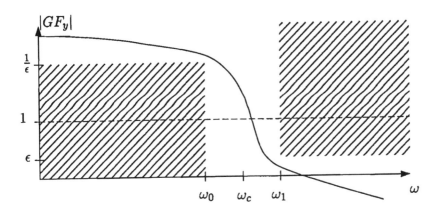

Figure 7.2: Shaded area: "forbidden region" for the singular values of the loop gain.

Amplitude and Phase Are Coupled

Pragmatically, we can describe the limitations for the change in loop gain in the following way: we know from elementary control theory that there has to be a positive phase margin, that is, the phase (argument) of GF must be greater than $-\pi$ when $|GF| = 1$. But if the slope of the amplitude curve is $-\alpha$ (i.e., $|GF|$ decreases with $20 \cdot \alpha$ dB per decade) over an interval, then the phase loss is at least $\approx -\alpha \cdot \frac{\pi}{2}$. This follows, e.g., from the rules for drawing Bode diagrams.

The amplitude curve thus cannot decrease faster than 40 dB/decade in an interval around the gain crossover frequency ω_c. This fact limits the transition from "small S" to "small T". From a formal mathematical point of view, these relations are described by *Bode's relationships*. A transfer function is an analytic function. It follows that there are certain relations between amplitude and phase. Let us define

$$A(\omega) = \log |G(i\omega)|, \quad \phi(\omega) = \arg G(i\omega) \tag{7.14}$$

A and ϕ are the quantities drawn in a Bode diagram (recall that the amplitude diagram normally is drawn in logarithmic scale). The properties of the complex logarithmic function ($\log z = \log |z| + i \arg z$) give the relation

$$\log G(i\omega) = A(\omega) + i\phi(\omega) \tag{7.15}$$

A problem is that ϕ is defined only up to a multiple of 2π. To handle that problem we will assume that $G(0) > 0$, set $\phi(0) = 0$ and define ϕ to be a continuous function. It is now possible to derive the following relation between A and ϕ.

Theorem 7.1 (Bode's relation) *Let $G(s)$ be a proper rational transfer function which has no poles in the right half plane and no poles or zeros on the imaginary axis. Assume that $G(0) > 0$ and let A and ϕ be defined as above. Then*

$$\phi(\omega) \le \frac{2\omega}{\pi} \int_0^\infty \frac{A(\nu) - A(\omega)}{\nu^2 - \omega^2} \, d\nu \tag{7.16}$$

for all $\omega \ge 0$. If the system has no zeros in the right half plane the inequality is replaced by an equality.

Proof: See Appendix 7A.   □

For stable minimum-phase systems the phase curve is consequently uniquely defined by the amplitude curve, provided the latter is given on the whole ω-axis. For systems with right half plane zeros, the phase curve will have a greater (negative) phase displacement. This is what motivates the terminology of minimum phase and non-minimum phase.

The result, which was first formulated by Bode, can be reformulated in many ways. The following one is perhaps most well-known. Let

$$f(\log \omega) = \log |G(i\omega)| \quad \text{i.e.,} \quad f(x) = \log |G(ie^x)| \tag{7.17}$$

(It is really the curve $f(x)$ which is drawn in a Bode diagram.) For a given amplitude curve $|G(i\omega)|$ the phase must then satisfy *Bode's relation.*

Theorem 7.2 (Bode's relation, variant) *Let $f(x)$ be defined by (7.17). Under the same conditions as in Theorem 7.1 the following holds.*

$$\arg G(i\omega) \le \frac{1}{\pi} \int_{-\infty}^\infty \frac{d}{dx} f(x) \cdot \psi(x - \log \omega) dx \tag{7.18}$$

where the weighting function ψ is given by

$$\psi(x) = \log \frac{e^x + 1}{|e^x - 1|} \tag{7.19}$$

see Figure 7.3. There is equality in (7.18) if there are no right half plane zeros.

Proof: The change of variables $\nu = e^x$ in the right hand side of (7.16), together with a partial integration, give (7.18). □

The phase curve must lie below a given curve, determined by the slope of the amplitude curve. Note that most of the contribution to the integral comes from the neighborhood of $x = \log \omega$ since ψ has the shape shown in Figure 7.3.

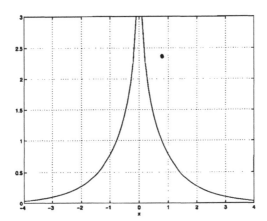

Figure 7.3: The weighting function $\psi(x)$ of Bode's relation.

It can be shown that

$$\int_{-\infty}^{\infty} \psi(x)dx = \frac{\pi^2}{2}$$

which means that Bode's relation gives the following approximation

$$\arg G(i\omega) \overset{<}{\approx} \frac{\pi}{2} \frac{d}{dx} f(x) \Big|_{x=\log \omega} = \frac{\pi}{2} \frac{d}{d\log \omega} \cdot \log |G(i\omega)| \qquad (7.20)$$

This is the same result as the pragmatic one, i.e., the phase decreases with $n \cdot 90°$ if the amplitude slope is $-n \cdot 20$ dB/decade.

7.4 S and T: Individual Limitations

The Sensitivity Cannot be Small Everywhere

We want the sensitivity function to be small over a large frequency range. On the other hand we have to weigh this desire against the need to keep T small, as discussed in the previous section. Since the gain of a physical system has to decrease for high enough frequencies, we must have

$$T(i\omega) \to 0, \quad S(i\omega) = 1 - T(i\omega) \to 1, \quad \text{when } \omega \to \infty$$

It would be natural to require $|S| \leq 1$ for all frequencies. The control action then does not amplify the disturbances at any frequency. In practical situations, however, one usually gets sensitivity functions similar to the ones shown in Figure 7.4. The figure shows the sensitivity

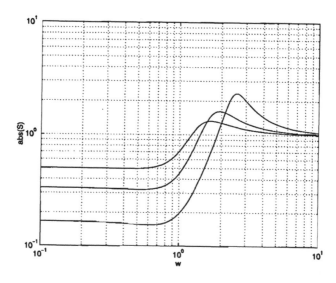

Figure 7.4: Sensitivity functions for $K = 1, 2, 5$ (growing peaks) when the loop gain is $K/(s^2 + s + 1)$.

functions obtained for P-control, $F_y = K$, of the system

$$G(s) = \frac{1}{s^2 + s + 1}$$

We note that an increase in loop gain decreases the magnitude of S at low frequencies, but at the price of larger peaks with $|S| > 1$ at higher frequencies. Experiments with other controllers and other systems give the same qualitative results. We are dealing with a general property of linear systems described by *Bode's Integral Theorem*.

Theorem 7.3 (Bode's Integral Theorem) *Assume that the magnitude of the loop gain, $|G(s)F_y(s)|$, decreases at least as fast as $|s|^{-2}$ as $|s|$ tends to infinity. Let the loop gain $G(s)F_y(s)$ have M poles in the right half plane, p_i, $i = 1, 2, \ldots, M$ (counted with multiplicity). Then the sensitivity function $S(s) = (1 + G(s)F_y(s))^{-1}$ satisfies the relation*

$$\int_0^\infty \log|S(i\omega)|\,d\omega = \pi \sum_{i=1}^M \mathrm{Re}(p_i) \qquad (7.21)$$

Proof: See Appendix 7A.   □

The theorem has several important interpretations and consequences.

1. The condition that $|GF_y|$ decreases as $1/|\omega|^2$ holds if neither the system G nor the regulator F_y lets through infinitely high frequencies, i.e., each of them decreases at least as fast as $1/\omega$. This is true for all physical systems described by rational transfer functions.

2. The integral expresses an *invariance*, i.e., no matter how F_y is chosen, the value of $\int \log|S(i\omega)|$ is preserved. (Provided unstable poles of F_y are not changed.)

3. If both the system and the controller are stable then

$$\int_0^\infty \log|S(i\omega)|\,d\omega = 0 \qquad (7.22)$$

Sensitivity $|S(i\omega)|$ less than 1 over a certain frequency interval must thus be paid for with a sensitivity greater than 1 somewhere else, as we saw in Figure 7.4. Expressing the sensitivity in dB, the surface below 0 dB equals the surface above 0 dB (with a linear frequency scale). In Figure 7.5 the sensitivity function of Figure 7.4 (for the case $K = 5$) is redrawn with a linear frequency scale.

4. If the loop gain is unstable the situation is worse. The surface corresponding to $|S| > 1$ then becomes larger than the one corresponding to $|S| < 1$. The faster an unstable pole is (the greater Re p_i is), the worse the sensitivity becomes. Note that the loop gain is unstable not only if the original system G is, but

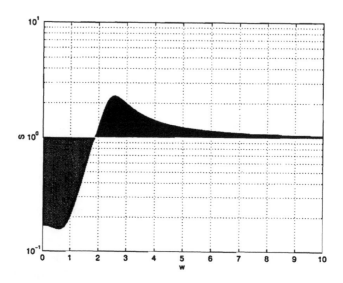

Figure 7.5: Illustration of Bode's integral theorem. The two areas are equal.

also if the controller F_y is unstable. There is thus good reason for avoiding unstable controllers. We also get a confirmation of our original conclusion of Section 7.2, that unstable systems are difficult to control.

5. The "repayment" in the form of sensitivity greater than 1 cannot take place at arbitrarily high frequencies. This can be seen by estimating the tail of the integral. If ω_1 is sufficiently large, we have the estimate

$$\int_{\omega_1}^{\infty} \log \frac{1}{|1 + GF_y|} d\omega \le \int_{\omega_1}^{\infty} \log \frac{1}{1 - |GF_y|} d\omega$$
$$\approx \int_{\omega_1}^{\infty} \log(1 + |GF_y|) d\omega \approx \int_{\omega_1}^{\infty} |GF_y| d\omega \quad (7.23)$$

Since GF_y has to decrease at least as ω^{-2} for the theorem to be valid, there is a constant C such that

$$|GF_y| \le \frac{C}{\omega^2} \quad (7.24)$$

for large enough ω. The tail of the integral can then be estimated with

$$\int_{\omega_1}^{\infty} \log |S(i\omega)| d\omega \stackrel{<}{\approx} \int_{\omega_1}^{\infty} C\omega^{-2} d\omega = \frac{C}{\omega_1} \tag{7.25}$$

The contribution to the integral from frequencies where GF_y is small (i.e., from frequencies well above the bandwidth) is thus very small.

Limitations of Specifications

In the previous chapter we discussed specifications of the sensitivity that have the form

$$\|W_S S\|_\infty \leq 1 \tag{7.26}$$

There are several limitations on the choice of W_S, to avoid specifications that are impossible to satisfy. Bode's integral gives some limitations. Other limitations arise from the fact that

$$S(z) = (1 + G(z)F_y(z))^{-1} = 1$$

if z is a zero of the loop gain. This relation is often referred to as an *interpolation condition*.

Theorem 7.4 *Let W_S be stable and minimum phase, and let S be the sensitivity function of a stable closed loop system. Then (7.26) can not be satisfied unless*

$$|W_S(z)| \leq 1 \tag{7.27}$$

for every right half plane zero z of the loop gain GF_y.

Proof: The proof is based on the fact that $W_S S$ is analytic in the right half plane and the fact that an analytic function achieves its maximum absolute value on the boundary of a set. Let the last fact be applied to a set enclosed by the imaginary axis and a half circle which is expanded to enclose all of the right half plane. Since the limit of $|W_S S|$ on the half circle equals the limit when $\omega \to \pm\infty$, we get

$$\sup_\omega |W_S(i\omega)S(i\omega)| \geq |W_S(s)S(s)|$$

for all s in the right half plane. If s is chosen as a zero of the loop gain, then (7.27) holds, since S becomes 1. □

Below we will use this fact to get a bandwidth limitation for non-minimum phase systems.

The Complementary Sensitivity Function

The complementary sensitivity function has to satisfy a number of conditions, similar to the ones restricting the sensitivity. The following condition is an analogy of Bode's integral theorem.

Theorem 7.5 *Let the loop gain $|G(s)F_y(s)|$ have a higher degree in the denominator than in the numerator and assume that $G(0)F_y(0) \neq 0$. Let $G(s)F_y(s)$ have M right half plane zeros (not counting the imaginary axis, but counted with multiplicity) z_i, $i = 1, 2, \ldots, M$. Then $T(s) = \frac{G(s)F_y(s)}{1+G(s)F_y(s)}$ satisfies the relation*

$$\int_0^\infty \log\left|\frac{T(i\omega)}{T(0)}\right| \frac{d\omega}{\omega^2} = -\frac{\pi}{2}\frac{e_1}{T(0)} + \pi\sum_{i=1}^{M}\mathrm{Re}\frac{1}{z_i} \qquad (7.28)$$

where e_1 is the error coefficient, cf (6.33):

$$e_1 = -\lim_{s\to 0}\frac{dT(s)}{ds} = \lim_{s\to 0}\frac{dS(s)}{ds}$$

Proof: The result is shown by computing

$$\int \log\left(\frac{T(s)}{T(0)}\right)\frac{ds}{s^2}$$

along an integration path, which in the limit encloses the right half plane, in analogy with the proofs of Theorems 7.3 and 7.2.

(For the case of a loop gain with at least two poles at the origin, the proof can be directly obtained from the proof of Bode's integral theorem. This is done by using the relation

$$T(s^{-1}) = \frac{1}{1 + F_y(s^{-1})^{-1}G(s^{-1})^{-1}}$$

which shows that $T(s^{-1})$ can be regarded as the sensitivity function of a loop gain $F_y(s^{-1})^{-1}G(s^{-1})^{-1}$ which has a numerator degree which is at least two units higher than that of the numerator. The result then follows from the fact that the poles of $F_y(s^{-1})^{-1}G(s^{-1})^{-1}$ are the inverses of the zeros of $F_y(s)G(s)$.)   □

We make some remarks.

1. For the case $F_r = F_y$ and $T(0) = 1$, the error coefficient e_1 gives the stationary error when the reference signal is a unit ramp.

2. If the loop gain has at least two poles at the origin, then $T(0) = 1$ and $e_1 = 0$. We then obtain an invariant integral for $|T|$ in analogy with the one for sensitivity of Theorem 7.3. If it desired to decrease $|T|$ at some frequencies then it has to increase somewhere else.

3. If the loop gain has precisely one pole at the origin then we still have $T(0) = 1$, but e_1 is nonzero. The integral of $\log|T|$ depends on e_1 (which normally is positive). It might therefore be easier to get a decrease in $|T|$ if a greater value of e_1 is accepted.

4. The restrictions on T of the two cases above become more severe if there are zeros in the right half plane, especially if they are close to the imaginary axis.

5. If the loop gain contains no pole at the origin, then $T(0) \neq 1$. In that case the ratio $|T(i\omega)/T(0)|$ satisfies an invariance condition. Equation (7.28) is then consistent with $|T|$ being small at all frequencies. There is, however, still the fact that S is close to 1 when T is small which has to be taken into account.

Limitations when Specifying T

In the previous chapter we discussed robustness specifications of the form

$$\|W_T T\|_\infty \leq 1 \qquad (7.29)$$

If the loop gain has a pole at p_1, then

$$T(p_1) = 1$$

This gives the following limitation in the choice of W_T.

Theorem 7.6 *Let W_T be stable and minimum phase. Then (7.29) can not be satisfied unless*

$$|W_T(p_1)| \leq 1 \qquad (7.30)$$

for every right half plane pole p_1 of the loop gain GF_y.

Proof: In analogy with Theorem 7.4 we get

$$\sup_\omega |W_T(i\omega)T(i\omega)| \geq |W_T(s)T(s)|$$

for all s in the right half plane. If s is chosen as a right half plane pole, the result follows. $\square$

7.5   Consequences for the System Performance

Consequences for Unstable Systems

We shall now use the results of the previous sections to get a deeper understanding of the difficulties with unstable systems. The intuitive reasoning of Section 7.2 suggested that a bandwidth larger than the unstable pole is required. We will first look at this question using Bode's integral theorem.

Suppose we control an unstable system G having a real pole at $s = p_1$, with a controller F_y. Let M be the maximum value of $|S|$. We then have from (7.21)

$$\pi p_1 = \int_0^{p_1} \log |S(i\omega)| d\omega + \int_{p_1}^{\infty} \log |S(i\omega)| d\omega$$

$$\leq p_1 \log M + \int_{p_1}^{\infty} \log |S(i\omega)| d\omega$$

Suppose the loop gain for angular frequencies above p_1 is negligible. We can use the same reasoning as with (7.25) to see that the contribution from the last integral is negligible and get

$$\log M \geq \pi \;\; \Rightarrow \;\; M \geq e^{\pi} \approx 23$$

This is normally quite unacceptable. It follows that the loop gain above p_1 cannot be too small. This agrees qualitatively with our intuitive conclusion that the bandwidth has to be greater than p_1.

A more precise quantitative estimate can be derived from Theorem 7.6. Suppose we require

$$\|W_T T\|_\infty \leq 1, \quad \text{i.e., } |T(i\omega)| \leq |W_T^{-1}(i\omega)|$$

with W_T chosen as

$$W_T = \frac{s}{\omega_o} + \frac{1}{T_o}$$

The Bode diagram of W_T^{-1} is shown in Figure 7.6. We thus specify a bound on the peak of $|T|$ parameterized by T_o, and an upper bound on the bandwidth, parameterized by ω_o. If the loop gain has a real pole at p_1, we get from Theorem 7.6

$$|W_T(p_1)| \leq 1 \;\; \Rightarrow \;\; \frac{p_1}{\omega_o} + \frac{1}{T_o} \leq 1 \;\; \Rightarrow \;\; \omega_o \geq \frac{p_1}{1 - 1/T_o} \qquad (7.31)$$

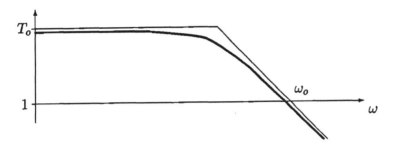

Figure 7.6: Bode diagram of W_T^{-1}.

If we accept $T_o = 2$, we have to choose $\omega_o > 2p_1$, a rule-of-thumb which is often given.

Example 7.1:   Balancing

We want to balance a stick on the finger, see Figure 7.7. The input is the acceleration of the finger and the output is the angle of the stick with respect to a vertical line. The length of the stick is ℓ and its mass is m. We consider

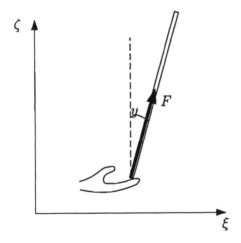

Figure 7.7: Balancing. The lower end point of the stick has the coordinates ξ and $\zeta = 0$.

horizontal motion by the finger along a line. The position is given by the coordinate ξ. The input is then

$$u = \ddot{\xi}.$$

The center of mass has the horizontal coordinate

$$\xi + \frac{\ell}{2}\sin y$$

and the vertical coordinate

$$\frac{\ell}{2}\cos y$$

We view the point of contact between the finger and the stick as a hinge without friction. The force F from the finger then must point in the direction of the stick. The Newtonian force equation gives

$$F\cos y - mg = m\frac{d^2}{dt^2}(\frac{\ell}{2}\cos y) = m\frac{\ell}{2}(-\ddot{y}\sin y - \dot{y}^2\cos y)$$

for the ζ-component and

$$F\sin y = m\ddot{\xi} + m\frac{d^2}{dt^2}(\frac{\ell}{2}\sin y) = mu + m\frac{\ell}{2}(\ddot{y}\cos y - \dot{y}^2\sin y)$$

for the ξ-component. Eliminating F and simplifying we obtain

$$\frac{\ell}{2}\ddot{y} - g\sin y = -u\cos y \tag{7.32}$$

For small angles we get the transfer function

$$G(s) = \frac{-2/\ell}{s^2 - \frac{2g}{\ell}}$$

The poles are $\pm\sqrt{\frac{2g}{\ell}}$ and the system is unstable. From our argument above we should have a bandwidth well in excess of $\sqrt{2g/\ell}$, say $2\pi\sqrt{2g/\ell}$ rad/s. The stabilization is thus more difficult to achieve for a short stick than for a long one. The practical limitation comes from the bandwidth of the hand-eye coordination. There are variations from one individual to another, but 10 Hz (time to react ≈ 0.1 second) is a good figure. This corresponds to a stick that is about 20 cm long. For a pole of one meter, a bandwidth of about $\sqrt{20} \approx 4.5$ Hz is enough. This can be managed by most people.

Consequences for Time Delays

A system with time delay can be written in the form

$$G(s) = G_1(s)e^{-sT_d}$$

Let us once more consider Equation (7.9) from Section 7.2:

$$u = G^{-1}(G_c r - (1 - S)w)$$

It is now impossible to achieve ideal control $G_c \to 1$, $S \to 0$, since it would give the non-causal controller

$$u = G_1^{-1}(s)e^{sT_d}(r - w)$$

To calculate $u(t)$, one would have to know $r(t + T_d)$ and $w(t + T_d)$. Thus a realistic controller can only be obtained if G_c and $1 - S = T$ contain a delay e^{-sT_d} which is cancelled when multiplying with G^{-1}. The best approximation of an ideal control then becomes

$$G_c = e^{-sT_d}, \quad T = e^{-sT_d}$$

The corresponding ideal sensitivity is

$$S = 1 - e^{-sT_d} \tag{7.33}$$

In Figure 7.8 we have given the ideal sensitivity function of (7.33) for $T_d = 1$. For small values of ω we get $S(i\omega) \approx i\omega T_d$ from (7.33).

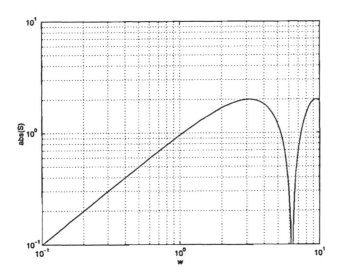

Figure 7.8: Ideal sensitivity function for a unit time delay.

The ideal sensitivity function thus has a magnitude less than 1 up to approximately $\omega = 1/T_d$. This is also approximately the bandwidth, ω_B, of S, so we have

$$\omega_B < 1/T_d \tag{7.34}$$

for a system containing a time delay.

Consequences for Non-Minimum Phase Systems

In Section 7.2 we discussed non-minimum phase systems and said on intuitive grounds that there would probably be an upper limit to the bandwidth. We will now use Theorem 7.4 to get a quantitative value.

Suppose we require S to satisfy

$$\|W_S S\|_\infty \le 1, \quad \text{i.e., } |S(i\omega)| \le |W_S^{-1}(i\omega)| \quad \forall \omega$$

where W_S^{-1} is given by

$$W_S^{-1} = \frac{S_o s}{s + \omega_o S_o}$$

with the Bode diagram of Figure 7.9. We require S to suppress

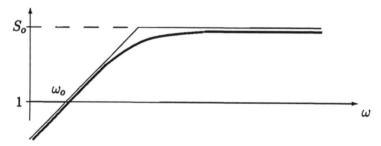

Figure 7.9: Bode diagram of W_S^{-1}.

disturbances in the frequency interval up to ω_o and we also set an upper bound on its magnitude, parameterized by S_o. If the loop gain contains a real zero $z > 0$, then Theorem 7.4 gives

$$|W_S(z)| \le 1 \; \Rightarrow \; \frac{z + \omega_o S_o}{S_o z} \le 1 \; \Rightarrow \; \omega_o \le (1 - 1/S_o)z \qquad (7.35)$$

If we accept $S_o = 2$ then we have to take $\omega_o \le z/2$. Even if we accept $S_o \to \infty$ we have to choose $\omega_o \le z$. The bandwidth of a system with non-minimum phase loop gain can realistically not be greater than approximately half the value of the zero.

Another point of view is the following. Assume that the loop gain has a real zero $z > 0$, which is the only right half plane zero. We can write the loop gain in the form

$$G(s)F(s) = \frac{z - s}{z + s}\tilde{G}(s)F(s)$$

where $\tilde{G}(s)$ is a minimum phase system. For $s = i\omega$ we get

$$|G(s)F(s)| = |\tilde{G}(s)F(s)|$$

The gain crossover ω_c is the same for GF as for the minimum phase system $\tilde{G}F$. The phase margin is, however, decreased by the factor $(z - i\omega_c)/(z + i\omega_c)$. For $\omega_c = z$ this gives a decrease of phase margin which is $\pi/2$. The phase shift of $\tilde{G}F$ can then be at most $-\pi/2$ to get any stability at all. From Bode's relation this is the phase shift given by having a slope of -20dB/decade, which is quite typical at the cross-over frequency. To get a reasonable stability margin we have to take a gain crossover frequency well below $\omega = z$.

It is interesting to compare the bandwidth limitations of non-minimum phase and time delay systems. Approximating the time delay with a rational function gives

$$e^{-sT} \approx \frac{1 - sT/2}{1 + sT/2}$$

with a zero in $2/T$. From (7.35) with $S_o = 2$ we get $\omega_o < 1/T$, which is the same limitation as we got when reasoning directly about the time delay.

Finally, we will also look at non-minimum phase system limitations in the time domain. Consider a system $G(s)$ with a right half plane zero:

$$G(z) = 0, \qquad \mathrm{Re}\, z > 0$$

Choose an arbitrary controller $F_r(s) = F_y(s) = F(s)$ giving a stable closed loop system with static gain equal to 1. The Laplace transform of the closed loop step response is

$$Y(s) = (I + G(s)F(s))^{-1}G(s)F(s)\frac{1}{s}$$

Hence $Y(s) = 0$ for the zero $s = z$. We get, using the definition of the Laplace transform

$$0 = Y(z) = \int_0^\infty y(t)e^{-zt}dt \tag{7.36}$$

(Here we have used the fact that the Laplace integral is convergent for $s > 0$.) Since the static gain is 1, $y(t)$ eventually tends to one, and in particular gets positive. For the integral to be zero, $y(t)$ then also has to take negative values.

We can estimate the magnitude of the negative values. Let T_s be the settling time to within 10% of the final value, i.e.,

$$|y(t) - 1| \le 0.1 \quad \text{for } t > T_s$$

and define the undershoot $M = -\min_{t \ge 0} y(t)$. From (7.36) we get

$$-\int_{y \le 0} y(t) e^{-zt} dt = \int_{y > 0} y(t) e^{-zt} dt$$

where we have divided the time axis into intervals where y has constant sign. The interval where $y \le 0$ must be contained in $[0, T_s]$ and the intervals where $y > 0$ must contain at least $[T_s, \infty]$. We then get the following estimates

$$M \int_0^{T_s} dt \ge -\int_{y \le 0} y(t) e^{-zt} dt$$

$$= \int_{y > 0} y(t) e^{-zt} dt \ge 0.9 \int_{T_s}^{\infty} e^{-zt} dt = \frac{0.9}{z} e^{-zT_s}$$

and conclude that

$$M \ge \frac{0.9}{zT_s} e^{-zT_s} \tag{7.37}$$

For small values of zT_s the undershoot will thus be very large. For a given value of z, there is therefore a lower limit on the settling time T_s if a reasonable step response is wanted. This agrees well with the bandwidth limitations we discussed above. Conversely, if z is much greater than the inverse of the desired settling time, then (7.37) will cause no problems.

7.6   Effects of Control Signal Bounds

Consider a control problem defined by

$$y = Gu + G_d d \tag{7.38}$$

where d is a disturbance. Assume that the variables are scaled as described in Section 7.1. We then have

$$|d(t)| \le 1, \quad |u(t)| \le 1$$

Let us focus on disturbance suppression and put $r = 0$. We see that the control signal

$$u = -G^{-1}G_d d \tag{7.39}$$

gives perfect disturbance rejection. The question is whether this control signal can be applied considering the bound $|u(t)| \leq 1$.

Theorem 7.7 *A necessary condition for all disturbances with $|d(t)| \leq 1$ to be perfectly rejected by control signals satisfying $|u(t)| \leq 1$ is that*

$$|G(i\omega)| \geq |G_d(i\omega)|, \quad all \ \omega \tag{7.40}$$

Proof: If d is a pure sinusoid, then from (7.39) the steady state u becomes a sinusoid amplified by the factor $|G(i\omega)^{-1}G_d(i\omega)|$. This factor then has to be less than 1, which gives the result. □

The condition given by (7.40) is obviously also sufficient if we consider steady state rejection of a single sinusoid. For more general disturbances it is not sufficient, as shown by the following example.

Example 7.2: Disturbance rejection

Let G and G_d be

$$G(s) = \frac{1}{2s+1}, \quad G_d = \frac{1.4}{(s+1)(s+2)}$$

In Figure 7.10 their amplitude curves are shown. We see that the requirement of Theorem 7.7 is satisfied, but with no margin. The control signal which gives perfect disturbance rejection is

$$u = -\frac{2.8s+1.4}{(s+1)(s+2)}d$$

If $d(t)$ is chosen as

$$d(t) = \begin{cases} 1 & 0 \leq t < 5 \\ -1 & t \geq 5 \end{cases}$$

we get the result of Figure 7.11. We see that the control signal has an amplitude greater than one during a short interval around $t = 6$.

The reasoning above assumed that we want perfect disturbance rejection. In practice we want to suppress the influence of disturbances

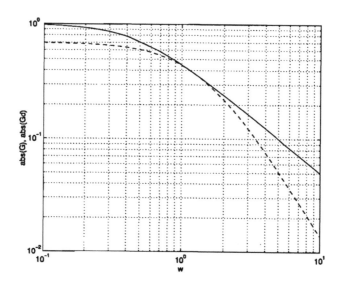

Figure 7.10: A comparison of $|G|$ (solid) and $|G_d|$ (dashed).

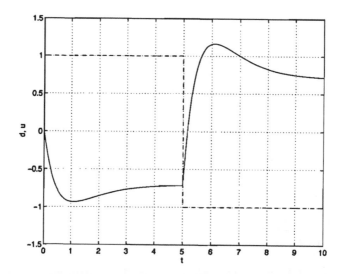

Figure 7.11: Disturbance (dashed) and control signal (solid) for perfect disturbance rejection.

below some threshold. Assume that the variables are scaled so that we accept an error less than 1. Consider the relation

$$e = -y = -Gu - G_d d$$

Let d be given by

$$d = \sin \omega t$$

where ω is an angular frequency where $|G(i\omega)| < |G_d(i\omega)|$. The best to do is to decrease the amplitude of e to the value

$$|G_d(i\omega)| - |G(i\omega)|$$

Requiring this amplitude to be less than 1 gives the following condition.

$$|G(i\omega)| \geq |G_d(i\omega)| - 1 \tag{7.41}$$

Note that there will be no requirement on G for frequencies where $|G_d(i\omega)| < 1$. For those frequencies the effect of the disturbance is so small that there is no need for control action.

7.7   The Multivariable Case

Most of the discussion of the previous sections can be generalized to the multivariable case, but the conclusions are often not as straightforward.

Bode's Integral

Theorem 7.3 has a generalization to multivariable systems.

Theorem 7.8 (Bode's Integral Theorem, MIMO) *Assume that every element of the loop gain $G(s)F_y(s)$ decreases at least as $1/|s|^2$ as $|s|$ tends to infinity. Let the loop gain have M poles in the right half plane: p_i, $i = 1, 2, \ldots, M$ (counted with multiplicity). Then the sensitivity function $S(s) = (1 + G(s)F_y(s))^{-1}$ satisfies*

$$\int_0^\infty \log |\det S(i\omega)| \, d\omega = \pi \sum_{i=1}^M \text{Re}(p_i) \tag{7.42}$$

Proof: See Appendix 7A. □

If S has the singular values $\sigma_1,..,\sigma_m$ then it can be shown that

$$|\det S| = \sigma_1 \cdots \sigma_m \tag{7.43}$$

From Theorem 7.8 we get:

Corollary 7.1 *Under the conditions stated in Theorem 7.8 the singular values of the sensitivity function satisfy*

$$\sum_{k=1}^{m} \int_0^{\infty} \log \sigma_k(S(i\omega))d\omega = \pi \sum_{i=1}^{M} \mathrm{Re}(p_i) \tag{7.44}$$

The largest singular value, $\bar{\sigma}$, must contribute at least $1/m$ of the sum, giving

Corollary 7.2 *Under the conditions stated in Theorem 7.8 the largest singular value of the sensitivity function satisfies*

$$\int_0^{\infty} \log \bar{\sigma}(S(i\omega))d\omega \geq \frac{\pi}{m} \sum_{i=1}^{M} \mathrm{Re}(p_i) \tag{7.45}$$

As in the SISO case the sensitivity function has an invariance property. The most useful quantity in practice, the largest singular value, only satisfies an inequality, however. The interpretation is therefore less straightforward. It is possible to get better estimates for the integral of the largest singular value, but at the price of more complicated calculations.

Right Half Plane Zeros

For simplicity we consider systems $G(s)$ which are square and have no zero coinciding with a pole. From Definition 3.5 we then get

$$\det G(z) = 0$$

for a zero $s = z$ of G. There is then a nonzero row vector a, such that

$$aG(z) = 0$$

Let us now assume that z lies in the right half plane. Since the closed loop system G_c and the complementary sensitivity function T satisfy (6.8), (6.10) i.e.,

$$(I + GF_y)G_c = GF_r, \quad (I + GF_y)T = GF_y$$

we get by multiplying from the left

$$aG_c(z) = 0, \quad aT(z) = 0 \tag{7.46}$$

(Since z is in the right half plane, it is not possible to cancel the zero by putting a pole in F_y or F_r without destroying the internal stability. All matrix elements are thus finite at $s = z$.) We conclude that the right half plane zero has to remain in the closed loop system and complementary sensitivity function. As in the SISO case we get *interpolation conditions* which have to be satisfied and limit the achievable performance. As a consequence we get a multivariable analogy of Theorem 7.4 of Section 7.4.

Theorem 7.9 *Assume that G has a zero z in the right half plane. Let the scalar transfer function W_S be stable and minimum phase. Then a necessary condition for*

$$\|W_S S\|_\infty = \sup_\omega \bar{\sigma}\left(W_S(i\omega)S(i\omega)\right) \le 1 \tag{7.47}$$

is that

$$|W_S(z)| \le 1 \tag{7.48}$$

Proof: The proof is analogous to the proof of Theorem 7.4, using the scalar function $W_S(s)vS(s)\bar{v}^T$, where v is a normed vector such that $vG(z) = 0$. $\Box$

The theorem shows that right half plane zeros give the same type of limitations as they do in the SISO case. Since the limitation lies in the requirements of the largest singular value there is a possibility of achieving better performance for certain linear combinations of signals. We illustrate this fact by considering Example 1.1.

Example 7.3:   Non-minimum Phase MIMO System

Consider the system (1.1):

$$G(s) = \begin{bmatrix} \frac{2}{s+1} & \frac{3}{s+2} \\ \frac{1}{s+1} & \frac{1}{s+1} \end{bmatrix}$$

From Example 3.2 we know that G has a zero at $s = 1$. We have

$$\begin{bmatrix} 1 & -2 \end{bmatrix} G(1) = \begin{bmatrix} 1 & -2 \end{bmatrix} \begin{bmatrix} 1 & 1 \\ 0.5 & 0.5 \end{bmatrix} = 0$$

Consider three different controllers.

Controller 1 With

$$F_y = F_r = \begin{bmatrix} K_1 \frac{s+1}{s} & -\frac{3K_2(s+0.5)}{s(s+2)} \\ -K_1 \frac{s+1}{s} & \frac{2K_2(s+0.5)}{s(s+1)} \end{bmatrix}$$

we get the diagonal loop gain

$$GF_y = \begin{bmatrix} \frac{K_1(-s+1)}{s(s+2)} & 0 \\ 0 & \frac{K_2(s+0.5)(-s+1)}{s(s+1)^2(s+2)} \end{bmatrix}$$

The sensitivity function, complementary sensitivity function and closed loop system then also become diagonal. We have achieved a decoupling of the control problem into two single loop problems. Since each diagonal element of the loop gain has a zero at $s = 1$, each loop has the restrictions due to right half plane zeros in the SISO case. None of the loops can thus have a bandwidth of more than approximately 0.5 rad/s. In Figure 7.12 the response of the closed loop system is shown for steps in the reference signals and $K_1 = 1$, $K_2 = 1$. Both loops show typical non-minimum phase behavior.

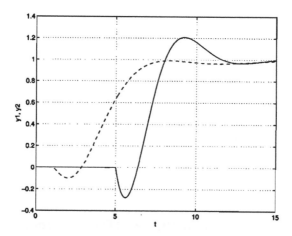

Figure 7.12: Output (y_1 solid, y_2 dashed) of a decoupled non-minimum phase system. There are reference steps at $t = 1$ (r_2) and $t = 5$ (r_1).

Controller 2 With the choice

$$F_y = F_r = \begin{bmatrix} K_1 \frac{s+1}{s} & K_2 \\ -K_1 \frac{s+1}{s} & K_2 \end{bmatrix}$$

we get the loop gain

$$GF_y = \begin{bmatrix} \frac{K_1(-s+1)}{s(s+2)} & \frac{K_2(5s+7)}{(s+2)(s+1)} \\ 0 & \frac{2K_2}{s+1} \end{bmatrix}$$

There is now a partial decoupling, output 2 is not affected by input 1. We also see that the right half plane zero has disappeared from the (2,2)-element but is still present in the (1,1)-element. The closed loop system is of the form

$$G_c = T = \begin{bmatrix} \frac{K_1(-s+1)}{d_{11}(s)} & g_{12}(s) \\ 0 & \frac{2K_2}{s+1+2K_2} \end{bmatrix}$$

By choosing K_2 large enough we can achieve an arbitrarily high bandwidth from reference 2 to output 2 despite the zero in $s = 1$. The non-minimum phase effect is present in the other output however. The response to steps in the reference signals with $K_1 = 1$ and $K_2 = 10$ is shown in Figure 7.13. We see that y_2 is very fast but at the price of a strong coupling to y_1. There is also a clear non-minimum phase effect in y_1.

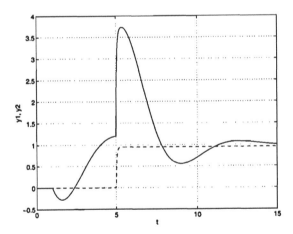

Figure 7.13: Output (y_1 solid, y_2 dashed) of a non-minimum phase system with u_2 and y_2 in a fast loop. The reference steps come at $t = 1$ (r_1) and $t = 5$ (r_2).

Controller 3 With

$$F_y = F_r = \begin{bmatrix} K_1 & -\frac{3K_2(s+0.5)}{s(s+2)} \\ K_1 & \frac{2K_2(s+0.5)}{s(s+1)} \end{bmatrix}$$

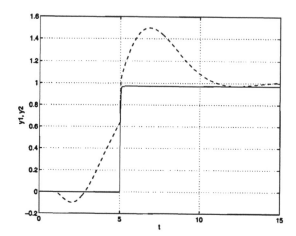

Figure 7.14: Output (y_1 solid, y_2 dashed) of a non-minimum phase system with u_1 and y_1 in a fast loop. The reference steps come at $t = 1$ (r_2) and $t = 5$ (r_1).

we get the loop gain

$$\begin{bmatrix} \frac{K_1(5\,s+7)}{(s+1)(s+2)} & 0 \\ \frac{2\,K_1}{s+1} & \frac{K_2(-1+s)(s+0.5)}{s(s+1)^2(s+2)} \end{bmatrix}$$

In this case y_1 is decoupled from u_2 and the bandwidth can be increased arbitrarily by increasing K_1. The response when the reference signals are steps and $K_1 = 10$, $K_2 = 1$ is shown in Figure 7.14. In this case the response in y_1 is fast and the bad behavior is seen in y_2.

The example suggests, as expected, that it is not possible to get around the effects of the right half plane zero. There is, however, considerable freedom in choosing where the effects become visible. We will analyze the situation further by considering a 2×2-system, with a real zero at z. We assume that both elements of the real vector $a = [a_1 \quad a_2]$ satisfying $aG(z) = 0$ are nonzero. To get a completely decoupled system we require T to be of the form

$$T(s) = \begin{bmatrix} t_{11}(s) & 0 \\ 0 & t_{22}(s) \end{bmatrix}$$

The interpolation condition $aT(z) = 0$ of (7.46) then gives

$$t_{11}(z) = 0, \quad t_{22}(z) = 0$$

The right half plane zero is thus present in both the decoupled loops in this case, as illustrated in the example. If we want perfect control from r_1 to y_1 we require

$$T(s) = \begin{bmatrix} 1 & 0 \\ t_{21}(s) & t_{22}(s) \end{bmatrix}$$

(Requiring the (1,1)-element to be 1 is of course an idealization.) The interpolation condition gives

$$t_{21}(z) = -a_1/a_2, \quad t_{22}(z) = 0$$

We see that there is no obstacle to the design in principle, but we have to accept the non-minimum phase effects from r_2 to y_2 and also the coupling from r_1 to y_2. Similarly we can see that it is possible to achieve the designs

$$\begin{bmatrix} t_{11}(s) & t_{12}(s) \\ 0 & 1 \end{bmatrix}, \quad \begin{bmatrix} 0 & 1 \\ t_{21}(s) & t_{22}(s) \end{bmatrix}, \quad \begin{bmatrix} t_{11}(s) & t_{12}(s) \\ 1 & 0 \end{bmatrix}$$

in principle. There is thus the possibility to achieve perfect control from one reference component to one output component, at the cost of having the undesirable behavior moved to the response of the other output component.

7.8 Some Examples

We will illustrate the previous analysis by presenting some physical examples.

Example 7.4: Temperature Control

Consider Example 5.5 describing temperature control. The system is described by the two equations

$$\dot{z}^f = K_1(x_1^f - z^f) - K_2(z^f - w^f)$$
$$\dot{x}_1^f = K_3(u^f - x_1^f) - K_4(x_1^f - z^f)$$

relating the room temperature z^f, the radiator temperature x_1^f, the outdoor temperature w^f and the water temperature u^f. With the numerical values

$$K_1 = K_2 = K_4 = 0.7, \quad K_3 = 35$$

(where the time scale is hours) a possible equilibrium is at

$$x_1^f = 50°C, \quad z^f = 20°C, \quad w^f = -10°C, \quad u^f = 50.6°C$$

In what follows we will interpret z^f, u^f, w^f and x_1^f as deviations from these values. The system model in transfer function form is the following (after some rounding of the numerical values)

$$z^f = \frac{0.5}{(0.03s+1)(0.7s+1)} u^f + \frac{0.01s+0.5}{(0.03s+1)(0.7s+1)} w^f$$

We note that the system has two time constants, 0.03 hours (the radiator) and 0.7 hours (the room).

Suppose the control system is required to keep room temperature within $\pm 1°C$ when the outdoor temperature varies $\pm 10°C$ and that u can be varied $\pm 20°C$.

To model the fact that w cannot change arbitrarily fast, we put

$$w^f = \frac{1}{s+1} d^f$$

where d^f also lies in the interval $\pm 10°C$. With the scaling

$$u = u^f/20, \quad z = z^f, \quad d = d^f/10$$

we get

$$z = \frac{10}{(0.03s+1)(0.7s+1)} u + \frac{0.1s+5}{(0.03s+1)(0.7s+1)(s+1)} d$$

In Figure 7.15 these transfer functions are shown. We see that the condition $|G(i\omega)| \geq |G_d(i\omega)|$ is satisfied at all frequencies. Since $|G_d(i\omega)|$ is greater than one for angular frequencies up to around 2 radians/hour, this is the frequency interval where control action is needed. It is thus natural to aim at a gain crossover frequency of about 2 radians/hour.

It might be interesting to investigate what happens if we take away the factor $1/(s+1)$ from G_d. This corresponds to a model of outdoor temperature which admits instantaneous changes in the interval $\pm 10°C$. A comparison of $|G|$ and $|G_d|$ is shown in Figure 7.16. With this disturbance model we can (as expected) not achieve perfect disturbance rejection over the whole frequency axis. However, at the point where the curves intersect, the gain of G_d is less than 0.1, so the disturbances that cannot be perfectly rejected will not affect the control error much. We also see that we require somewhat higher bandwidth of the regulator, since $|G_d(i\omega)|$ is greater than one in a larger frequency interval.

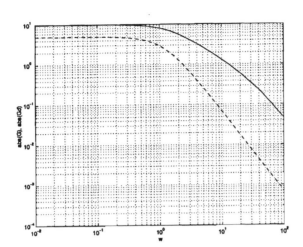

Figure 7.15: $|G|$ (solid) and $|G_d|$ (dashed) for temperature control. Since $|G| > |G_d|$ at all frequencies, complete disturbance rejection is in principle possible with the available input amplitude.

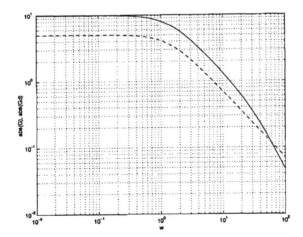

Figure 7.16: $|G|$ (solid) and $|G_d|$ (dashed) for temperature control without time constant for the outdoor temperature. Since $|G| > |G_d|$ except where $|G_d|$ is very small, sufficient disturbance suppression can, in principle, be obtained.

Example 7.5:   A Reason for Using Flywheels in Rotating Machinery

Consider a motor turning a shaft. Let J be the moment of inertia of the rotating parts, ω the angular velocity, D the applied torque, B the braking torque from the load and b the coefficient of viscous friction. Then the system is described by

$$J\dot{\omega} = -b\omega + D - B \tag{7.49}$$

Regarding D as control signal and B as a disturbance we get the transfer functions

$$G(s) = -G_d(s) = \frac{1}{Js + b} \tag{7.50}$$

Let D_0 be the maximum torque from the motor, B_0 the maximum disturbance and ω_0 the greatest allowed error in controlling the speed of rotation. The scaled transfer functions are then

$$G(s) = \frac{D_0/\omega_0}{Js + b}, \quad G_d = -\frac{B_0/\omega_0}{Js + b} \tag{7.51}$$

In principle, the Bode diagram is the one shown in Figure 7.17 if we assume that D_0 and B_0 are considerably larger than $b\omega_0$. The condition (7.40) gives

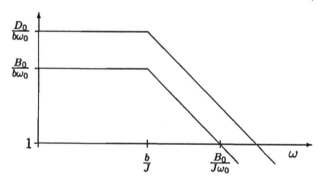

Figure 7.17: Asymptotic Bode diagrams for G and G_d for the flywheel example.

the obvious requirement

$$D_0 \geq B_0 \tag{7.52}$$

The gain crossover frequency of G_d is

$$\omega = \frac{B_0}{J\omega_0} \tag{7.53}$$

This is the smallest possible bandwidth of the system if sudden disturbances are to be rejected. From the Bode diagram we see that G automatically gets a higher gain crossover if (7.52) is satisfied. We have, however, not modeled G completely. For an electric motor we also have the inductance and resistance of the windings and power amplifier dynamics. Other types of machinery will similarly have additional dynamics. All this unmodeled dynamics has to have a sufficiently high bandwidth in order to be negligible. The same has to be true for sensors and the digital equipment used for control. This means that the bandwidth requirement (7.53) can be unrealistic. One way of moderating the requirement is to increase the moment of inertia, J, e.g., by putting a flywheel on the axis. Note that this also lowers the gain crossover of G. This is acceptable as long as it is not in conflict with other requirements, e.g., the speed of response to reference signals.

Example 7.6: A Zero in the Right Half Plane

Consider the system

$$G(s) = \frac{-s+1}{s(s+1)} \tag{7.54}$$

It has a zero at $s = 1$. We use a second order controller F_y whose coefficients can be chosen to place the four poles of the closed loop system as double poles at $-a \pm ia$. The parameter a is used to change the speed of response of the closed loop system. In Figure 7.18 the sensitivity function and the complementary sensitivity function are shown for the a-values 0.5, 1, 2, 5 och 10. We see that the bandwidth of T increases approximately in proportion to the distance to the origin of the poles, which is expected. For S, however, the crossing of $|S| = 1$ cannot be pushed higher than approximately $\omega = 1$, at the price of a high peak. This agrees entirely with (7.35).

Example 7.7: A Pole in the Right Half Plane

Let us change the previous example so that the zero is in the left half plane but a pole is in the right half plane:

$$G(s) = \frac{s+1}{s(s-1)} \tag{7.55}$$

As in the previous example, we use a control law where F_y is a second order system whose coefficients have been chosen so that the four poles of the closed loop system are double poles at $-a \pm ia$. The parameter a is chosen so that the closed loop system has varying speed of response: $a = 0.1, 0.2, 0.5, 1$ and 2. The results are shown in Figure 7.19. We see that the bandwidth of T never drops below 1 even if the closed loop system has very slow poles. This is in agreement with (7.31). We also note that both S and T have large peaks when this limit is approached. The gain crossover of the sensitivity function, however, is decreased when the system becomes slower.

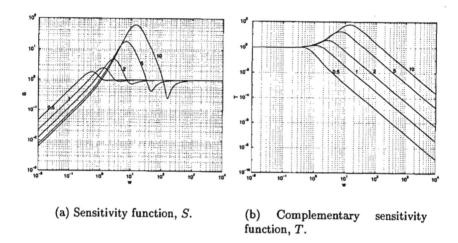

(a) Sensitivity function, S.   (b) Complementary sensitivity function, T.

Figure 7.18: Sensitivity function and complementary sensitivity function for non-minimum phase system. The four poles are placed as double poles at $a(-1 \pm i)$ where $a = 0.5, 1, 2, 5, 10$ (shown at the curves).

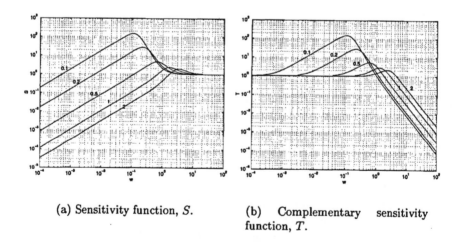

(a) Sensitivity function, S.   (b) Complementary sensitivity function, T.

Figure 7.19: Sensitivity function and complementary sensitivity function for a system with unstable loop gain. The four closed loop poles are placed as double poles at $a(-1 \pm i)$ where $a = 0.1, 0.2, 0.5, 1, 2$.

Example 7.8: A System with Both a Zero and a Pole in the Right Half Plane

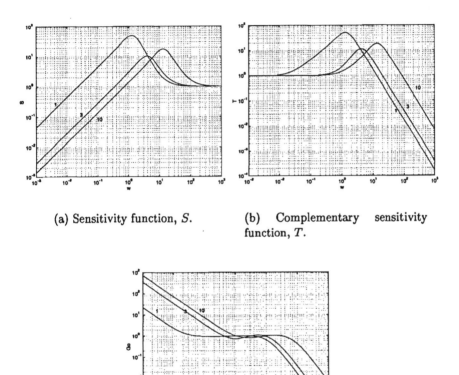

(a) Sensitivity function, S.   (b) Complementary sensitivity function, T.

(c) Loop gain.

Figure 7.20: Sensitivity function, complementary sensitivity function and loop gain for the system (7.56) with a controller placing the four closed loop poles as double poles at $a(-1 \pm i)$ with $a = 1, 3, 10$.

Now consider the system

$$G(s) = \frac{10(-s + 1)}{s(s - 10)} \qquad (7.56)$$

It has both a zero at $s = 1$ and a pole at $s = 10$. From (7.31) the bandwidth of

T should be more than 20 rad/s, while the bandwidth of S from (7.35) should be less than 0.5 rad/s. Translating into gain crossover frequency we get the conflicting demands $\omega_c \leq 0.5$ and $\omega_c \geq 20$ respectively. With a controller in analogy with the previous examples, with four closed loop poles as double poles at $-a \pm ia$ and $a = 1, 3, 10$, we get the result of Figure 7.20.

We see that the conflict between the requirements $\omega_c \leq 0.5$ and $\omega_c \geq 20$ are "solved" by having the loop gain close to 1 in an interval which includes 0.5–10. We also note that both S and T are bad in this interval. As shown by (7.31) and (7.35) no good control design exists in this case.

7.9   Comments

Main Points of the Chapter

The examples, together with the theoretical results, show that there are performance limitations stemming from fundamental principles as well as from practical considerations. For systems with one input and one output, much of the discussion can be summarized in a Bode diagram, see Figure 7.21. We assume that the transfer functions are scaled in the

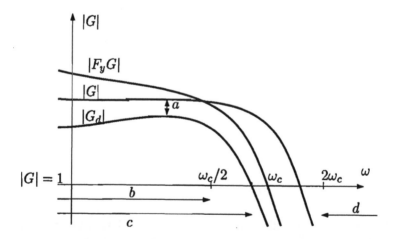

Figure 7.21: Specifications and limitations in a Bode diagram. G, G_d and F_y are the transfer functions of the process, the disturbances and the controller respectively after scaling. a: the margin for disturbance suppression. b: poles in the right half plane. d: zeros in the right half plane. c: in this interval disturbance suppression is needed.

way discussed in Section 7.1. In the figure "c" denotes the frequencies where $|G_d| > 1$. It is in this interval that disturbances cause enough problems to motivate control action. The loop gain has to large enough in this interval. In particular, the gain crossover frequency ω_c should be to the right of "c". To suppress disturbances with acceptable control signal magnitude, we know from Section 7.6 that $|G|$ has to be greater than $|G_d|$. The margin for this is given by "a". We derived two rules-of-thumb: that the gain crossover should be at least twice the value of an unstable pole and at most half the value of a right half plane zero. The gain crossover of the figure satisfies these rules for unstable poles in "b" and non-minimum phase zeros in "d".

Literature

Bode's relation and Bode's integral theorem were formulated for the SISO case already around 1940. These results and other related ones are described in Bode (1945). The interest in fundamental limitations grew strong in the 1980s, inspired in part by the work on controllers for unstable aircraft. Excellent reviews of limitations are presented in several books on controller design, e.g., Zhou et al. (1996), Skogestad & Postlethwaite (1996), Maciejowski (1989). A thorough discussion of frequency properties of multivariable systems is given in Freudenberg & Looze (1988). The fundamental theorems on performance limitations are derived in Seron, Braslavsky & Goodwin (1997), which presents a number of extensions and also discusses the nonlinear case.

Software

Several functions in the CONTROL SYSTEMS TOOLBOX of MATLAB are relevant when investigating fundamental limitations e.g.,

   eig Calculation of eigenvalues (poles).
   tzero Calculation of zeros.
   bode Bode diagrams.
   sigma Plots of singular values.

Appendix 7A Proofs

Proof of Theorem 7.1 Define $H(s) = \log G(s)$. Assume to begin with that G has no zeros in the closed right half plane. Then H is an analytic function there. Now consider the integral

$$\int_C \frac{H(s) - A(\omega)}{s^2 + \omega^2} \, ds = 0 \tag{7.57}$$

where C is the curve given by Figure 7.22. The integral is zero from Cauchy's

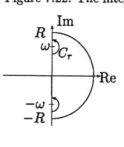

Figure 7.22: Integration path used to derive the relation between amplitude and phase.

Integral Theorem since the integrand is analytic inside the curve. The indentations around $\pm i\omega$ are made to avoid the poles of the integrand. Since $|H(s)|/|s| \to 0$ as $|s| \to \infty$ we conclude that the integral along the half circle with radius R will tend to zero as $R \to \infty$. Let C_r be the small half circle around $i\omega$, and let its radius be r. As r tends to zero, all parts of the integrand, except the factor $s - i\omega$ of the denominator, can be replaced by their values at $s = i\omega$. The integral along C_r is thus given by

$$\frac{i\phi(\omega)}{2i\omega} \int_{C_r} \frac{ds}{s - i\omega} = \frac{\phi(\omega)}{2\omega} \int_{-\pi/2}^{\pi/2} \frac{rie^{i\theta} \, d\theta}{re^{i\theta}} = i\pi \frac{\phi(\omega)}{2\omega}$$

as r tends to zero. Analogous calculations give the same value for the integral around $-i\omega$. As r tends to zero and R tends to infinity we thus get from (7.57)

$$\int_{-\infty}^{\infty} \frac{H(i\nu) - A(\omega)}{\omega^2 - \nu^2} \, d\nu + \frac{\phi(\omega)\pi}{\omega} = 0$$

Since $H(i\nu) = A(\nu) + i\phi(\nu)$, where A is even and ϕ is odd, it follows that (7.16) is satisfied with equality.

Now assume that G has a zero $s = z$ in the right half plane. G can then be written in the form

$$G(s) = G_m(s)(z - s) = G_m(s)(z + s)\frac{z - s}{z + s}$$

where G_m has no zeros in the right half plane. For $G_m(s)(z + s)$ the relation (7.16) holds with equality, according to the argument above. For the factor

$(z - s)/(z + s)$ we have

$$\frac{|z - i\omega|}{|z + i\omega|} = 1, \quad \arg\frac{z - i\omega}{z + i\omega} = -2\arctan(\omega/z)$$

We conclude that this factor does not alter the amplitude curve, but adds a negative contribution to the phase curve. The result is that (7.16) is satisfied with inequality. The argument is easily extended to systems having several zeros in the right half plane. ◻

Proof of Theorem 7.3: The proof is based on properties of the complex logarithm. The basic formula is

$$\log z = \log|z| + i\arg z$$

By restricting z to a region where $\arg z$ can be defined uniquely as a continuous function, $\log z$ can be made analytic. Now consider the sensitivity function

$$S(s) = \frac{1}{1 + F_y(s)G(s)}$$

Assume to begin with that the loop gain GF_y has no poles in the right half plane or on the imaginary axis. S is then an analytic function which is nonzero in the right half plane and on the imaginary axis. It is then possible to define $\log S$ as an analytic function there. If C is the closed contour of Figure 7.23, then from Cauchy's Integral Theorem it follows that

$$0 = \int_C \log S(s)\,ds = \int_{C_I} \log S(s)\,ds + \int_{C_R} \log S(s)\,ds \qquad (7.58)$$

Here C_I is the part of C which lies along the imaginary axis and C_R the half circle. We have

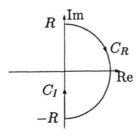

Figure 7.23: Path for calculating the integral of $\log S$.

$$\int_{C_I} \log S(s)\,ds = \int_{C_I} \log|S(s)|\,ds + i\int_{C_I} \arg S(s)\,ds$$

$$= i\int_{-R}^{R} \log|S(i\omega)|\,d\omega = 2i\int_{0}^{R} \log|S(i\omega)|\,d\omega \qquad (7.59)$$

where the second equality follows from $\arg S$ being an odd function and the third equality from $\log |S|$ being an even function. Along the half circle with radius R we have $|F_y G| \leq K_1/R^2$ for some constant K_1. If R is big enough we then have

$$\log S = \log(1 + F_y G)^{-1} \approx -F_y G$$

implying that

$$|\log S| \leq K_2/R^2$$

for some constant K_2. We then get

$$\left| \int_{C_R} \log S \, ds \right| \leq \pi R \frac{K_2}{R^2} \to 0, \quad R \to \infty \tag{7.60}$$

Together (7.58), (7.59) and (7.60) imply, by letting R tend to infinity, that

$$\int_0^\infty \log |S(i\omega)| \, d\omega = 0$$

which is the theorem for the case $M = 0$.

Now assume that GF_y has a single pole at $s = p_1 > 0$. This implies that

$$S(p_1) = 0$$

with S being nonzero in the rest of the right half plane. We can then perform the integration along the curve C of Figure 7.24. Since $\log S$ is analytic inside

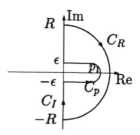

Figure 7.24: Integration path when S has a zero at p_1.

the curve, we get from Cauchy's Integral Theorem

$$\int_{C_I} \log S(s) \, ds + \int_{C_p} \log S(s) \, ds + \int_{C_R} \log S(s) \, ds = 0 \tag{7.61}$$

where C_I is that part of C which lies on the imaginary axis, C_p is the segment which goes from $-\epsilon i$ around p_1 to ϵi and C_R is the half circle with radius R.

In analogy with the previous case above we get

$$\int_{C_I} \log S(s)\, ds = 2i \int_{\epsilon}^{R} \log |S(i\omega)|\, d\omega \tag{7.62}$$

Let us consider the integral along C_p. Since S has a zero at $s = p_1$ we can write it in the form

$$S(s) = S_m(s)(s - p_1)$$

where S_m has no zero in the closed right half plane. The presence of the factor $s - p_1$ makes the argument of S discontinuous when passing the real axis to the left of p_1. The limit from above is π and the limit from below $-\pi$. One can show that this is the only contribution one gets in

$$\int_{C_p} \log S(s)\, ds = \int_{C_p} \log S_m(s)(s - p_1)\, ds$$

which results in

$$\lim_{\epsilon \to 0} \int_{C_p} \log S(s)\, ds = -i\pi p_1 - i\pi p_1 = -2\pi p_1 i \tag{7.63}$$

As above the integral along C_R tends to zero as R tends to infinity. Letting $\epsilon \to 0$ and $R \to \infty$ in the equations (7.61), (7.62) and (7.63) then gives

$$\int_0^{\infty} \log |S(i\omega)|\, d\omega = \pi p_1$$

The theorem is now proved for $M = 1$. The proof is extended to the general case by enclosing every zero in the right half plane in an integration path similar to the one around p_1. $\qquad \square$

Proof of Theorem 7.8: The proof is carried out along the lines of the proof of Theorem 7.3 with S replaced by $\det S$ (note that $\det S$ is a rational function of s). From

$$(I + G(s)F_y(s))S(s) = I$$

it follows that $S(s)$ must be singular for those s-values where GF_y has a pole, since the product otherwise could not be a finite value. We thus have $\det S(p_i) = 0$ for all right half plane poles p_i. We can then write

$$\det S(s) = S_m(s)(s - p_1) \cdots (s - p_{M_p})$$

where S_m has no zero in the right half plane. The handling of right half plane zeros is therefore the same as in the proof of Theorem 7.3. From the formulas for computing the determinant we get

$$|\det S| = |\det(I + G(s)F_y(s))^{-1}| \le 1 + C_2 R^{-2}$$

for some constant C_2, and

$$|\log \det S(s)| \le C_3 R^{-2}, \quad |s| = R$$

for some other constant C_3. This means that the integral along the big half circle will tend to zero as the radius goes to infinity. The proof can then be carried out in analogy with the proof of Theorem 7.3.   □

Chapter 8

Controller Structures and Control Design

The process to construct a control system is often complex and time consuming. For large systems, like an aircraft, several man-years of work are required. Much of the work involves *configuration*, *specification*, and *structuring*. It concerns questions of which signals should, or must, be measured and decisions about the quality of the sensors and actuators to be acquired. In particular, the process involves how to break down the whole control problem into pieces of cascaded loops, possibly with feedforward action; which sensor signals to use to determine a certain control input, etc.

Experience and intuition play very important roles in this process, and it is not easy to formulate systematic methods for how to proceed. The basic limitations, described in Chapter 7 may, however, give useful guidance.

We shall in this, and the next two chapters, discuss the design of a single control loop (which very well could be multivariable). This means that we focus on the choice of F_r and F_y in the linear controller (6.1). We then start from a given model G of the system, a given description of the reliability of this model, and given information about what disturbances influence the system.

8.1   Main Ideas

Model Based Control

The foundation for all modern control design is a model of the system to be controlled. The model should include information about typical disturbances, and typical reference signals (set points) that the controlled signals should follow. The model need not be – and cannot be – an exact and correct description of the real system. Feedback is forgiving in the sense that one can achieve good control even with approximate models. It is in any case important to have an understanding of how reliable the model is.

The model allows us to "experiment" with the system, both by analytically calculating properties of the closed loop system obtained by a certain controller, and by simulating a number of typical situations. In simulations one may also include all those practical limitations, like saturating control signals, that may be difficult to treat analytically.

Controller Structures: Degrees of Freedom

We have described the controller in transfer function form as

$$u = F_r r - F_y y \tag{8.1}$$

The controller (8.1) is a *two-degree-of-freedom* controller (or *2-DOF-controller*). It is determined by two independently chosen transfer functions F_r and F_y. See Figure 8.1. We know from Chapter 6 that

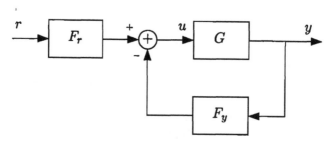

Figure 8.1: Linear controller with two degrees of freedom.

the loop properties depend only on F_y, while F_r affects the response to the reference input.

A common special case is $F_r = F_y$ which gives

$$u = F_y(r - y) \tag{8.2}$$

i.e., feedback directly from the estimated control error $r - y$. Such a controller is a *one-degree-of freedom controller* (or *1-DOF controller*). Often such a controller is first designed, and, if necessary, modified to two degrees of freedom by

$$u = F_y(\tilde{F}_r r - y) \tag{8.3}$$

This corresponds to $F_r = F_y \tilde{F}_r$ in (8.1) and can also be seen as a 1-DOF controller, where the reference signal has been prefiltered: $\tilde{r} = \tilde{F}_r r$. See Figure 8.2.

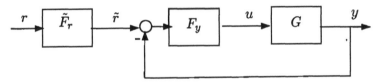

Figure 8.2: Prefiltering of the reference signal, followed by a 1-DOF controller.

Signal or Loop Based Design

There are two main approaches to controller design:

1. **Signal based design.** The focus is on the sizes of the signals of interest, control error and control signals. Typically, criteria are formed which weigh together desired properties. The model is used to calculate the size of the control error, as a function of the control signals.

2. **Loop based design.** The focus is on the crucial transfer functions that describe the closed loop. These can all be derived from the loop gain GF_y (see (6.8)–(6.10)). The loop gain is formed either "by hand" or by computing the F_y that satisfies criteria and constraints that have been specified for the loop gain. The model is used as an explicit part of the loop gain as well as to formulate requirements to yield desired sensitivity and robustness properties.

Sometimes, these two approaches more or less coincide. This is due to the close relationship (via Parseval's equality) between signal size and transfer functions. See Example 6.5.

A Split into Servo and Loop Properties

It is useful to treat the closed loop properties (i.e., sensitivity and robustness) separately from the servo properties (i.e., the transfer function G_c from r to z). The key to this separation is that the sensitivity function S and the complementary sensitivity function T depend only on F_y, and not on F_r in (8.1) or $\tilde{F}_r$ in (8.3).

A natural approach is thus to first choose F_y so that S and T get desired properties, according to the discussion in Chapter 6. With F_y chosen, the transfer function from $\tilde{r}$ to y is given (and equals T). If this does not give acceptable servo properties for the closed loop, the filter $\tilde{F}_r$ in (8.3) is modified. This filter can have different characters:

1. $\tilde{F}_r$ can improve the bandwidth in the transfer function from r to z by increasing the gain at frequencies where T has been forced down due to model uncertainty. The limiting factor is in this case the size of the control input. Increasing bandwidth requires larger inputs.

2. $\tilde{F}_r$ can be used to "soften" fast changes in the reference signal. $\tilde{F}_r$ then is of low pass character which means that the peak values of the control input decrease.

The Work Flow

The steps leading to a complete controller can be summarized as follows:

1. *Modeling.* Construct a model of the system. Include disturbances and measures of the model's reliability. This may be quite a time consuming process, but in many case a crude model will do.

2. *Overall specification considerations.* Scale the system signals according to Section 7.1 and make some back-of-an-envelope calculations to check if the actuator capacities will be sufficient for the desired control accuracy. (See Example 7.4.) Also relate the requirements to fundamental design limitations, according to Chapter 7.

3. *Configuration and design.* Determine a controller structure and its parameters using some of the methods that will be treated in this and the following chapters.

4. *Testing and validation.* Test the candidate controller by simulation, with all practical constraints, like non-linearities, at hand, which may have been neglected in step 3. If necessary, iterate from step 3.

8.2 Configuration of Multivariable Controllers

Input–Output Couplings

The cross coupling between the inputs and outputs is the root of the difficulties in multivariable control: If one input is changed, several outputs are affected. A simple example is to control both flow and temperature of tap water using the hot and cold water taps. The stronger the couplings, the more difficult is the control problem.

RGA: A Measure of Interaction

To measure the degree of cross coupling or interaction in a system, the concept of *Relative Gain Array (RGA)* can be used. For an arbitrary square matrix A, it is defined as

$$\text{RGA}(A) = A. * (A^{-1})^T \tag{8.4}$$

Here, ".*" denotes – like in MATLAB – element-wise multiplication. For

$$A = \begin{bmatrix} 1 & 2 \\ 3 & 4 \end{bmatrix}, \quad \text{we obtain} \quad (A^{-1})^T = \begin{bmatrix} -2 & 1.5 \\ 1 & -0.5 \end{bmatrix}$$

and

$$\text{RGA}(A) = \begin{bmatrix} 1 & 2 \\ 3 & 4 \end{bmatrix} . * \begin{bmatrix} -2 & 1.5 \\ 1 & -0.5 \end{bmatrix} = \begin{bmatrix} -2 & 3 \\ 3 & -2 \end{bmatrix}$$

RGA for a matrix has a number of interesting properties:

- Summing the elements in any row or column always gives +1.

- RGA is independent of any scaling, i.e., $\text{RGA}(A) = \text{RGA}(D_1 A D_2)$ if D_i are diagonal matrices.

- The sum of the absolute values of all elements in RGA(A) is a good measure of A's true condition number, i.e., the best condition number than can be achieved in the family D_1AD_2, where D_i are diagonal matrices.

- Permutation of the rows (columns) of A leads to permutation of the corresponding rows (columns) of RGA(A).

Clearly RGA of a diagonal matrix is the unit matrix. The deviation of RGA from the unit matrix can be taken as a measure of cross coupling between x and y in the relation $y = Ax$.

For a non-square or singular matrix, the RGA is defined by replacing the inverse in (8.4) by the pseudo-inverse. The general definition is:

Definition 8.1 (Relative Gain Array – RGA.) *For an arbitrary, complex-valued matrix A, the RGA is defined as*

$$RGA(A) = A. * (A^\dagger)^T \tag{8.5}$$

In MATLAB-*notation this means* A.*pinv(A).'.

Note that the matrix is to be transposed without being complex-conjugated. (This is what .' does in MATLAB.)

Remark. The pseudoinverse $A^\dagger$ is defined as

$$(A^*A)^{-1}A^* \text{ or } A^*(AA^*)^{-1}$$

provided the indicated inverse exists. For a full rank matrix with more rows than columns we have $A^\dagger A = I$, and $AA^\dagger = I$ if the matrix has more columns than rows.

A Control Interpretation of RGA

It is not immediately clear in what sense RGA measures the cross coupling between inputs and outputs. A control oriented interpretation of RGA can be given as follows:

Consider the square transfer function matrix $y = G(p)u$. Study the entry from input u_j to output y_k: how is y_k affected by changes in u_j? Well, that will depend on how the other inputs behave. Let us consider two extreme cases:

1. *"Open loop control"*: The other inputs are all zero. Then the transfer function from u_j to y_k is given by the (k, j) element of G:

$$y_k = G_{kj}(p)u_j. \tag{8.6}$$

2. *"Closed loop control"*: The remaining inputs are chosen so that the other outputs $(y_i, i \neq k)$ are forced to zero. We have the inverse relation $u = G^{-1}(p)y$. Since $y_i, i \neq k$ equal zero, the inputs u_j will relate to y_k as $u_j = (G^{-1}(p))_{jk}y_k$, i.e.,

$$y_k = \frac{1}{(G^{-1}(p))_{jk}}u_j \tag{8.7}$$

Here $(G^{-1}(p))_{jk}$ is the (j, k) element of $G^{-1}(p)$. The ratio between the gains in these two cases is thus given by $G_{kj}(p)(G^{-1}(p))_{jk}$, i.e., the (k, j) element of $RGA(G(p))$. If the RGA-element equals 1 we have the same gain from u_j to y_k regardless whether the other inputs are zero or chosen by feedback to keep the other outputs zero. This indicates that the cross couplings in the system are quite innocent. Conversely, RGA-elements that differ much from 1 indicate substantial cross coupling effects.

Decentralized Control: Pairing

A simplistic approach to control of multivariable systems is to disregard cross couplings and treat one loop at at time, as if other inputs and outputs did not exist. That is, construct a controller for u_i that is based only on the measured signal y_j and its setpoint r_j:

$$u_i = F_r^i r_j - F_y^i y_j \tag{8.8}$$

With obvious renumbering of the inputs and outputs, the feedback matrix F_y will be *diagonal*. Such a controller is said to be *decentralized*.

Decentralized controllers are always square: if the number of inputs and outputs are not the same, we simply do not use some of the signals. It is easy to realize that such a control structure has problems. The actual loop gain is GF_y – a $p \times p$ matrix – and we have only designed its diagonal elements. The cross couplings from the off-diagonal elements may of course affect both stability and performance of the closed loop system.

It is reasonable to expect that decentralized control will work better the less cross couplings there are between the different inputs and outputs. It is thus important to link the signals y_j and u_i in (8.8) which have the strongest interactions. This is *the pairing problem.* This problem includes deciding which control inputs (measurements) to use, if these exceed the number of measurements (control inputs).

It has proved useful to measure the degree of coupling and cross coupling between inputs and outputs using the RGA of the transfer function matrix. There are two formal results:

- If a decentralized controller is stable in each SISO loop, and if $\text{RGA}(G(i\omega)) = I$, $\forall \omega$, then the closed loop system is also stable. (The requirement that $\text{RGA} = I$ is quite restrictive: it essentially means that G is triangular.)

- If some diagonal element of $\text{RGA}(G(0))$ is negative, and a diagonal controller F_y is used, then either the closed loop system is unstable, or it will become unstable if one of the SISO loops are broken.

It is not very difficult to prove these results, see the books in the list of references. The first result really relates to the properties of RGA close to the cross-over frequency. The second result is natural, since it indicates that we need different signs of the gain in the studied loop, depending on whether the other loops are open or closed.

Two main pairing rules follow from these results: ($G(s)$ here denotes the square transfer function matrix obtained when the inputs have been paired with the outputs and renumbered accordingly.)

1. Try to make the diagonal elements of $\text{RGA}(G(i\omega_c))$ as close as possible to the point 1 in the complex plane. Here ω_c is the intended cross-over frequency, or bandwidth, of the closed loop system.

2. Avoid pairings that give negative diagonal elements of $\text{RGA}(G(0))$.

If it turns out to be difficult to reach these goals, and if the RGA at the cross-over frequency differs much from the unit matrix, one may expect that decentralized control will not give good performance of the closed loop system

Note that the elements of $\text{RGA}(A)$ are permuted in the same way as the rows and columns of A itself. This means that it is sufficient to compute RGA for the original (not renumbered) system to find good pairings.

Example 8.1: A System With Two Inputs and Two Outputs

In Examples 1.1 and 7.3 we studied the system

$$G(s) = \begin{bmatrix} \frac{2}{s+1} & \frac{3}{s+2} \\ \frac{1}{s+1} & \frac{1}{s+1} \end{bmatrix}$$

We have

$$G(0) = \begin{bmatrix} 2 & 1.5 \\ 1 & 1 \end{bmatrix} \quad \text{and} \quad RGA(G(0)) = \begin{bmatrix} 4 & -3 \\ -3 & 4 \end{bmatrix}$$

and

$$G(i5) = \begin{bmatrix} 0.0769 - 0.3846i & 0.2069 - 0.5172i \\ 0.0385 - 0.1923i & 0.0385 - 0.1923i \end{bmatrix}$$

$$RGA(G(i5)) = \begin{bmatrix} -1.7692 + 1.1538i & 2.7692 - 1.1538i \\ 2.7692 - 1.1538i & -1.7692 + 1.1538i \end{bmatrix}$$

Obviously it is difficult to find a pairing that gives a desirable RGA. This confirms the difficulties we had in the earlier examples. Note in particular the problems we had in Example 1.1 with the pairing $u_1 \rightarrow y_2$ and $u_2 \rightarrow y_1$. These can be traced to the negative values that $RGA(G(0))$ has for this pairing.

Example 8.2: A Distillation Column

Distillation columns are the most common processes in the chemical industry. We shall study an example from the petrochemical industry, viz. a heavy oil fractionator. This process normally follows an operation such as catalytic cracking. The feed to this column consists of a hot mixture of gasoils of varying molecular weights. The objective of the column is to systematically remove heat from this feedstream and in the process distill it into heavy to light fractions. The example is provided by Shell Development, and a basic sketch of the process is shown in Figure 8.3.

The essentials of the process are as follows: Hot gasoil is entered at the lower part, at FEED and is cooled in the column. It is condensed at various levels at so called trays. The composition of the condensed substance depends on the level. At the trays, part of the product is drawn from the column and at some of the trays part of it is fed back into the column as reflux. The control signals are the exit flows (DRAW) from the different trays, as well as the temperature control at the bottom of the column. The controlled variables (the outputs) are the product qualities (compositions) at some of the trays. Moreover, the bottom reflux temperature should be seen as an output, since it must be kept between certain values for safe operation.

The disturbances are variations in the heat fluxes (INTERMEDIATE and UPPER REFLUX). These are coupled to other parts of the process due to integrated energy control, and may therefore vary with changes in other columns.

In this example we choose to consider the *outputs*

y_1 = top draw composition

y_2 = side draw composition

y_3 = bottom reflux temperature

and the *control inputs*

u_1 = top draw flowrate

u_2 = side draw flowrate

u_3 = bottom temperature control input

as well as the *disturbance*

v_1 = heat flow at UPPER REFLUX

The transfer function matrix for this system is

$$\begin{bmatrix} Y_1(s) \\ Y_2(s) \\ Y_3(s) \end{bmatrix} = G(s) \begin{bmatrix} U_1(s) \\ U_2(s) \\ U_3(s) \end{bmatrix} + G_d(s)V_1(s) \tag{8.9}$$

where

$$G(s) = \begin{bmatrix} \frac{4.05e^{-27s}}{50s+1} & \frac{1.77e^{-28s}}{60s+1} & \frac{5.88e^{-27s}}{50s+1} \\ \frac{5.39e^{-18s}}{50s+1} & \frac{5.72e^{-14s}}{60s+1} & \frac{6.90e^{-15s}}{40s+1} \\ \frac{4.38e^{-20s}}{33s+1} & \frac{4.42e^{-22s}}{44s+1} & \frac{7.20}{19s+1} \end{bmatrix} \tag{8.10}$$

$$G_d(s) = \begin{bmatrix} \frac{1.44e^{-27s}}{40s+1} \\ \frac{1.83e^{-15s}}{20s+1} \\ \frac{1.26}{32s+1} \end{bmatrix} \tag{8.11}$$

The time unit is minutes, and a typical desired response time is about 50 minutes.

The problem is to select two out of the possible three control inputs to control the first two outputs with a decentralized controller. The third output, y_3, is not a quality variable and need not be kept constant. It must however not drop below certain values for safe operation. Which control inputs shall be chosen, and how shall they be paired with the outputs?

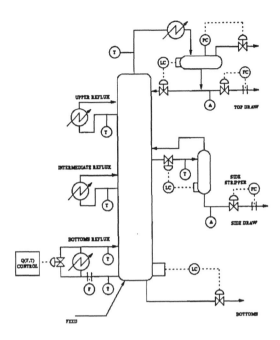

Figure 8.3: A heavy oil fractionator.

Let $\tilde{G}(s)$ be the first two rows of $G(s)$. We compute the RGA-matrices for $\tilde{G}(0)$ and $\tilde{G}(i/50)$ as

$$\begin{bmatrix} 0.3203 & -0.5946 & 1.2744 \\ -0.0170 & 1.5733 & -0.5563 \end{bmatrix}$$

and

$$\begin{bmatrix} 0.4794 - 0.3558i & -0.6325 - 0.0158i & 1.1532 + 0.3716i \\ -0.1256 + 0.3558i & 1.5763 + 0.0158i & -0.4507 - 0.3716i \end{bmatrix}$$

respectively. To find which control signal shall be determined by y_1, we look at the first row in each of these matrices. Obviously, u_2 must be avoided, since this gives negative RGA-values. Also, we see that u_3 is the best choice at both frequencies. Similarly we see, from the second row, that y_2 shall affect u_2. Consequently, we should use u_3 (temperature control) determined from y_1 (top tray composition), and u_2 (side draw flowrate), determined from y_2 (side composition).

Decoupled Control

Successful decentralized control requires natural pairs of inputs and outputs that account for the dominating dynamics. If such pairs do not exist, change variables $\tilde{y} = W_2 y$ and $\tilde{u} = W_1^{-1} u$, so that the transfer function matrix from $\tilde{u}$ to $\tilde{y}$, i.e.,

$$\tilde{G}(s) = W_2(s)G(s)W_1(s) \tag{8.12}$$

becomes as diagonal as possible. With the new inputs and outputs a decentralized (diagonal) controller could be constructed:

$$\tilde{u} = -F_y^{\text{diag}} \tilde{y} \tag{8.13}$$

Expressed in the original variables we then obtain the controller

$$u = -W_1 F_y^{\text{diag}} W_2 y \tag{8.14}$$

Such a controller structure is said to be *decoupled*.

The question is how to choose W_1 and W_2. To obtain a truly decoupled (diagonal) system $\tilde{G}$ in (8.12), requires dynamical, i.e., s-dependent, matrices W_i, that typically are related to the system inverse G^{-1}. Such *dynamic decoupling* is closely connected to internal model control, to be discussed in Section 8.3. It clearly requires good knowledge of the plant model and may be quite sensitive to model errors.

A simpler case is to use *steady-state decoupling*, i.e., to make $\tilde{G}(0)$ diagonal. That can be achieved by real-valued, constant matrices W. Another possibility is to use constant real-valued decoupling matrices to make $\tilde{G}(i\omega)$ as diagonal as possible for a certain selected frequency ω, often picked as the cross-over frequency ω_c.

Example 8.3:   Distillation Column

Consider the distillation column from Example 8.2. If the control inputs are chosen according to that example, the transfer function matrix from $[u_3, u_2]^T$ to $[y_1, y_2]^T$ will be

$$G(s) = \begin{bmatrix} \frac{5.88e^{-27s}}{50s+1} & \frac{1.77e^{-28s}}{60s+1} \\ \frac{6.90e^{-15s}}{40s+1} & \frac{5.72e^{-14s}}{60s+1} \end{bmatrix} \tag{8.15}$$

Disregarding the cross couplings (non-diagonal elements) gives a reasonable *decentralized* PI-controller

$$F_y^{\text{dec}}(s) = \begin{bmatrix} \frac{63s+1}{50s} & 0 \\ 0 & \frac{67s+1}{50s} \end{bmatrix}$$

(This design is based on a common rule of thumb: the ratio between the I-term and the P-term is the time constant of the system + half the dead time;

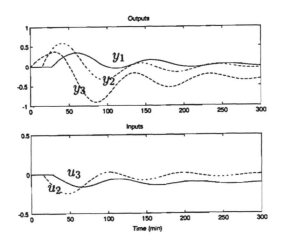

(a) Decentralized control.

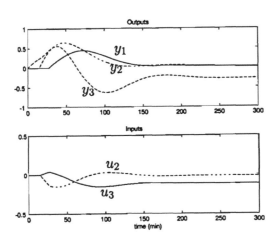

(b) Decoupled control.

Figure 8.4: PI-control of the distillation column. At time 0 a step disturbance of size 0.5 occurs in v_1. All setpoints are zero.

see Section 8.3.) This controller applied to (8.15) gives the result shown in Figure 8.4a.

Let us now consider *decoupled control*, where the decoupling is made at stationarity. We have

$$G(0) = \begin{bmatrix} 5.88 & 1.77 \\ 6.90 & 5.72 \end{bmatrix}$$

In (8.12) we take $W_1 = G^{-1}(0)$ and $W_2 = I$. This gives

$$\tilde{G}(s) = \begin{bmatrix} \frac{1.57e^{-27s}}{50s+1} - \frac{0.57e^{-28s}}{60s+1} & X \\ X & \frac{1.57e^{-14s}}{60s+1} - \frac{0.57e^{-15s}}{40s+1} \end{bmatrix}$$

where we ignore the off-diagonal elements X (we know that they are zero at frequency zero), and design a decentralized controller for the diagonal elements, e.g., the PI-regulators

$$F_y^{\text{diag}}(s) = \begin{bmatrix} \frac{63s+1}{50s} & 0 \\ 0 & \frac{55s+1}{50s} \end{bmatrix}$$

This leads to the controller

$$F_y(s) = W_1 F_y^{\text{diag}}(s) = \begin{bmatrix} \frac{0.267(63s+1)}{50s} & -\frac{0.0826(55s+1)}{50s} \\ -\frac{0.3221(63s+1)}{50s} & \frac{0.2745(55s+1)}{50s} \end{bmatrix}$$

which applied to (8.15) gives the results of Figure 8.4b. A clear improvement in error amplitude as well as resolution time can be seen, compared to the decentralized case.

8.3 Internal Model Control

The Idea: Only Feed Back the New Information.

Consider control of a stable system with model G. The model may very well differ from the true system G_0. If there are no model errors and no disturbances, there would be no reason for feedback. To pinpoint this, it is natural to focus the feedback on the information in the output that is new, i.e., the part that originates from disturbances, measurement errors, and model errors:

$$y - Gu$$

This new information is fed back to the control input by a transfer function Q:

$$u = -Q(y - Gu) \tag{8.16}$$

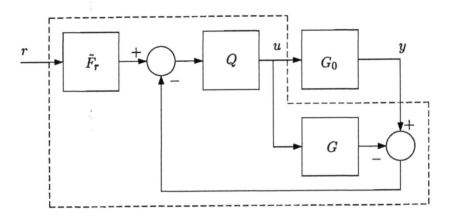

Figure 8.5: Feedback from model error.

See Figure 8.5. Including the link from the reference signal we obtain

$$u = -Q(y - Gu) + Q\tilde{r}, \quad \tilde{r} = \tilde{F}_r r \qquad (8.17)$$

By *Internal Model Control, IMC*, is meant methods to describe and design controllers in terms of feedback via Q of the *new information* $y - Gu$. The model G evidently plays an explicit role in this control structure.

Expressions for Some Important Transfer Functions

Based on (8.17) we can compute the essential transfer functions that describe the closed loop system:

The transfer function from y to u is obtained as

$$u = -(I - QG)^{-1}Qy, \quad \text{i.e.,} \quad F_y = (I - QG)^{-1}Q \qquad (8.18)$$

This controller is parameterized by the – as yet – arbitrary transfer function Q.

The closed loop system in the nominal case $G_0 = G$ is

$$G_c = (I + GF_y)^{-1}GF_y\tilde{F}_r = GQ\tilde{F}_r \qquad (8.19)$$

To realize this, note that in the absence of disturbances and model errors, the signal in the feedback path is zero in Figure 8.5. The transfer function from r to y is then the product of the terms in the forward path. Moreover, the complementary sensitivity function is

$$T = (I + GF_y)^{-1}GF_y = GQ \qquad (8.20)$$

(see (8.19)) hence the sensitivity function

$$S = I - GQ \qquad (8.21)$$

The transfer function from the output disturbance w to u will be

$$G_{wu} = -(I + F_y G)^{-1} F_y = -Q \qquad (8.22)$$

(this is actually how Q was chosen).

All these expressions refer to the case $G_0 = G$. This gives the nominal system and the nominal sensitivity and robustness functions. These are also the expressions used in the results of Section 6.3. With a model error the closed loop system becomes

$$G_c = G_0(I + Q(G_0 - G))^{-1} Q \tilde{F}_r \qquad (8.23)$$

Parameterization of All Stabilizing Controllers

From equations (8.19)–(8.22) it follows that all relevant transfer functions for the closed loop system are stable *if and only if Q is stable*. Recall the assumption that G itself is stable.

The class of all stabilizing controllers is thus given by (8.18) where Q is an *arbitrary stable transfer function*. This is a very simple description of the complicated conditions on the feedback function F_y to assure closed loop stability.

The parameterization we have treated here is only valid for stable G. With a similar technique, also unstable G can be dealt with. The general parameterization of all stabilizing regulators is known as *Youla-parameterization* or *Q-parameterization*.

Design Based on IMC

A very important and interesting side-effect of the IMC controller structure is that *all the transfer functions of interest S, T, G_c and G_{wu} are linear in Q*. This makes it particularly easy to discuss controller design in terms of Q. With Q determined, F_y is calculated from (8.18).

How to choose Q? The "ideal" choice $Q = G^{-1}$ – which would make $S \equiv 0$ and $G_c \equiv I$ – is not possible (it corresponds to $F_y \equiv \infty$), but is still important guidance. Let us discuss, in the SISO case, what makes this choice impossible, and how it can be modified to a possible and good choice:

1. G has more poles than zeros, and the inverse cannot be physically realized

   - Use $Q(s) = \frac{1}{(\lambda s+1)^n} G^{-1}(s)$ with n chosen so that $Q(s)$ can be realized (numerator degree $\leq$ denominator degree). The number λ is a design parameter that can be adjusted to desired bandwidth of the closed loop system.

2. G has an unstable zero (it is non-minimum phase) which would be canceled when GG^{-1} is formed. This would give an unstable closed loop system (see Example 6.1). There are two possibilities if $G(s)$ has the factor $(-\beta s + 1)$ in the numerator:

   (a) Ignore this factor when $Q \approx G^{-1}$ is formed.

   (b) Replace this factor with $(\beta s + 1)$ when $Q \approx G^{-1}$ is formed. This only gives a phase error and no amplitude error in the model.

   In both cases the original G is used when the controller is formed according to (8.18).

3. G has a time-delay (dead-time), i.e., the factor $e^{-s\tau}$. Two possibilities:

   (a) Ignore this factor when Q is formed, but not in (8.18) (see Example 8.5 below).

   (b) Approximate this factor by

   $$e^{-s\tau} \approx \frac{1 - s\tau/2}{1 + s\tau/2}$$

   and use step 2. Possibly use the same approximation of $e^{-s\tau}$ in (8.18).

In Chapter 7 we saw that non-minimum phase and time delays limit the performance of a system. Here we see how these properties explicitly show up in the control design.

Example 8.4: Some Simple Controllers

Consider first the system

$$G(s) = \frac{1}{\tau s + 1}, \quad \text{which gives} \quad Q(s) = \frac{\tau s + 1}{\lambda s + 1}$$

i.e.,

$$F_y(s) = \frac{\frac{\tau s+1}{\lambda s+1}}{1 - \frac{1}{\lambda s+1}} = \frac{\tau s + 1}{\lambda s}$$

This is a PI-regulator

$$F_y(s) = \frac{\tau}{\lambda}\left(1 + \frac{1}{s\tau}\right)$$

where τ/λ corresponds to the gain and the relation between the P-part and the I-part is determined by the model time constant τ.

Another simple example that illustrates how to deal with non-minimum phase models is

$$G(s) = \frac{-\beta s + 1}{\tau s + 1}.$$

Approach 2a in the recipe above gives $Q_1(s) = (\tau s + 1)/(\lambda s + 1)$ while approach 2b gives $Q_2(s) = (\tau s + 1)/(\beta s + 1)$. The corresponding controllers are computed as

$$F_y(s) = \frac{\tau s + 1}{(\lambda + \beta)s}, \quad \text{for } Q_1, \text{ while } Q_2 \text{ gives } \quad F_y(s) = \frac{\tau s + 1}{2\beta s}$$

Example 8.5:   Dead-time Compensation

Suppose the model is given by $G(s) = G_1(s)e^{-s\tau}$. Approach 3a for handling time delays – to ignore them – implies that Q should be chosen as a suitable approximation of G_1^{-1}. This gives the controller

$$F_y(s) = \frac{Q(s)}{1 - Q(s)G_1(s)e^{-s\tau}}$$

Let $F_y^0 = Q/(1-QG_1)$ be the controller designed for the system without time delays. Simple manipulations then give

$$F_y(s) = \frac{F_y^0(s)}{1 + (1 - e^{-s\tau})F_y^0(s)G_1(s)}$$

Such a controller is called *dead-time compensation according to Otto Smith*. See, e.g., Franklin et al. (1994), Section 7.10.

In the same way, we can generate more examples of how the general IMC rules of thumb give good controller designs. The PID-regulator, for example, is the "answer" to second order process models, where like in Example 8.4, λ is a design parameter and the relationships between the gains of the P, I and D-terms are determined by the process model.

A typical feature of IMC controllers is that they cancel the dominating poles when $QG \approx I$ is formed. Load disturbances at the input (w_u in Figure 6.1) affect the output via the transfer function $G_{w_u y} = S_u G$ (see (6.7)). Here the system poles are still present, and IMC design may therefore give poorly damped response to load disturbances.

Finally, we may remark that the IMC rules very well can be applied also to multivariable systems.

8.4   Feedback from Reconstructed States

One of the most important controller structures is obtained as feedback from reconstructed/estimated state variables. Some general optimal control problems also turn out to have solutions that naturally can be determined in this structure. This will be shown in the following two chapters. In this section we shall first give the expressions for a number of important transfer functions relating to this structure.

The starting point is the state space description

$$px = Ax + Bu + Nv_1 \qquad (8.24a)$$
$$z = Mx \qquad (8.24b)$$
$$y = Cx + v_2 \qquad (8.24c)$$

Here it is assumed that the system does not have any direct term Du. By straightforward modifications of the expressions below such a term can be included. Moreover, all said in this section also applies to discrete time systems, by replacing p by q.

The Basic Expressions

The essence of the concept of state is that it summarizes all that is worth knowing about the past behavior of the system. It is therefore natural to base the choice of control signal on a good estimate of the current state, computed as in Section 5.7. The controller structure is then set up as feedback from reconstructed states as depicted in Figure 8.6:

$$u(t) = -L\hat{x}(t) + \bar{r}(t) \qquad (8.25a)$$
$$p\hat{x}(t) = A\hat{x}(t) + Bu(t) + K(y(t) - C\hat{x}(t)) \qquad (8.25b)$$

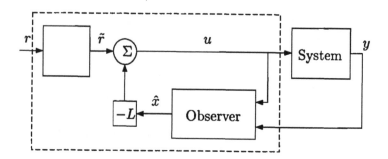

Figure 8.6: State feedback from reconstructed states.

This is a parameterization of the controller in terms of the $m \times n$ matrix L and the $n \times p$ matrix K. Here $\tilde{r}$ is a scaled or filtered version of the reference signal. We shall later discuss how $\tilde{r}$ is best formed.

To find the transfer function matrix for the controller we note that

$$p\hat{x}(t) = (A - BL - KC)\hat{x}(t) + B\tilde{r}(t) + Ky(t)$$

This means that

$$u(t) = -F_y(p)y(t) + F_r(p)\tilde{r}(t) \tag{8.26a}$$
$$F_y(p) = L(pI - A + BL + KC)^{-1}K \tag{8.26b}$$
$$F_r(p) = I - L(pI - A + BL + KC)^{-1}B \tag{8.26c}$$

The loop gain is

$$C(pI - A)^{-1}BF_y(p) \tag{8.27}$$

To determine the closed loop system, we assume that the control object is described by (8.24). Let $\tilde{x} = x - \hat{x}$. This gives

$$p\tilde{x} = px - p\hat{x} = (A - KC)\tilde{x} + Nv_1 - Kv_2,$$

and with the state vector

$$X(t) = \begin{bmatrix} x(t) \\ \tilde{x}(t) \end{bmatrix}$$

we obtain

$$pX = \begin{bmatrix} A - BL & BL \\ 0 & A - KC \end{bmatrix} X + \begin{bmatrix} B \\ 0 \end{bmatrix} \tilde{r}$$
$$+ \begin{bmatrix} N \\ N \end{bmatrix} v_1 + \begin{bmatrix} 0 \\ -K \end{bmatrix} v_2 \tag{8.28a}$$
$$z = \begin{bmatrix} M & 0 \end{bmatrix} X \tag{8.28b}$$

or

$$z = \begin{bmatrix} M & 0 \end{bmatrix} \begin{bmatrix} pI - A + BL & -BL \\ 0 & pI - A + KC \end{bmatrix}^{-1}$$

$$\times \left(\begin{bmatrix} B \\ 0 \end{bmatrix} \tilde{r} + \begin{bmatrix} N \\ N \end{bmatrix} v_1 + \begin{bmatrix} 0 \\ -K \end{bmatrix} v_2 \right)$$

Using the matrix equality

$$\begin{bmatrix} R & S \\ 0 & T \end{bmatrix}^{-1} = \begin{bmatrix} R^{-1} & -R^{-1}ST^{-1} \\ 0 & T^{-1} \end{bmatrix}$$

we find that

$$\begin{bmatrix} M & 0 \end{bmatrix} \begin{bmatrix} pI - A + BL & -BL \\ 0 & pI - A + KC \end{bmatrix}^{-1}$$
$$= \begin{bmatrix} M(pI - A + BL)^{-1} & M(pI - A + BL)^{-1}BL(pI - A + KC)^{-1} \end{bmatrix}$$

This means that the closed loop system can be written

$$z = G_c(p)\tilde{r} - \tilde{T}v_2 + \tilde{S}v_1$$

where

$$G_c = M(pI - A + BL)^{-1}B \qquad (8.29a)$$
$$\tilde{T} = M(pI - A + BL)^{-1}BL(pI - A + KC)^{-1}K \qquad (8.29b)$$
$$\tilde{S} = M(pI - A + BL)^{-1}\left(BL(pI - A + KC)^{-1}N + N\right) \quad (8.29c)$$

Note that $\tilde{T}$ is the complementary sensitivity function only if $M = C$, since v_2 equals the additive output disturbance only in that case. Similarly, $\tilde{S}$ is not the usual sensitivity functions, since the state disturbance v_1 in (8.24) does not equal the system disturbance w in (6.2). Instead we obtain the sensitivity function from $S = (I + GF_y)^{-1}$ as

$$S = \left(I + \left(C(pI - A)^{-1}B\right) L(pI - A + BL + KC)^{-1}K\right)^{-1}$$

$$(8.30)$$

Direct Term in the Controller

The observer (8.25b) has no direct term, i.e., in discrete time $\hat{x}(t)$ will only depend on $y(t-1)$ and earlier terms. In continuous time this means that the frequency function from y to $\hat{x}$ – and hence to u – has a gain that tends to zero as the frequency tends to infinity. Sometimes this might not be desirable, but a controller that has a quicker response to measurements is desired. Then (8.25b) has to be replaced by an observer with a direct term according to (5.67). This leads to straightforward adjustments of the expressions above.

The Reference Signal

A reference signal r for the controlled output z can be introduced into the structure (8.25) in several ways. A systematic approach is to include the properties (typical frequency contents) of the reference signal in the basic model (8.24). Then z has to represent the control error and the measured signals y should include r. That way r is introduced into the controller (8.25) via y. This somewhat laborious approach may be necessary if the relationship between y and z is complicated, or if the typical reference signal has very specific properties.

If the underlying model for the reference signal is simple, e.g., piecewise constant with random jumps, the result is however straightforward: Let $\tilde{r} = L_r r$ where the matrix L_r is chosen so that the closed loop system has the desired steady state gain. This means that the controller (8.25b) will be

$$p\hat{x}(t) = (A - BL)\hat{x}(t) + BL_r r(t) + K(y(t) - C\hat{x}(t)) \qquad (8.31)$$

See Theorem 9.2 in the next chapter.

Otherwise it may be suitable to follow the recipe (8.3) to introduce the reference signal. This means that we choose a prefilter

$$\tilde{r} = \tilde{F}_r r$$

and then replace y in (8.25b) with $y - \tilde{r}$. This gives the controller

$$p\hat{x}(t) = (A - BL)\hat{x}(t) - K\tilde{F}_r(p)r(t) + K(y(t) - C\hat{x}(t)) \qquad (8.32)$$

Control Design: Pole Placement, etc.

Control design for the structure (8.25) consists of determining the matrices K and L. The choice of L will directly affect the closed loop system $G_c(p)$ according to (8.29a). The poles of $G_c(s)$ are the eigenvalues of $A - BL$. According to Theorem 3.4 these can be chosen arbitrarily using L if the system (A, B) is controllable.

The system matrix for the complete closed loop system is given by (8.28a). Since the matrix is block triangular, its eigenvalues are given by those of the two diagonal blocks. The eigenvalues of $A - KC$ are thus also poles of the complete closed loop system according to (8.28a). Together with the poles of G_c, these determine how disturbances are damped and fade out. Again, according to Theorem 3.4 the eigenvalues of $A - KC$ can be chosen arbitrarily by K if (A, C) is observable. We may consequently handle the design criterion 6.3 using K and L if the system is controllable and observable. This design method is known as *pole placement*.

For SISO systems this is straightforward: every desired pole configuration for G_c and S leads to unique choices of K and L. This approach is treated in a comprehensive manner in e.g., Franklin et al. (1994), Chapter 7. For multivariable systems there are many K and L that give the same poles, and it may be difficult to choose particular values. It is then better to determine these matrices indirectly from more explicit design criteria. We shall in the two following chapters discuss the most important methods for this:

- Signal based optimization according to design criterion 6.4. (Chapter 9.)

- Loop based specifications according to the design criteria 6.5 and 6.6. (Chapter 10.)

8.5 Comments

Main Points of the Chapter

- There are two ways to perform multivariable control design in terms of single loops:

  - *Decentralized control:* Treat the multivariable systems as a number of independent single loops. Use, e.g., the RGA method to pair measurements and control inputs.

> – *Decoupled control:* Make the system more diagonal by
> creating new measurements and control inputs as linear
> combinations of the original ones. Then apply decentralized
> control to these new signals.

- *The IMC method* is a way of both configuring a controller and
  determining its parameters.

- *State feedback from reconstructed states* is a fundamental control
  structure that is well suited to compute and implement multivariable controllers that are more complex than PID-regulators.

Literature

To configure multivariable controllers is a problem long discussed.
Traditional solutions are often based on combinations of feedforward
and cascade controllers. The RGA method was introduced by Bristol
(1966) and is described in many books on process control, e.g.,
Skogestad & Postlethwaite (1996). This book also treats decentralized
and decoupled control.

The IMC principle has a long history: that a good controller must
build upon a model of the system and its disturbances is natural.
The approach has been put into a systematic framework by Morari
and is treated comprehensively in Morari & Zafiriou (1989) with many
examples. The related, but more general Youla-parameterization has
its name from the article Youla, Jabr & Bongiorno (1976) but had
as such been used before that. It has played an important role in
particular for the methods treated in Chapter 10.

The controller structure in Section 8.4 is the basis of modern control
theory and it is treated in all recent text books. See e.g., Åström &
Wittenmark (1997) and Franklin et al. (1990).

Software

MATLAB has the commands `connect`, `feedback`, `series`, `parallel`
to link various partial systems to build up a controller. The control
structure of Section 8.4 is generated by the command `reg`, while `place`
computes the feedback matrices L and K from specified desired pole
locations.

Chapter 9

Minimization of Quadratic Criteria: LQG

9.1 The Criterion and the Main Ideas

The controlled object is described as before, (6.2):

$$z = Gu + w \tag{9.1a}$$
$$y = z + n \tag{9.1b}$$

and the controller can be written in general terms as (6.3):

$$u = F_r r - F_y y \tag{9.2}$$

For the control design we choose the criterion 6.4, i.e., to minimize a trade-off between the size of the control error $e = z - r$ and the size of the control input u:

$$\min \left(\|e\|_{Q_1}^2 + \|u\|_{Q_2}^2 \right) = \min \int e^T(t) Q_1 e(t) + u^T(t) Q_2 u(t) dt \tag{9.3}$$

Recall that the left hand side can also be interpreted as the time average ("power") of the integrand in the right hand side, as well as mathematical expectations (variances) of stationary stochastic processes, see (5.26). (It might be a good idea to go back and check Sections 5.2 and 5.3 at this point.) We seek the optimal linear controller, so the minimization should be carried out over all linear transfer functions F_y and F_r. The formal solution to this problem is most easily found if the system (9.1) is represented in state space form. We shall compute this solution in the next section. Before that, some general reflections are in order.

What will influence the solution?

1. The optimal controller will of course depend on the system G, which we regard as given.

2. The controller will also depend on the weighting matrices Q_1 and Q_2, i.e., how we value the trade-off between control effort and control error. In the multivariable case, these matrices also determine how the different components of the control errors are weighted relative to each other. Similarly, the relative efforts in the different control inputs are weighted together in Q_2. Normally these weights are not given *a priori*, but it is natural to view Q_1 and Q_2 as *design variables*.

3. The sizes of e and u will depend on the disturbance model, that is the spectra of w and n. Moreover, they are affected by the properties of the reference signal. The disturbance model and the reference signal model will therefore affect the solution. Even if these, in some sense, are given by the problem, they are seldom unique in an objective manner. It is natural to view also *disturbance model and reference signal model as design variables*.

4. Moreover, the solution will depend on what we measure, that is y. Often the measured signals are just the controlled outputs, disturbed by measurement noise as in (9.1). The relationship may however be more complicated. The measured signals should include any measurable disturbances. In that way, feedforward from these disturbances will automatically be included in the controller (9.2). See Examples 5.7 and 9.2. The conclusion is that also a good description of the measured signals is part of the design work.

Is the optimal solution good?

- In Chapters 6 and 7 we listed a number of desired properties of the closed loop system. Sensitivity and robustness issues were of particular interest. It is important to realize that a solution that is optimal in the sense that (9.3) is minimized, need not be good in the sense that it has all our desired features.

- This means that we will have to *twist and turn* the solution that the machinery provides us with as the answer to the question (9.3). This entails plotting all relevant frequency curves (sensitivity, complementary sensitivity, G_{ru}, etc.), and also simulating a number of time responses to changes in set-points and disturbances.

- The *design variables* listed above will have to be used to generate a good solution, typically after having tried out several preliminary ones.

- The essence of the optimization approach is that if there is a good solution for the problem, it will also be an optimal solution to (9.3) for suitably chosen design variables. It is typically not so difficult to understand how to modify these design variables to improve non-desired features of a candidate solution.

State Space Description of the System

In this chapter we shall start from the general system description (9.1), given in standard state space form according to Chapter 5. That is:

$$\dot{x} = Ax + Bu + Nv_1 \tag{9.4a}$$
$$z = Mx \tag{9.4b}$$
$$y = Cx + v_2 \tag{9.4c}$$

Here v_1 and v_2 are white noises (or impulses; see Section 5.3) that correspond to the disturbance sources. z is the controlled variable. All relevant measurements are contained in y, including any measured disturbances. When the goal is to follow a reference signal $r \neq 0$, the model may be completed as described below in connection with Theorem 9.2.

Remark. In (9.4) a direct term Du has been omitted, for simplicity. Such a term in the expression for y will lead to straightforward modifications in the Kalman filter below. A direct term in the expression for z leads to a cross term in the criterion as in (9.14)–(9.15).

9.2   The Optimal Controller: Main Results

In this section we shall collect the formal results about the optimal controller in some different cases. We shall in the following sections discuss how to use these results in practice. First the case with no reference input, $r \equiv 0$ will be treated.

Theorem 9.1 (Linear-Quadratic Optimization.) *Consider the system (9.4) where* $\begin{bmatrix} v_1 \\ v_2 \end{bmatrix}$ *is white noise with intensity* $\begin{bmatrix} R_1 & R_{12} \\ R_{12}^T & R_2 \end{bmatrix}$. *We seek the linear, causal feedback law* $u(t) = -F_y(p)y(t)$ *that minimizes the criterion*

$$V = \|z\|_{Q_1}^2 + \|u\|_{Q_2}^2 \tag{9.5}$$

for a positive definite matrix Q_2 *and a positive semidefinite matrix* Q_1. *Suppose that* (A, B) *is stabilizable and that* $(A, M^T Q_1 M)$ *is detectable, (see Definition 3.3) and that the assumptions about* A, C, R_1, R_2 *and* R_{12} *in Theorem 5.4 are satisfied.*

Then the optimal controller is given by

$$u(t) = -L\hat{x}(t) \tag{9.6a}$$
$$\dot{\hat{x}}(t) = A\hat{x}(t) + Bu(t) + K(y(t) - C\hat{x}(t)) \tag{9.6b}$$

Here (9.6b) is the Kalman filter for the system, so K *is determined by Theorem 5.4 as*

$$AP + PA^T + NR_1 N^T - (PC^T + NR_{12})R_2^{-1}(PC^T + NR_{12})^T = 0$$
$$K = (PC^T + NR_{12})R_2^{-1}$$

while L *is given by*

$$L = Q_2^{-1} B^T S \tag{9.7a}$$

where S *is the unique, positive semidefinite, symmetric solution to the matrix equation*

$$A^T S + SA + M^T Q_1 M - SBQ_2^{-1}B^T S = 0 \tag{9.7b}$$

Moreover, it follows that $A - BL$ *has all eigenvalues inside the stability region.*

The proof is given in Appendix 9A.

Remark. If we omit the assumption that $(A, M^T Q_1 M)$ is detectable, the equation (9.7b) may have several positive semidefinite solutions. There is only one, though, that makes the corresponding $A - BL$ stable. The theorem still holds if this solution to (9.7) is chosen. Compare with Remark 2 after Lemma 5.1.

Corollary 9.1 (LQG: Gaussian Disturbances) *Suppose that the signal* $[v_1, v_2]$ *in (9.4) is Gaussian (Normal) distributed white noise. Then the controller that minimizes (9.5) among all possible, causal controllers (including nonlinear ones) is linear and given by (9.6).*

Proof: The proof of the main theorem uses that the controller is linear, and that there is no correlation among the innovations. For general controllers it is required that the innovations are independent. This is easily shown in the discrete time case. See Corollary 9.3 in Section 9.5. □

To design controllers according to Theorem 9.1 is known as the *Linear Quadratic* approach or *LQ Control*, since linear controllers and quadratic criteria are used. With Gaussian disturbances, the method is known as *Linear Quadratic Gaussian Control, LQG control*. Theorem 9.1 is also called the *separation theorem*, since it shows that the optimization problem is split into two separate parts: to compute the optimal state estimator (the Kalman filter) and to compute the optimal feedback law from these estimated states. The key equation for the computations is the *algebraic Riccati equation, ARE* (9.7b).

State Feedback

If all states are measurable, so that $y = x$, we have $\hat{x} = x$ and the Kalman filter (9.6b) can be dispensed with. The regulator is then pure *state feedback* $u = -Lx$. The same is true if the system is given in innovations form (5.84),

$$\dot{\hat{x}}(t) = A\hat{x}(t) + Bu(t) + Kv(t)$$
$$y(t) = C\hat{x}(t) + v(t) \tag{9.8}$$

with $A - KC$ stable. Then $\hat{x}$ can be directly computed and the controller takes the form

$$\dot{\hat{x}} = (A - KC)\hat{x} + Bu + Ky, \quad u(t) = -L\hat{x}(t) \tag{9.9}$$

Of course, this is quite consistent with Theorem 9.1, since (9.8) is "its own Kalman filter".

The Servo Problem

The problem that the controlled variable z shall follow a reference signal r is already contained in the main result of Theorem 9.1. The formally correct way to treat the servo problem as an optimization problem, is to include a model for the frequency contents of the reference signal in the state space model. This is done, analogously to the disturbance case, by describing r as the output of a linear system driven by white noise. The signal v_1 will then also contain the driving source of r, and the state vector will have to be extended to be able to describe also the reference signal. If the controlled output is $z = M_1 x$ and the reference signal is $r = M_2 x$, the quantity to penalize is $e = r - z = \begin{bmatrix} -M_1 & M_2 \end{bmatrix} x = Mx$. Since r is known, it is to be included in the measured outputs

$$\bar{y} = \begin{bmatrix} y \\ r \end{bmatrix}$$

and the controller $u = -F_{\bar{y}}\bar{y}$ can then be written as $u = F_r r - F_y y$.

This is quite a laborious process and is worthwhile only if the reference signal has some very specific properties (like a typical resonance frequency). The most common case is that r is modeled as piecewise constant or a random walk. Then the result is much simpler:

Theorem 9.2 (The Servo Problem) *Let the system from control input and driving noise source to controlled output z and measured output y be given by (9.4). Suppose that the number of control inputs is at least as large as the number of controlled outputs, and assume that $u^*(r)$ is a well-defined constant control signal that in stationarity and absence of disturbances would give $z(t) \equiv r$. The criterion to be minimized is*

$$V = \|z - r\|_{Q_1}^2 + \|u - u^*(r)\|_{Q_2}^2 \tag{9.10}$$

and the reference signal is described as integrated white noise, that is as piecewise constant or a random walk.

The optimal regulator then is

$$u(t) = -L\hat{x}(t) + L_r r(t) \tag{9.11}$$

Here L and $\hat{x}$ are given as in Theorem 9.1, while

$$L_r = [M(BL - A)^{-1}B]^\dagger \tag{9.12}$$

Here † denotes pseudoinverse (see (8.5)), and this choice of L_r means that the static gain of the closed loop system is the identity matrix.

Proof: From the proof of the main theorem we know that it is sufficient to consider the impulse response case (i.e., r constant, without random jumps) with the complete state vector known. Let $x^*(r)$ be the state in stationarity for the control input $u^*(r)$. Then

$$0 = Ax^*(r) + Bu^*(r)$$
$$r = Mx^*(r) \qquad (9.13)$$

For the differences $\tilde{x} = x - x^*(r)$ and $\tilde{u} = u - u^*(r)$ we have

$$\dot{\tilde{x}} = \dot{x} = Ax + Bu = A(x - x^*(r)) + B(u - u^*(r)) = A\tilde{x} + B\tilde{u}$$
$$\tilde{z} = z - r = Mx - Mx^*(r) = M\tilde{x}$$

From this follows that the optimal feedback is $\tilde{u} = -L\tilde{x}$, i.e., $u = -Lx + (u^* + Lx^*)$, with L computed as in Theorem 9.1. Consider the term $u^* + Lx^*$. Using (9.13) we obtain

$$M(BL - A)^{-1}B(u^*(r) + Lx^*(r)) = M(BL - A)^{-1}(Bu^*(r) + BLx^*(r))$$
$$= M(BL - A)^{-1}(BL - A)x^*(r) = Mx^*(r) = r$$

Multiplying from the left by (9.12) now gives the desired result. □

Cross Term between x and u in the Criterion

It is sometimes of interest to modify (9.5) to contain a cross term between u och x, that is

$$V(L) = \int z^T(t)Q_1 z(t) + 2x^T(t)Q_{12}u(t) + u^T(t)Q_2 u(t) dt \qquad (9.14)$$

Since $z = Mx$ the criterion can also be written

$$\begin{bmatrix} x^T & u^T \end{bmatrix} \tilde{Q} \begin{bmatrix} x \\ u \end{bmatrix} = \left\| \begin{bmatrix} x \\ u \end{bmatrix} \right\|_{\tilde{Q}}^2 \quad \text{where} \quad \tilde{Q} = \begin{bmatrix} M^T Q_1 M & Q_{12} \\ Q_{12}^T & Q_2 \end{bmatrix}$$

For the minimization to be meaningful, it is required that $\tilde{Q}$ is positive semidefinite. The controller that minimizes this criterion is given by (9.6), with the feedback law L computed as

$$L = Q_2^{-1}(B^T S + Q_{12}^T) \qquad (9.15)$$
$$A^T S + SA + M^T Q_1 M - (SB + Q_{12})Q_2^{-1}(SB + Q_{12})^T = 0$$

The proof of this is based on the change of input variable $\tilde{u} = u + Q_2^{-1}Q_{12}^T x$. In terms of $\tilde{u}$ and x the problem has now been transformed to the standard formulation and Theorem 9.1 can be applied.

The Riccati Equation

Th computational key to the optimal control law (9.6b) is the equation (9.7b). This equation is knows as the (algebraic or stationary) *Riccati Equation*. In this section we shall discuss how to solve it numerically. Introduce the matrix

$$H = \begin{bmatrix} A & -BQ_2^{-1}B^T \\ -M^TQ_1M & -A^T \end{bmatrix} \tag{9.16}$$

This matrix is intimately related to (9.7b). Let the number of system states be n. Then H is a $2n \times 2n$ matrix. Suppose that we find an $n \times n$ matrix Λ and W_1, W_2, such that

$$H \begin{bmatrix} W_1 \\ W_2 \end{bmatrix} = \begin{bmatrix} W_1 \\ W_2 \end{bmatrix} \Lambda \tag{9.17a}$$

In longhand we then have

$$AW_1 - BQ_2^{-1}B^TW_2 = W_1\Lambda \tag{9.17b}$$
$$-M^TQ_1MW_1 - A^TW_2 = W_2\Lambda \tag{9.17c}$$

Multiply the first equation by W_1^{-1} from the left and substitute the thus obtained expression for Λ in the second equation. Then multiply this equation by W_1^{-1} from the right. Introduce the notation

$$S = W_2W_1^{-1} \tag{9.18}$$

This gives

$$-M^TQ_1M - A^TS = SA - SBQ_2^{-1}B^TS$$

The matrix S defined by (9.18) will thus solve the Riccati equation (9.7b). Moreover, by multiplying (9.17b) by W_1^{-1} from the right we obtain

$$A - BQ_2^{-1}B^TS = W_1\Lambda W_1^{-1}$$

The right hand side is a similarity transformation, so $A - BQ_2^{-1}B^TS$ and Λ have the same eigenvalues. To assure that the conditions in Theorem 9.1 are satisfied, Λ must have all its eigenvalues in the left half plane. It remains to be proven that it is possible to achieve (9.17a) with such a matrix Λ. This is summarized as a theorem:

Theorem 9.3 (Solution of the Riccati Equation) *Let the matrix H be defined by (9.16). Suppose that the system $(A, B, M^T Q_1 M)$ is stabilizable and detectable. Then there exists a stable $n \times n$ matrix Λ such that (9.17a) holds with an invertible W_1. Let the matrix S be defined by (9.18). It is a real, positive definite, symmetric matrix, satisfying the Riccati equation (9.7b). Moreover, $A - BQ_2^{-1}B^T S$ has all eigenvalues strictly in the left half plane.*

Proof: See Appendix 9A. □

Solving the Riccati equation as an eigenvalue problem for the matrix (9.16) is a numerically sound method, and is the preferred approach today.

9.3 Some Practical Aspects

To determine the linear quadratic control law is computationally demanding and only simple examples can be treated by hand. Computer support is thus necessary for this design method. On the other hand, efficient numerical methods have been developed for matrix equations like (9.7b). As a consequence there are now many interactive software packages for LG control design.

On the Use of Quadratic Criteria

One cannot expect that a quadratic criterion like (9.5) corresponds to an actual cost or demand. Instead, the matrices Q_1 and Q_2 must be seen as design variables, which via Theorem 9.1 generate good controller candidates. Once the controller has been computed, all relevant properties of the closed loop system have to be tested. Typically, this means that one calculates and plots the transfer functions G_c, T and S in the frequency domain. Time responses to reference signals and disturbances also have to be simulated.

If the results are not satisfactory, the problem formulation has to be changed, i.e., other penalty matrices Q_1 and Q_2, as well as other disturbance descriptions have to be chosen, and the procedure is repeated. If the state vector is measurable, the disturbance description does not play any role for the regulator, but the design process can be focused on the trade-off between control accuracy and control effort. In that case, it is rather straightforward to adjust the sizes of Q_1 and

Q_2 until a good balance is struck. Often diagonal penalty matrices are used. It turns out that rather large changes in the elements of the Q-matrices have to be made – a factor of ten or so – to achieve essential effects in the closed loop system.

Example 9.1:   Wind Gust Disturbances on an Aircraft.

In this example we shall design a lateral control law to the Gripen aircraft, based on the model in Example 2.3.

The problem is to control the aileron angle δ_a and the rudder angle δ_r so that the aircraft shows an acceptable response to disturbances of wind gust character. The control signal u consists of commanded control surface angles, as pointed out in Example 2.3. In this simple example, the roll angle and heading angle are chosen as controlled outputs z:

$$M = \begin{bmatrix} 0 & 0 & 0 & 1 & 0 & 0 & 0 \\ 0 & 0 & 0 & 0 & 1 & 0 & 0 \end{bmatrix}$$

We shall work with the wind gust model presented in Example 5.4. As a test case we select the situation where the wind speed w_1 varies over time as depicted in Figure 9.1. The wind acceleration w_4 is then a staircase function. Furthermore, we assume that the wind motion is linear, so that the rotational speeds w_2 and w_3 are zero.

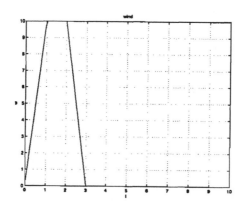

Figure 9.1: Wind gusts in a test case. Wind speed as a function of time.

We start by a linear quadratic criterion (9.5) with

$$Q_1 = I, \quad Q_2 = I$$

The optimal feedback is $u = -Lx$ with

$$L = \begin{bmatrix} -0.0022 & 0.17 & 0.12 & 0.98 & 0.31 & 0.76 & 0.018 \\ -0.0038 & 0.028 & -0.25 & 0.16 & -0.95 & 0.072 & 0.10 \end{bmatrix}$$

The eigenvalues of the closed loop system $A - BL$ are

$$-20.28, -20.00, -6.02 \pm 6.19i, -1.22 \pm 3.21i, -0.13$$

The response to the wind gust in Figure 9.1 is shown in Figure 9.2. It also shows the control surface rates $(\dot{x}_i; i = 6, 7)$. Since in practice there are rate limitations of the control signals, it is important to check that the regulator gives reasonable control inputs also in this respect. We see that the amplitude of the roll angle is substantially reduced, while the heading angle still shows a clear impact from the wind gust. To improve the disturbance response, several options are possible. One may, e.g., increase the penalty on the heading angle

$$Q_1 = \begin{bmatrix} 1 & 0 \\ 0 & 10 \end{bmatrix}$$

or reduce the weight on the control inputs to one tenth

$$Q_1 = I, \quad Q_2 = 0.1I$$

The latter case results in

$$L = \begin{bmatrix} -0.0036 & 0.39 & 0.21 & 3.11 & 0.63 & 1.54 & 0.046 \\ -0.0073 & 0.085 & -0.78 & 0.57 & -3.10 & 0.20 & 0.30 \end{bmatrix}$$

with eigenvalues in

$$-21.9, -20.03, -9.54 \pm 11.3i, -2.69 \pm 4.07i, -0.19$$

The response is shown in Figure 9.3. Here the disturbance attenuation on the heading angle is much better, at the price of larger control signals.

In this way, different penalty matrices in the criterion (9.5) can be tried out to find a controller that gives desired behavior of the closed loop system. The controller for the fighter aircraft Gripen has been designed in this way. Good controllers have been determined by LQ technique for a large number of flight cases. The actual controller then uses an L which is linearly interpolated from the precomputed flight cases, closest to the current one.

It might be added that in Gripen, the feedback from the heading angle ψ (the elements 0.63 and -3.10 in our L) is left to the pilot, and the controller is primarily modifying the dynamics to the angular velocities. This is achieved by choosing criteria that penalize roll- and yaw angular velocities, and possibly also v_y.

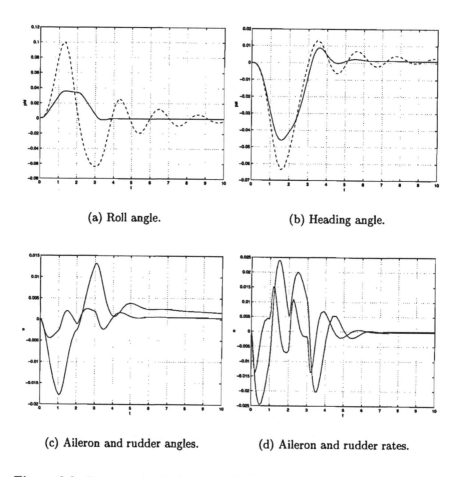

(a) Roll angle.

(b) Heading angle.

(c) Aileron and rudder angles.

(d) Aileron and rudder rates.

Figure 9.2: Response to a wind gust disturbance, $Q_2 = I$. For comparison the response of the open loop system is shown as dashed lines.

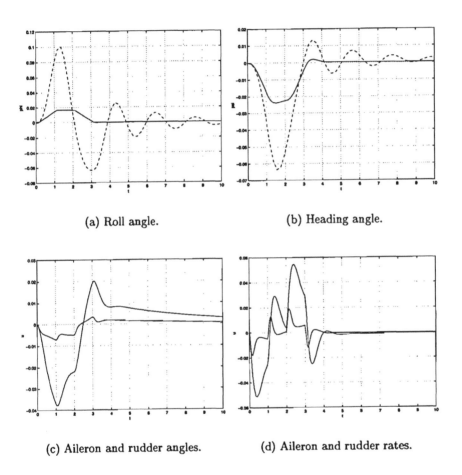

(a) Roll angle.   (b) Heading angle.

(c) Aileron and rudder angles.   (d) Aileron and rudder rates.

Figure 9.3: Response to wind gust disturbances, $Q_2 = 0.1I$. For comparison, the response of the open loop system is shown as dashed lines.

We noted in Example 3.3 that Gripen is non-minimum phase, and this is a potential problem according to Example 7.3. However, the zeros in 77 and 57 are so far to the right in the right half plane that they would pose a problem only if a bandwidth of more than *ca* 50 rad/sec was desired (see Figure 7.21). This is not at all the case for Gripen, as seen from the computed eigenvalues of the closed loop system.

Example 9.2:   Room Temperature Control

Let us return to Example 5.7. (See also Example 7.4.) The most common case is that heating the radiator has fast dynamics (K_1 is a rather large number) while the influence of the outdoor temperature on the indoor one is rather slow (K_2 is a rather small number). Here, however, we shall consider a house with the opposite conditions. (Such houses are typical in, e.g., the Stanford area, where the outer walls are rather thin -- the risk of earthquakes! -- and the heating is achieved by water pipes, set into the concrete floor. At the same time, the outdoor temperature shows substantial variation over the 24-hour period, which makes it quite a challenge to keep a constant room temperature.)

We let the time constant from outdoor to indoor temperature be 1 hour ($K_1 = 1$) and the time constant for heating the floor from the hot water be 8 hours ($K_3 = 1/8$). Moreover, we assume that the ratio between the heat capacities of the floor and the room air is 5, and choose $K_2 = 1$ and $K_4 = 0.2$. These values are inserted into the state space model (5.57)–(5.58). Recall that the states are

$$
\begin{aligned}
x_1 &= \text{the radiator (floor) temperature} \\
x_2 = z &= \text{the room temperature} \\
x_3 = \dot{w} &= \text{the time derivative of the outdoor temperature} \\
x_4 = w &= \text{the outdoor temperature} \\
x_5 &= \text{the constant bias in thermometer 1}
\end{aligned}
$$

Moreover, we have the measured signals and the control signal

$$
\begin{aligned}
y_1 &= \text{noiseless measurement of the indoor temperature,} \\
&\quad\text{with a constant but unknown calibration error (bias)} \\
y_2 &= \text{correctly calibrated, but noisy measurement} \\
&\quad\text{of the indoor temperature} \\
y_3 &= \text{outdoor temperature} \\
u &= \text{water temperature in the pipes}
\end{aligned}
$$

We consider the case that the mean outdoor temperature is 18° and that the setpoint for the indoor temperature is 20°. This corresponds to an equilibrium temperature for the radiator (floor) of 22° and for the heating water of 38°. In the sequel, all temperatures are measured as deviations from this equilibrium.

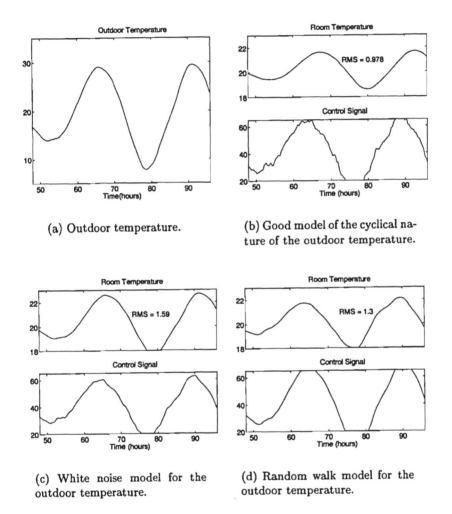

(a) Outdoor temperature.

(b) Good model of the cyclical nature of the outdoor temperature.

(c) White noise model for the outdoor temperature.

(d) Random walk model for the outdoor temperature.

Figure 9.4: Temperature control when the outdoor temperature is measured, and different models of its character are used. The root mean square (RMS) value of the deviation from the indoor temperature setpoint is marked in each figure.

The Kalman filter for this model is now determined for the covariance matrices

$$R_1 = \begin{bmatrix} 10 & 0 \\ 0 & 1 \end{bmatrix}, \quad R_2 = \begin{bmatrix} 0.0001 & 0 & 0 \\ 0 & 0.1 & 0 \\ 0 & 0 & 0.001 \end{bmatrix}$$

This gives

$$K = \begin{bmatrix} -0.0039 & 0.0003 & 0.0035 \\ 0.0463 & 0.0027 & 0.9676 \\ 8.5224 & 0.0148 & 99.7543 \\ 1.1426 & 0.0097 & 14.1198 \\ 99.9634 & -0.0026 & -0.8533 \end{bmatrix}$$

For the criterion (9.5) we choose $Q_1 = 1$ and $Q_2 = 0.01$. This gives

$$L = \begin{bmatrix} 2.8106 & 1.4071 & 6.8099 & 3.7874 & 0 \end{bmatrix} \qquad (9.19)$$

We thus introduce feedback from (estimated values of) floor temperature, indoor temperature, as well as from the outdoor temperature and its derivative. The latter feedback is really feedforward from the disturbance signal w. Figure 9.4b illustrates the performance of this controller.

Let us consider some variants of this controller. Figure 9.4c,d illustrates the importance of a model for the outdoor temperature w. If this is assumed to vary as white noise, we obtain

$$L = \begin{bmatrix} 24.8672 & 46.7304 & 0 & 4.8531 & 0 \end{bmatrix}$$

(In the models that are based on a white noise model for the outdoor temperature, we use $Q_2 = 0.001$ to obtain similar variations in u.) This controller does not have any feedback from x_3, the estimated value of the time derivative of the outdoor temperature. This is natural, since the temperature is not believed to change in any predictable fashion. The control is considerably worse, since future load disturbances from the outdoor temperature are not predicted. A random walk model for w (Figure 9.4d) gives a clearly better result.

If the outdoor temperature is not measured, but we use a good model for its variations, we have the case (5.57b). This gives

$$K = \begin{bmatrix} 0.3346 & 0.1839 \\ 29.5766 & 2.2505 \\ 218.6779 & 6.7601 \\ 315.4723 & 7.2870 \\ 71.1826 & -2.2209 \end{bmatrix}$$

and (of course) the same L as in (9.19). The performance of this controller is shown in Figure 9.5a. The somewhat unexpected result is that the control

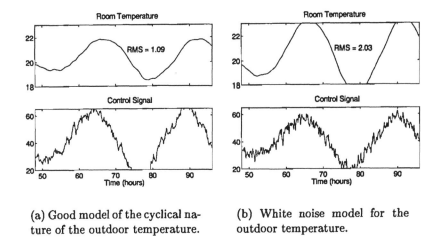

(a) Good model of the cyclical nature of the outdoor temperature.

(b) White noise model for the outdoor temperature.

Figure 9.5: Temperature control when the outdoor temperature is not measured, and different models of its character are used. The root mean square (RMS) value of the deviation from the indoor temperature setpoint is marked in each figure.

is almost as good as when the outdoor temperature is measured. The explanation is that, with the reliable model, and from the measurements of y_1, y_2 and u, the outdoor temperature can be calculated with good accuracy. This is obviously what the Kalman filter does. If there are significant model errors, the value of measuring y_3 increases. (This can be illustrated by simulating the same controllers for a system with an 80% higher value of K_3. Then the RMS value of the control error is twice as big without outdoor temperature measurement as with it.)

The model for the variations of outdoor temperature is essential. Figure 9.5b shows the result when the outdoor temperature is not measured, and believed to be white noise.

Ensuring Integration and Other Loop Properties

It is often desired to have an integrator in the feedback loop, so that constant disturbances can be eliminated in steady state, i.e., to ensure that the sensitivity function is zero at zero frequency. There may also be other requirements on the sensitivity function and the complementary sensitivity function. The LQG-technique produces solutions that reflect the problem formulation: if very small sensitivity is desired at low frequencies, we have to "tell the machinery" that there are very large system disturbances there. This is done by giving w in the disturbance model (9.1a) much power at low frequencies, i.e.,

$$w(t) = \frac{1}{p+\delta}v(t) \qquad (9.20)$$

where v is white noise and δ is a small number.

Remark. The choice $\delta = 0$ is not possible, since those states in (9.4) that are introduced to describe the noise properties are normally not controllable from u. They must then be stable, in order to secure that (A, B) is stabilizable, which is required by Theorem 9.1.

In the same way one can ensure that the complementary sensitivity function will be small at certain frequencies by giving much power to the measurement noise n in those ranges. The loop gain properties of the LQG-solution can thus be manipulated by adjusting the character of measurement and system disturbances. See Example 6.5. With specific requirements on the loop properties, it is however easier and better to directly use the LQ-technique for loop shaping according to Section 10.3.

Example 9.3:   A Multivariable System

Consider again the system (1.1):

$$y(t) = \begin{bmatrix} \frac{2}{p+1} & \frac{3}{p+2} \\ \frac{1}{p+1} & \frac{2}{p+1} \end{bmatrix} u(t) + w(t)$$

(See Examples 3.1, 3.2 and 7.3.) In state space form it is given by (3.14). To ensure that the sensitivity function tends to zero at zero frequency, i.e., to have an integrator in the feedback loop, we let the system disturbance w be of the form (9.20) with $\delta = 0.001$. This gives the state space form

$$\dot{x} = \begin{bmatrix} -1 & 0 & 0 & 0 & 0 \\ 0 & -1 & 0 & 0 & 0 \\ 0 & 0 & -2 & 0 & 0 \\ 0 & 0 & 0 & -0.001 & 0 \\ 0 & 0 & 0 & 0 & -0.001 \end{bmatrix} x + \begin{bmatrix} 1 & 0 \\ 0 & 1 \\ 0 & 1 \\ 0 & 0 \\ 0 & 0 \end{bmatrix} u + \begin{bmatrix} 0 & 0 \\ 0 & 0 \\ 0 & 0 \\ 1 & 0 \\ 0 & 1 \end{bmatrix} v$$

$$y = \begin{bmatrix} 2 & 0 & 3 & 1 & 0 \\ 1 & 1 & 0 & 0 & 1 \end{bmatrix} x$$

We assume the measurements to have a small amount of noise and solve the LQ-problem according to Theorem 9.1 with $R_1 = I, R_{12} = 0$, $R_2 = 0.001$ and $M = C, Q_1 = I, Q_2 = 0.1I$. This gives

$$L = \begin{bmatrix} 4.05 & 0.81 & 4.23 & 2.31 & -0.055 \\ 5.05 & 1.01 & 5.96 & 1.65 & 2.03 \end{bmatrix}$$

$$K = \begin{bmatrix} 0 & 0 \\ 0 & 0 \\ 0 & 0 \\ 31.62 & 0 \\ 0 & 31.62 \end{bmatrix}$$

Moreover, L_r in (9.11) is chosen according to(9.12) as

$$L_r = \begin{bmatrix} 4.24 & -3.43 \\ 0.12 & 4.82 \end{bmatrix}$$

The step responses for this controller are shown in Figure 9.6. We see that we can handle the speed-up of the response better than in Example 1.1. However, we see also that there are some problems in the step responses. According to Examples 3.2 and 7.3, the system is non-minimum phase with a zero in the point 1. This means that the maximum achievable bandwidth is about 0.5 rad/sec. (Rise time about 2 sec.) The difficulties that were shown in (7.35) are thus fundamental and cannot be eliminated by the LQ-technique.

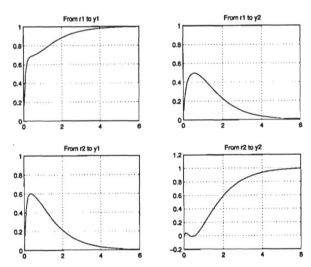

Figure 9.6: Step responses for the non-minimum phase system in Example 9.3.

9.4   Robustness for LQG Controllers

In this section we shall study the robustness aspects of the closed loop system, that are achieved by LQ and LQG design. All sensitivity and robustness properties are direct consequences of the loop gain GF_y (see (6.7)–(6.12)), so it is a matter of studying the properties of this gain.

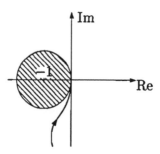

Figure 9.7: A Nyquist curve, subject to (9.24) is always outside the indicated circle.

An Inequality

First consider, from a formal point of view, the transfer function

$$H(s) = L(sI - A)^{-1}B \tag{9.21}$$

Suppose that L has been determined from (9.7a)–(9.7b) as an optimal feedback law for some penalty matrices Q_1 och Q_2, that is

$$
\begin{aligned}
L &= Q_2^{-1}B^T S \\
A^T S &+ SA + Q_1 - SBQ_2^{-1}B^T S = 0
\end{aligned}
\tag{9.22}
$$

It follows from Lemma 5.2 that

$$(I + H(-i\omega))^T Q_2(I + H(i\omega)) \geq Q_2 \tag{9.23}$$

(In (5.82) we associate the matrices as follows: $P \mapsto S, A \mapsto A^T, R_2 \mapsto Q_2, N \mapsto M^T, C \mapsto B^T, K \mapsto L^T$. Then we use that $R_1 \mapsto Q_1$ is positive semidefinite.) In the SISO case the conclusion is that

$$|1 + H(i\omega)| \geq 1 \tag{9.24}$$

or, in words, that $H(i\omega)$ has at least the distance 1 to the point -1. The Nyquist curve for $H(i\omega)$ is thus always outside the circle marked in Figure 9.7.

In the multivariable case, the conclusion is that $Q_2 = \rho I$ implies that all singular values of $I + H(i\omega)$ will be larger or equal to 1 for all frequencies.

The State Feedback Loop Gain

If all states are measurable, the optimal state feedback is

$$u(t) = -Lx(t)$$
$$\dot{x}(t) = Ax(t) + Bu(t)$$

The corresponding controller is depicted in Figure 9.8. The loop gain

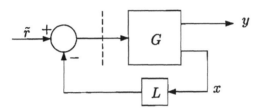

Figure 9.8: Feedback from measured states.

for this closed loop system is

$$G_k(s) = L(sI - A)^{-1}B \tag{9.25}$$

To realize this, assume that the loop is broken at the dashed line in Figure 9.8. Starting to the right of this line with the input u_1 gives $x = (sI - A)^{-1}Bu_1$ and coming back to the left of the line, the signal is $Lx = L(sI - A)^{-1}Bu_1$, which gives the loop gain.

If L is chosen as the solution to a linear quadratic problem, the loop gain (9.25) will thus have the property (9.24); illustrated in Figure 9.7. The consequences are as follows for SISO systems:

1. *The phase margin for the system is at least 60°.* This follows from the fact that the unit circle cuts the circle in Figure 9.7 in two points that correspond to the phase angles (arguments) $-120°$ and $-240°$. The Nyquist curve therefore cannot be inside the unit circle with a phase angle between these values. Then the phase margin must be at least $180 - 120 = 60$ degrees.

2. *The gain margin is infinite,* since the Nyquist curve for the loop gain does not cut the negative real axis between -2 and 0.

3. *The sensitivity function S (measured at Lx) is always less than 1,* since $S(s) = 1/(1 + G_k(s))$ and the denominator is always larger

than 1 according to (9.24). (This apparently contradicts Bode's integral, Theorem 7.3, but feedback from measured states is an "unrealistic" controller that does not satisfy the assumption that the loop gain decreases faster than $1/\omega$ as the frequency ω tends to infinity.)

4. *The complementary sensitivity function T is always less than 2,* since $T = 1 - S$.

In the multivariable case, (3) and (4) are to be interpreted so that all singular values satisfy the inequalities in question, and (1) and (2) so that each control input can suffer a phase loss of up to 60° or an arbitrary gain increase without the system becoming unstable.

Optimal feedback from perfectly measured states thus gives a loop gain with very good robustness and sensitivity properties. The question is whether these good properties are retained in the more realistic case of feedback from observer-reconstructed states.

Feedback from Reconstructed States: Robustness

When reconstructed states are fed back, the transfer function of the controller is given by (8.26b). Multiplying (8.26b) by the transfer function of the open loop system gives the loop gains

$$F_y G = L(sI - A + BL + KC)^{-1}KC(sI - A)^{-1}B \tag{9.26}$$

and

$$GF_y = C(sI - A)^{-1}BL(sI - A + BL + KC)^{-1}K \tag{9.27}$$

In the SISO case there is of course no difference between these, but in the multivariable case, S and T are determined by (9.27) according to (6.9) and (6.10), while (9.26) determines the input properties (6.11), (6.14)–(6.16).

It can be shown that even if K and L are chosen as in Theorem 9.1 there are no guarantees for the robustness of the closed loop system; the stability margin may in principle be arbitrarily poor.

Suppose that L has been chosen as the solution of a linear quadratic control problem. The question then is if K can be chosen so that the loop gain (9.26) resembles the ideal loop gain (9.25), that is, if

$$L(sI - A + BL + KC)^{-1}KC(sI - A)^{-1}B \approx L(sI - A)^{-1}B \tag{9.28}$$

We shall show that (9.28) holds if K is chosen as

$$K = \rho B \qquad (9.29)$$

with a sufficiently large value of ρ. Note that this choice can be made only if the number of inputs equals the number of outputs.

Lemma 9.1 *As ρ tends to infinity, the following limit hold:*

$$
\begin{aligned}
F_y(s)G(s) &= L(sI - A + BL + \rho BC)^{-1}\rho BC(sI - A)^{-1}B \\
&\to L(sI - A)^{-1}B
\end{aligned}
$$

$$(9.30)$$

Proof: The proof will only be given in the SISO case. Without loss of generality it then can be assumed that (A, B) is given in controller canonical form. (The basis can always be changed to this form, without affecting the loop gain.)

From (2.17) we have

$$
A = \begin{bmatrix}
-a_1 & -a_2 & \cdots & -a_n \\
1 & 0 & \cdots & 0 \\
0 & 1 & \cdots & 0 \\
\vdots & \vdots & & \\
0 & 0 & \cdots 1 & 0
\end{bmatrix}, \qquad
B = \begin{bmatrix}
1 \\
0 \\
0 \\
\vdots \\
0
\end{bmatrix}
$$

$$C = \begin{bmatrix} c_1 & \cdots & c_n \end{bmatrix}$$

$$C(sI - A)^{-1}B = \frac{\tilde{C}(s)}{\tilde{A}(s)}$$

where

$$
\begin{aligned}
\tilde{C}(s) &= c_1 s^{n-1} + c_2 s^{n-2} + \ldots + c_n \\
\tilde{A}(s) &= s^n + a_1 s^{n-1} + \ldots + a_n
\end{aligned}
$$

In our case also $(A - BL - \rho BC, B)$ will be given in controller canonical form (the elements of the L- och C-matrices are shifted into the first row of A.) Hence

$$L(sI - A + BL + \rho BC)^{-1}\rho B = \frac{\rho \tilde{L}(s)}{\tilde{A}(s) + \tilde{L}(s) + \rho \tilde{C}(s)}$$

where

$$\tilde{L}(s) = \ell_1 s^{n-1} + \ldots + \ell_n, \qquad L = \begin{bmatrix} \ell_1 & \cdots & \ell_n \end{bmatrix}$$

This means that the loop gain is subject to

$$F_y G = \frac{\tilde{L}(s)}{\frac{1}{\rho}\tilde{A}(s) + \frac{1}{\rho}\tilde{L}(s) + \tilde{C}(s)} \cdot \frac{\tilde{C}(s)}{\tilde{A}(s)} \rightarrow \frac{\tilde{L}(s)}{\tilde{A}(s)} = L(sI - A)^{-1}B$$

as $\rho \rightarrow \infty$ □

Note that, as the loop gain is formed, the factor $\tilde{C}(s)$ is cancelled in the limit. This factor corresponds to the open loop zeros, and according to Example 6.1 it corresponds to non-observable modes in the closed loop system. The zeros must therefore be stable, i.e., the system must be minimum phase for $K = \rho B$ to be a reasonable choice.

Where will the poles of the observer move? From the proof of Lemma 9.1, it follows that the characteristic equation is

$$\det(sI - A + KC) = \det(sI - A + \rho BC) = \tilde{A}(s) + \rho\tilde{C}(s)$$

When $\rho \rightarrow \infty$, the poles of the observer will thus tend partly to the zeros of the system, and partly to infinity.

Loop Transfer Recovery

The procedure to regain the ideal loop gain properties of $L(sI - A)^{-1}B$ is known as Loop Transfer Recovery, LTR. It requires an equal number of inputs and outputs.

To apply LTR means that the state feedback law L is first determined as the solution to a linear quadratic problem. Then the observer gain K is taken as $K = \rho B$ and ρ is increased until the system has desired sensitivity and robustness properties. According to Lemma 9.1 a sufficiently large ρ will give properties arbitrarily close to the ideal ones as listed above, provided the system is minimum phase.

There is another way to achieve

$$K \approx \rho B \qquad \text{with large} \quad \rho$$

Consider (5.79)–(5.80):

$$K = PC^T R_2^{-1}$$
$$PA^T + AP - PC^T R_2^{-1}CP + NR_1N^T = 0 \tag{9.31}$$

Suppose we choose

$$R_1 = \alpha \cdot R_2 \quad \text{and} \quad N = B \tag{9.32}$$

with a large α. Then (9.31) gives

$$PA^T + AP - KR_2K^T + \alpha BR_2B^T = 0$$

As α increases, the two last terms will dominate, so

$$K \approx \sqrt{\alpha}B$$

must hold, provided this K yields a stable $A - KC$. This is the case if the zeros of the open loop system are inside the stability region, as shown above.

Note that the conditions (9.32) correspond to assuming that there are system disturbances with large power that enter together with the input

$$\dot{x} = Ax + B(u + v_1)$$

We can now summarize the LTR procedure as follows

1. Choose L as optimal LQ-state feedback for penalty matrices Q_1 and Q_2 which give a suitable dynamics for the closed loop system. Verify that the ideal loop gain

   $$L(sI - A)^{-1}B$$

   gives good sensitivity and robustness.

2. Choose the observer gain K as optimal (Kalman) gain for the disturbance matrix $N = B$ and $R_1 = \alpha R_2$. Let the scalar α increase until the loop gain (sensitivity functions) has acceptable properties.

To choose K via a Kalman filter instead of directly using $K = \rho B$ gives the advantage that $A - KC$ is guaranteed to be stable for any value α.

Output-LTR

The technique just described is more precisely known as *input-LTR*. The reason is that this method recovers the ideal loop gain F_yG in (9.26), which determines the loop properties at the input side. By completely analogous reasoning, there is also an *output-LTR* procedure. Then a Kalman filter with good loop gain $C(sI - A)^{-1}K$ is first determined. Then, L is determined as LQ-feedback, with $M = C$, and a large penalty $Q_1 = \alpha Q_2$. As α increases to infinity, the loop gain at the output, that is GF_y, will tend to $C(sI - A)^{-1}K$. For a SISO system there is clearly no difference between input- and output-LTR.

Note that for multivariable systems, input-LTR does not give any guarantees for the sensitivity at the output and vice versa.

9.5   Discrete Time Systems

The LQ theory for discrete time systems is entirely parallel to the continuous time case, but the formulas are different. In this section we shall describe the necessary modifications.

The system description is

$$x(t+1) = Ax(t) + Bu(t) + Nv_1(t) \tag{9.33a}$$
$$z(t) = Mx(t) \tag{9.33b}$$
$$y(t) = Cx(t) + v_2(t) \tag{9.33c}$$

The criterion takes the form

$$\min \left(\|e\|_{Q_1}^2 + \|u\|_{Q_2}^2 \right) = \min \sum_{t=1}^{\infty} e^T(t)Q_1 e(t) + u^T(t)Q_2 u(t) \tag{9.34}$$

The interpretation of the left hand side can also be the time average of the sum in the right hand side or expected values (variances) of stationary stochastic processes.

The Optimal Controller: Main Results

Theorem 9.4 (Linear Quadratic Optimization)   *Consider the system (9.33) where* $\begin{bmatrix} v_1 \\ v_2 \end{bmatrix}$ *is white noise with intensity* $\begin{bmatrix} R_1 & R_{12} \\ R_{12}^T & R_2 \end{bmatrix}$. *The linear controller* $u(t) = -F_y(q)y(t)$ *that minimizes the criterion*

$$V = \|z\|_{Q_1}^2 + \|u\|_{Q_2}^2 \tag{9.35}$$

is to be found. The controller is assumed to be causal and contain a time delay, (i.e., $u(t)$ may depend on $y(t-1), u(t-1), y(t-2), u(t-2), \dots$). Suppose that (A, B) is stabilizable and that $(A, M^T Q_1 M)$ is detectable and that the assumptions on A, C, R_1, R_2 and R_{12} in Theorem 5.6 are satisfied.

Then the optimal linear controller is given by

$$u(t) = -L\hat{x}(t) \tag{9.36a}$$
$$\hat{x}(t+1) = A\hat{x}(t) + Bu(t) + K(y(t) - C\hat{x}(t)) \tag{9.36b}$$

Here (9.36b) is the Kalman filter for the system, so K is determined by Theorem 5.6 as

$$P = APA^T + NR_1N^T - (APC^T + NR_{12})(CPC^T + R_2)^{-1}$$
$$\times (APC^T + NR_{12})^T$$
$$K = (APC^T + NR_{12})(CPC^T + R_2)^{-1}$$

while L is given by

$$L = (B^TSB + Q_2)^{-1}B^TSA \tag{9.37a}$$

where S is the unique, positive semidefinite, symmetric solution to the matrix equation

$$S = A^TSA + M^TQ_1M - A^TSB(B^TSB + Q_2)^{-1}B^TSA \tag{9.37b}$$

Moreover, it follows that A − BL has all its eigenvalues inside the stability region.

The proof is given in Appendix 9A. The key equation is the *discrete time Riccati equation* (9.37b).

Corollary 9.2 (Controller without Delay) *Suppose a direct term is allowed in the controller, i.e., u(t) may depend on*

$$y(t), y(t-1), u(t-1), y(t-2), \ldots$$

Then the feedback is given by

$$u(t) = -L\hat{x}(t|t) \qquad\qquad \text{if } R_{12} = 0 \tag{9.38a}$$
$$u(t) = -L\hat{x}(t|t) - L_v\hat{v}_1(t|t) \quad \text{if } R_{12} \neq 0 \tag{9.38b}$$
$$L_v = (B^TSB + Q_2)^{-1}B^TS \tag{9.38c}$$

where $\hat{x}(t|t)$ is the measurement updated state estimate (5.102) and $\hat{v}_1(t|t)$ is given by (5.102d). L is the same matrix as in (9.37).

Proof: In the calculations in the proof $\hat{x}(t|t)$ is used instead of $\hat{x}(t|t-1)$. Also, the expression (5.102) must be used at the time update. $\square$

Corollary 9.3 (LQG: Gaussian Noise) *Suppose that the signal* $[v_1, v_2]$ *in (9.4) is Gaussian white noise. Then the controller that minimizes the criterion (9.5) among all possible causal controllers (including nonlinear ones) is linear and is given by (9.6).*

Proof: If the noises are Gaussian, then also $\nu(t)$ and $y(s)$, $s < t$ are Gaussian. Since these are uncorrelated, they are also independent. This means that $u(t)$ and $\nu(t)$ will be independent for any causal controller, and the step marked in italics in the proof of Theorem 9.4 is valid also for nonlinear controllers. $\square$

The Servo Problem

The treatment of the servo problem is entirely analogous to the time continuous case and Theorem 9.2 is valid without changes.

Cross Term Between x and u in the Criterion

A criterion with a cross term

$$V(L) = \sum_{t=1}^{\infty} \left(z^T(t) Q_1 z(t) + 2x^T(t) Q_{12} u(t) + u^T(t) Q_2 u(t) \right)$$

(9.39)

is minimized by

$$
\begin{aligned}
L &= (B^T S B + Q_2)^{-1} (A^T S B + Q_{12})^T \\
S &= A^T S A + M^T Q_1 M - (A^T S B + Q_{12}) \\
&\quad \times (B^T S B + Q_2)^{-1} (A^T S B + Q_{12})^T
\end{aligned}
$$

(9.40)

This follows as in Theorem 9.4.

Minimization of Continuous Time Criteria with Sampled Data Control

To specify the behavior of the control variables only at the sampling instants may be questionable. Problems may occur, as illustrated in Figure 4.3. Even in the case of sampled data control it would be natural to use a time continuous criterion of the type

$$V_c(L) = \int_0^\infty z^T(t) Q_1^c z(t) + u^T(t) Q_2^c u(t) \, dt$$

(9.41)

This is the same expression as (9.5) in the time continuous case, and is thus minimized by a continuous time controller (9.7a). Now, we seek instead a sampled-data controller with sampling interval T, that minimizes (9.41). Note that this controller is not given by (9.37a), since the latter minimizes a criterion that just looks at the values of z at the sampling instants. On the other hand, we can exactly compute the state $x(t), kT \leq t < (k+1)T$ from $x(kT)$ och $u(kT)$, using (3.2). In that way, (9.41) can be exactly translated to a criterion that is just based on the signals at the sampling instants. The result is as follows:

Theorem 9.5 (Continuous Time Criterion in Sampled Form)
To minimize (9.41) with a sampled data controller with sampling interval T is the same as minimizing the discrete time criterion

$$V_d(L) = \sum_{t=1}^{\infty} \left(x^T(t)Q_1 x(t) + 2x^T(t)Q_{12}u(t) + u^T(t)Q_2 u(t) \right)$$

(9.42a)

where

$$Q_1 = \int_0^T (e^{At})^T M^T Q_1^c M e^{At} dt$$

(9.42b)

$$Q_{12} = \int_0^T \Gamma^T(t) M^T Q_1^c M e^{At} dt$$

(9.42c)

$$Q_2 = \int_0^T \Gamma^T(t) M^T Q_1^c M \Gamma(t) dt + T Q_2^c$$

(9.42d)

Here

$$\Gamma(t) = \int_0^t e^{As} B ds$$

(9.42e)

Note that (9.42a) penalizes the whole state vector x. This corresponds to $M = I$ in the discrete time model.

Proof: The proof is carried out by inserting the time function $z(t)$ as a function of $x(kT)$ and $u(kT)$ and then cleaning up the expression by straightforward calculations. $\square$

The sampled data feedback that minimizes (9.41) is thus given by (9.40) with the Q-matrices given above.

Example 9.4:   Optimal Inventory Control – White Sales

Consider the simple inventory control model of Example 5.10 where the daily sales are described as white noise: (5.87). Suppose that we would like to avoid irregular orders by introducing a penalty on the deviation from the normal daily order ($u = 0$). This is captured by state feedback, such that the criterion

$$V = Ey^2(t) + \mu Eu^2(t)$$

is minimized.

With the state space description (5.87) we have

$$A = \begin{bmatrix} 1 & 1 \\ 0 & 0 \end{bmatrix}, \quad B = \begin{bmatrix} 0 \\ 1 \end{bmatrix}, \quad M = \begin{bmatrix} 1 & 0 \end{bmatrix}, \quad Q_1 = 1, \quad Q_2 = \mu$$

The system of equations (9.37a)–(9.37b) is to be solved for these matrices. Introduce

$$\bar{S} = \begin{bmatrix} s_1 & s_2 \\ s_2 & s_3 \end{bmatrix}$$

With the current values we obtain

$$L = \left(\mu + \begin{bmatrix} 0 & 1 \end{bmatrix} \begin{bmatrix} s_1 & s_2 \\ s_2 & s_3 \end{bmatrix} \begin{bmatrix} 0 \\ 1 \end{bmatrix} \right)^{-1} \begin{bmatrix} 0 & 1 \end{bmatrix} \begin{bmatrix} s_1 & s_2 \\ s_2 & s_3 \end{bmatrix} \begin{bmatrix} 1 & 1 \\ 0 & 0 \end{bmatrix}$$

$$= \frac{s_2}{\mu + s_3} \begin{bmatrix} 1 & 1 \end{bmatrix}$$

Moreover, we have (9.37b):

$$\begin{bmatrix} s_1 & s_2 \\ s_2 & s_3 \end{bmatrix} = \begin{bmatrix} 1 & 0 \\ 1 & 0 \end{bmatrix} \begin{bmatrix} s_1 & s_2 \\ s_2 & s_3 \end{bmatrix} \begin{bmatrix} 1 & 1 \\ 0 & 0 \end{bmatrix} + \begin{bmatrix} 1 \\ 0 \end{bmatrix} \begin{bmatrix} 1 & 0 \end{bmatrix}$$
$$- \frac{s_2^2}{\mu + s_3} \begin{bmatrix} 1 & 1 \\ 1 & 1 \end{bmatrix}$$

This gives

$$s_1 = s_1 + 1 - \frac{s_2^2}{\mu + s_3}, \quad s_2 = s_1 - \frac{s_2^2}{\mu + s_3}, \quad s_3 = s_1 - \frac{s_2^2}{\mu + s_3}$$

with the solution

$$s_2 = s_3 = \frac{1 + \sqrt{1 + 4\mu}}{2}, \quad s_1 = 1 + s_2$$

The optimal feedback law is

$$L = \frac{\sqrt{1 + 4\mu} - 1}{2\mu} \begin{bmatrix} 1 & 1 \end{bmatrix}$$

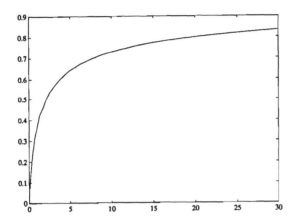

Figure 9.9: One of the poles of the closed loop system as a function of the input weight μ. The other pole is in the origin.

which gives the feedback

$$u(t) = -\frac{\sqrt{1+4\mu}-1}{2\mu}(y(t) + u(t-1))$$

Introduce the notation $K(\mu) = (\sqrt{1+4\mu}-1)/2\mu$.The system matrix of the closed loop system then is

$$A - BL = \begin{bmatrix} 1 & 1 \\ -K(\mu) & -K(\mu) \end{bmatrix}$$

The poles are thus located in 0 and $1 - K(\mu)$. See Figure 9.9. Note that for $\mu = 0$ (no input penalty) both poles are in the origin. Such a controller is called *dead-beat control*.

Example 9.5:   Optimal Inventory Control – Colored Sales

Suppose now that the sales are described as correlated random variables. In Example 2.7 we gave the equations for this description. With the state vector

$$x(t) = \begin{bmatrix} y(t) \\ u(t-1) - 0.9v(t-1) \end{bmatrix}$$

we have

$$x(t+1) = \begin{bmatrix} 1 & 1 \\ 0 & 0 \end{bmatrix} x(t) + \begin{bmatrix} 0 \\ 1 \end{bmatrix} u(t) + \begin{bmatrix} -1 \\ -0.9 \end{bmatrix} v(t) \qquad (9.43a)$$

$$y(t) = \begin{bmatrix} 1 & 0 \end{bmatrix} x(t). \qquad (9.43b)$$

As in the previous example, let us minimize

$$Ey^2(t) + \mu Eu^2(t)$$

Since the matrix N and the covariance matrix R_1 do not influence the solution we have the same *state feedback matrix* L as in Example 9.4, i.e.,

$$u(t) = -\frac{\sqrt{1+4\mu}-1}{2\mu}\begin{bmatrix}1 & 1\end{bmatrix}x(t). \tag{9.44}$$

Note that it is still not the same *control law* as in Example 9.4, since we have a different state vector here. The state vector in this example,

$$x(t) = \begin{bmatrix} y(t) \\ u(t-1) - 0.9v(t-1) \end{bmatrix}$$

is not directly measurable due to the noise term $0.9v(t-1)$. On the other hand, this term can be computed (i.e., not only estimated) from past observations $y(s)$, $u(s)$, $s \le t$. From (5.89) we have

$$-v(t-1) = \frac{1-q^{-1}}{1+0.9q^{-1}}y(t) - \frac{1}{1+0.9q^{-1}}u(t-2) \tag{9.45}$$

that is

$$-v(t-1) = y(t) - y(t-1) - u(t-2) + 0.9v(t-2) \tag{9.46}$$

This value is known, or can be computed at time t. With the notation

$$K(\mu) = \frac{\sqrt{1+4\mu}-1}{2\mu}$$

(9.44) can be written

$$u(t) = -K(\mu)(y(t) + u(t-1) - 0.9v(t)) \tag{9.47}$$

The control is based on the expression within parentheses. This has a very natural interpretation. The term $y(t)$ is today's inventory level. The term $-0.9v(t)$ is the best estimate of tomorrow's sales. The term $u(t-1)$ is what we know will be delivered tomorrow. The expression

$$y(t) + u(t-1) - 0.9v(t)$$

consequently is our best estimate of tomorrow's inventory level. The control action is based on this prediction. Using (9.45) we can rewrite (9.47) without the term $v(t)$. We have

$$u(t) = -K(\mu)\left(y(t) + u(t-1) + \frac{0.9(1-q^{-1})}{1+0.9q^{-1}}y(t)\right.$$
$$\left. - \frac{0.9q^{-2}}{1+0.9q^{-1}}u(t)\right) = -K(\mu)\frac{1.9y(t) + u(t-1)}{1+0.9q^{-1}}$$

which can be written

$$u(t) + 0.9u(t-1) = -K(\mu) \cdot 1.9y(t) - K(\mu)u(t-1)$$

that is

$$u(t) = -K(\mu) \cdot 1.9y(t) - (K(\mu) + 0.9)u(t-1) \qquad (9.48)$$

For $\mu = 0$, we have $K(0) = 1$, which gives *minimum variance control*, i.e., a control law that minimizes the variance of the control error, without any penalty on the size of the control input.

9.6 Comments

Main Points of the Chapter

The optimization technique described in this chapter actually gives a whole family of methods for control design. Three different viewpoints can be distinguished

1. **The Optimization Approach.** Theorem 9.1 and the expressions for the Kalman filter give the optimal solution to quite a realistic optimization problem. With this approach one would select the noise matrices R_i based on physical insight into the nature of the disturbances. The criterion matrices Q_i reflect some kind of cost, but must be adjusted until simulations of the closed loop system are acceptable. The starting point for this view is thus the design criterion 6.4.

2. **The Loop Shaping Approach.** The starting point is specifications like design criterion 6.5. The requirements concern the frequency functions for S, T, G_{ru}, G_c etc. Since the control error and the input are subject to (6.7) and (6.14), respectively,

$$e = (I - G_c)r - Sw + Tn$$
$$u = G_{ru}r + G_{wu}(w + n)$$

the optimization criterion is directly influenced by the spectra of r, w and n. By using these three spectra as design variables (as well as the relative weighting of u and e), the LQG-machinery will generate controllers, whose G_c, S and T-functions directly reflect the spectra of r, w and n. This approach will be discussed in more detail in the next chapter.

3. **The LTR Approach.** In the LTR approach, first a desired loop gain $L(sI - A)^{-1}B$ is constructed using the LQ method. This loop gain will automatically have very good loop properties. The second step is to use a Kalman filter that corresponds to substantial noise disturbances at the control input.

It should be added that regardless of the approach, any candidate controller must be scrutinized by simulation, by studying all relevant frequency functions, etc., before it can be accepted.

Literature

The signal optimization approach received considerable attention with the book Newton, Gould & Kaiser (1957), which shows how controller parameters could be determined by optimization methods. This approach in turn goes back to Wiener's work, e.g., Wiener (1949).

From a formal point of view, optimal control is a special case of classical calculus of variations, which is further dealt with in Chapter 18. The methods therefore have a long history. The real breakthrough in the control area came with Kalman's fundamental work, Kalman (1960a), which gave the solution in a form that suited the growing state space theory in an excellent manner. LQ and LQG techniques are treated extensively in many text books, like Åström & Wittenmark (1997), Anderson & Moore (1989) and Kwakernaak & Sivan (1972).

Robustness of the LQG controllers is treated in Safonov (1980) and Doyle & Stein (1981), while the LTR technique was introduced in Stein & Athans (1987). This technique is also treated comprehensively in many text books, e.g., Maciejowski (1989). Numerical solution of the Riccati equation is studied in, for example, Van Dooren (1981). A summary of the properties of the Riccati equation is given in Zhou et al. (1996).

Software

In MATLAB the LQG-solution is configured by the command lqgreg. The calculation of L is done by lqr in continuous time and by dlqr in discrete time. The commands lqe and dlqe compute K in a corresponding manner. The continuous time criterion is minimized by a sampled-data controller according to Theorem 9.5, using lqrd. The

LTR technique is implemented as ltru and ltry. The singular values of the loop gains are computed and plotted by sigma and dsigma.

Appendix 9A: Proofs

Proof of Theorem 9.1

Proof: Lemma 5.1 proves that (9.7b) has a unique solution, such that $A - BL$ is stable. (Take $A = A^T, C = B^T, R_1 = M^T Q_1 M$ etc.) We can write $x(t) = \hat{x}(t) + \tilde{x}(t)$ where $\hat{x}$ is given by the Kalman filter, so that

- $\tilde{x}(t)$ does not depend on the control input u. (See (5.66).)

- $Ex(t)x^T(t) = E\hat{x}\hat{x}^T + E\tilde{x}\tilde{x}^T$ (See (5.85.)

- $\nu(t) = y(t) - C\hat{x}(t)$ (the innovations) is white noise with intensity R_2 (see Theorem 5.5).

This gives

$$\|z\|_{Q_1}^2 = \|x(t)\|_{M^T Q_1 M}^2$$
$$= \|\hat{x}(t)\|_{M^T Q_1 M}^2 + \|\tilde{x}(t)\|2_{M^T Q_1 M}$$

The last term will be denoted $\tilde{V}$. It is entirely independent of the control input, and the criterion can now be rewritten as

$$V = \|\hat{x}(t)\|_{M^T Q_1 M}^2 + \|u(t)\|_{Q_2}^2 + \tilde{V} = V_1 + \tilde{V} \tag{9.49}$$

where only the first two terms, i.e., V_1, have to be considered. Moreover,

$$\dot{\hat{x}} = A\hat{x} + Bu + KR_2^{1/2}\tilde{\nu} \tag{9.50}$$

where $\tilde{\nu}$ is white noise with intensity I.

The criterion V depends on the controller $u = -F_y y$, and we now only consider stabilizing control laws. Then all signals will be stationary processes. It is rather difficult to treat continuous time white noise, and we shall therefore make use of (5.31)–(5.32) to compute the expected values. According to this expression we have

$$V_1 = V_1^{(1)} + \ldots + V_1^{(p)}$$

where

$$V_1^{(k)} = \int_0^\infty (\hat{x}_{(k)}^T(t)M^T Q_1 M\hat{x}_{(k)}(t) + u_{(k)}^T(t)Q_2 u_{(k)}(t))dt$$

with index (k) indicating the signal obtained from the system (9.50) when $\tilde{v}(t)$ is a unity impulse (Dirac function) in component k, i.e., $\hat{x}_{(k)}(t)$ is the solution to

$$\dot{\hat{x}}_{(k)}(t) = A\hat{x}_{(k)}(t) + Bu_{(k)}(t), \quad x(0) = \tilde{K}_k$$

where $\tilde{K}_k$ is the k:th column of the matrix $KR_2^{1/2}$.

Introduce

$$W^{(k)}(t) = \int_t^\infty \hat{x}_{(k)}^T(\tau)M^TQ_1M\hat{x}_{(k)}(\tau) + u_{(k)}^T(\tau)Q_2u_{(k)}(\tau)d\tau$$
$$- \hat{x}_{(k)}(t)S\hat{x}_{(k)}(t)$$

where S is the solution to (9.7b). This means that we have

$$V_1^{(k)} = W^{(k)}(0) + \hat{x}_{(k)}(0)S\hat{x}_{(k)}(0) = W^{(k)}(0) + \tilde{K}_k^T S\tilde{K}_k$$

In the sequel the index (k) will be suppressed. We have

$$\dot{W}(t) = -\hat{x}^T(t)M^TQ_1M\hat{x}(t) - u^T(t)Q_2u(t) - (A\hat{x}(t) + Bu(t))^T S\hat{x}(t)$$
$$- \hat{x}^T(t)S(A\hat{x}(t) + Bu(t)) = -\hat{x}^T(M^TQ_1M + A^TS + SA$$
$$- SBQ_2^{-1}B^TS)\hat{x} - (u + Q_2^{-1}B^TS\hat{x})^TQ_2(u + Q_2^{-1}B^TS\hat{x})$$
$$= -(u + Q_2^{-1}B^TS\hat{x})^TQ_2(u + Q_2^{-1}B^TS\hat{x})$$

The first step follows by differentiation and the equation for $\hat{x}$. The next step consists of algebraic manipulations collecting all terms containing u in a quadratic expression. The last step follows by choosing S as the solution to (9.7b). We now have

$$\int_0^\infty \dot{W}(t)dt = W(\infty) - W(0) = -W(0)$$

$W(\infty) = 0$ since the closed loop system is stable and hence $\hat{x}(\infty) = 0$. This gives

$$V_1^{(k)} = \tilde{K}_k^T S\tilde{K}_k + \int_0^\infty (u(t) + Q_2^{-1}B^TS\hat{x}(t))^TQ_2(u(t) + Q_2^{-1}B^TS\hat{x}(t))dt$$

We see that the control input that minimizes this expression is given by $u(t) = -Q_2^{-1}B^TS\hat{x}(t)$ regardless of the component (k). Therefore V_1 and hence V is minimized by (9.6)–(9.7).

$\square$

Proof of Theorem 9.3

Proof: It is easy to verify the following: If λ is an eigenvalue to H with eigenvector $\begin{bmatrix} w_1 \\ w_2 \end{bmatrix}$, then $-\lambda$ is an eigenvalue to H^T with eigenvector $\begin{bmatrix} w_2 \\ -w_1 \end{bmatrix}$.

Since a matrix has the same eigenvalues as its transpose, we have then shown that the eigenvalues of H are placed symmetrically around the imaginary axis. It follows in particular that there are n eigenvalues strictly in the left half plane if there are no eigenvalues on the imaginary axis. It can be shown (see e.g., Zhou et al. (1996)) that if the underlying system is detectable and stabilizable, then H has no purely imaginary eigenvalues.

If H is diagonalizable, Λ can be chosen as a diagonal matrix with the stable eigenvalues and $W = \begin{bmatrix} W_1 \\ W_2 \end{bmatrix}$ as the eigenvectors for these eigenvalues. If H cannot be diagonalized, it can at least be transformed to Jordan normal form (or any block diagonal form) with the stable eigenvalues collected in one block.

Since H is real, complex eigenvalues form complex conjugated pairs. This means that $\overline{W} = W\Pi$ for some permutation matrix Π, which implies that $\overline{W}_1 = W_1\Pi, \overline{W}_2 = W_2\Pi$ and hence $\overline{S} = S$, that is, S is real.

It now only remains to prove that S is symmetric and positive semidefinite. Form the difference between

$$A^T S + SA + M^T Q_1 M - SBQ_2^{-1}B^T S = 0$$

and the transposed equation

$$A^T S^T + S^T A + M^T Q_1 M - S^T BQ_2^{-1}B^T S^T = 0$$

which gives

$$(A - BQ_2^{-1}B^T S)^T(S - S^T) + (S - S^T)(A - BQ_2^{-1}B^T S) = 0$$

Since $A - BQ_2^{-1}BS$ is stable, Theorem 5.3 can be applied, which proves that $(S - S^T) = 0$.

Moreover, with $L = BQ_2^{-1}B^T S$ we have that

$$(A - BL)^T S + S(A - BL) + M^T Q_1 M + L^T Q_2 L = 0$$

which according to Theorem 5.3 (again using that $A - BL$ is stable) shows that S is a covariance matrix, and hence positive semidefinite.

We should also prove that W_1 is invertible, so that S is well defined. We know from Theorem 9.1 that the Riccati equation has a unique, positive definite solution S if the underlying system is detectable and stabilizable. This can be utilized to prove that W_1 must be invertible, see Zhou et al. (1996). $\square$

Proof of Theorem 9.4

Proof: Existence, uniqueness and stability of the solution, as well as the expression

$$V = E\hat{x}^T(t)M^T Q_1 M\hat{x}(t) + Eu^T(t)Q_2 u(t) + \tilde{V} \tag{9.51}$$

follow just as in the proof of Theorem 9.1.

We now only consider control laws that are stabilizing. Then all signals will be stationary processes. Let S be the solution to (9.37b) and introduce

$$W(t) = E\hat{x}^T(t)S\hat{x}(t) + tV \tag{9.52}$$

Denote the innovation by ν and represent $\hat{x}$ as

$$\hat{x}(t+1) = A\hat{x}(t) + Bu(t) + K\nu(t)$$

Now let

$$
\begin{aligned}
V &= W(t+1) - W(t) = E[\hat{x}(t+1)^T S\hat{x}(t+1) - \hat{x}^T(t)S\hat{x}(t)] + V \\
&= E[\hat{x}(t)^T(A^T SA - S)\hat{x}(t) + u^T(t)B^T SA\hat{x}(t) \\
&\quad + \nu^T(t)K^T S(A\hat{x}(t) + Bu(t)) + \nu(t)^T K^T SK\nu(t) + u(t)^T B^T SBu(t) \\
&\quad + \hat{x}^T(t)A^T SBu(t) + (\hat{x}^T(t)A^T + u^T(t)B^T)K\nu(t) \\
&\quad + \hat{x}^T(t)M^T Q_1 M\hat{x}(t) + u^T(t)Q_2 u(t)] + \tilde{V} \\
&= E[\hat{x}^T\{A^T SA - S + M^T Q_1 M - A^T SB^T(B^T SB + Q_2)^{-1}BSA\}\hat{x} \\
&\quad + (u(t) + (B^T SB + Q_2)^{-1}B^T SA\hat{x}(t))^T(B^T SB + Q_2)(u(t) \\
&\quad + (B^T SB + Q_2)^{-1}B^T SA\hat{x}(t))] + \tilde{V} + E\nu^T(t)K^T SK\nu(t) \\
&= E|u + (B^T SB + Q_2)^{-1}B^T SA\hat{x}|^2_{B^T SB + Q_2} \\
&\quad + \tilde{V} + \text{tr}(K^T SKR_\nu)
\end{aligned}
$$

The first two equalities follow from the definition of W. (Note that the first term of (9.52) is independent of t, due to the stationarity.) In the third equality, the Kalman filter (9.36b) is used, plus the expression (9.51). The fourth equality follows from $\nu(t)$ being uncorrelated with $\hat{x}(t)$ and with $u(t)$ ($u(t)$ may only depend on past $y(s)$, $s < t$. *Here we used that the controller is linear: it is not true in general that $\nu(t)$ is uncorrelated with non-linear transformations of past $y(s)$.*) Moreover, the terms were regrouped. The fifth equality follows from S being the solution to (9.37b).

In this last expression the two last terms are independent of the controller. Moreover, $B^T SB + Q_2$ is a positive semidefinite (typically also positive definite) matrix. V will thus be minimized by $u(t) = -L\hat{x}(t) = -(B^T SB + Q_2)^{-1}B^T SA\hat{x}(t)$.

□

Chapter 10

Loop Shaping

10.1 Direct Methods

Classical Control

Directly shaping the loop gain to obtain desired properties is really at the heart of what is now called "classical control". Bode, Nichols and others developed graphical methods, and special diagrams to handle this in an efficient way for SISO systems. The means for directly manipulating the loop gain are the lead- and lag-compensators in basic control theory.

Horowitz' Method

Quantitative Feedback Design Theory (QFT) was developed in the same spirit by Horowitz. This method explicitly ascertains that desired loop properties are maintained under given model uncertainty. The basic principle is to describe the controlled system by a set of transfer functions, which at each frequency defines a set in a Nyquist or Nichols diagram. These sets are called *templates* or *value sets*. The requirement is that, at each frequency, the specifications for the closed loop must be satisfied for all elements in the value set. This leads to constraints on the controller's frequency response which are easily translated to constraints on the nominal loop gain GF_y. In this way, the design problem is transformed from a simultaneous synthesis problem (i.e., one where the control object is described by several models) to a classical, conventional control problem with *one* process model with constraints. The synthesis cannot be carried out in a Bode diagram, though, since

277

the constraints are phase dependent. A Nichols diagram is more appropriate. This technique also works for multivariable systems.

Modern Development

During the '80s and '90s considerable research has been carried out to develop more systematic and efficient methods for multivariable control design, based on loop shaping. The methods and the theory behind them are quite advanced, mathematically. On the other hand, there are now a number of widely used commercial software packages (several MATLAB-based) which implement the methods. This means that there is reason to understand and master the methods even without a complete mathematical foundation. In this chapter we shall describe some examples of such a systematic theory.

10.2   Formalization of the Requirements

The Basic Specifications

A system $y = Gu + w$ with p measured outputs and m control inputs is given. The task is to find a feedback law $u = -F_y y$, that gives a good controlled system. Much of the design work concerns shaping the loop gain GF_y so that the sensitivity function S, the complementary sensitivity function T, and the transfer function from system disturbance (or reference signal) to control input, G_{wu}, have desired properties. Recall the definitions of these (see (6.9)–(6.16)):

$$S = (I + GF_y)^{-1} \tag{10.1a}$$

$$T = I - S = (I + GF_y)^{-1}GF_y = GF_y(I + GF_y)^{-1} \tag{10.1b}$$

$$G_{wu} = -(I + F_yG)^{-1}F_y = -F_y(I + GF_y)^{-1} = -F_yS \tag{10.1c}$$

It is desirable that all these transfer functions are "small". Since this is impossible to achieve for all frequencies simultaneously, we have to modify the requirements by considering frequency weighted transfer functions. We thus introduce three weighting matrices W_S, W_T (both $p \times p$ matrices) and W_u ($m \times m$-matrix) and require that

$$W_S(i\omega)S(i\omega) \tag{10.2a}$$

$$W_T(i\omega)T(i\omega) \tag{10.2b}$$

$$W_u(i\omega)G_{wu}(i\omega) \tag{10.2c}$$

"be small for all frequencies". See Section 6.6. We shall later discuss how to formalize these requirements.

We assume that G and W_S are strictly proper, i.e., state space representations of these transfer function matrices have the D-matrix equal to zero.

An Extended System

Starting from the given system G, we form an extended system G_e with $m + p$ inputs:

$$\begin{bmatrix} u \\ w \end{bmatrix} \tag{10.3}$$

and $m + 3p$ outputs:

$$z_1 = W_u u \tag{10.4a}$$
$$z_2 = W_T G u \tag{10.4b}$$
$$z_3 = W_S (G u + w) \tag{10.4c}$$
$$y = G u + w \tag{10.4d}$$

This system depends only on the three weighting matrices, W, and the nominal system G. No noise description is used here.

The system (10.4) has two kinds of outputs. First

$$z = \begin{bmatrix} z_1 \\ z_2 \\ z_3 \end{bmatrix} \tag{10.5}$$

which reflects our requirements and, secondly, y, which describes the measurements. See Figure 10.1. The point with this extended model is that if u is chosen as the feedback from y, $u = -F_y y$, then the transfer function from w to z will be

$$z = \begin{bmatrix} -W_u G_{wu} \\ -W_T T \\ W_S S \end{bmatrix} w = G_{ec} w \tag{10.6}$$

This can be seen by inserting $u = -F_y y$ into (10.4d), solving for $u = -F_y (I + G F_y)^{-1} w$ and using this expression for u in (10.4a,b,c). The outputs z will thus reflect our requirements (10.2). The extended, closed loop system is shown in Figure 10.2.

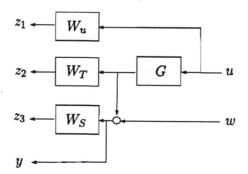

Figure 10.1: How the z-variables are generated from u and w.

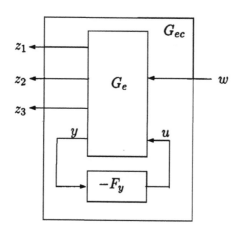

Figure 10.2: The extended open loop system G_e and the closed loop system G_{ec}.

Now, represent the open loop system G_e from u and w to z and y in state space form:

$$\dot{x} = Ax + Bu + Nw \tag{10.7a}$$
$$z = Mx + Du \tag{10.7b}$$
$$y = Cx + w \tag{10.7c}$$

The absence of w and u, respectively, in the last equations is a consequence of the assumption that G and W_S are strictly proper. To obtain simpler expressions, we also assume that

$$D^T \begin{bmatrix} M & D \end{bmatrix} = \begin{bmatrix} 0 & I \end{bmatrix} \tag{10.8}$$

(see Remark 2 below). Note that (10.7) is a state space representation of standard type (5.55). If $A - NC$ is a stable matrix, this representation is also in innovations form (see (5.84)). This means that the problem to reconstruct the state, that is the observer, is trivial.

Remark 1. The controllers discussed in this chapter simplify considerably if (10.7) is in innovations form, that is that $A - NC$ is stable. It is however often the case that $A - NC$ has eigenvalues on the boundary of the stability region, typically at the origin. This happens if it is required that the sensitivity function is zero at steady state – then W_S must contain an integrator. Another reason may be that the system itself contains an integration. Formally, this can be handled by modifying the model somewhat, so that the eigenvalue in question is pulled slightly into the stability region. Such a small modification has no practical consequences. It often turns out, however, that the machinery that we are about to describe will work even if $A - NC$ has eigenvalues on the boundary of the stability region.

Remark 2. The condition (10.8) may seem restrictive, but it can always be satisfied if the matrix $D^T D$ is invertible. We make a change of variables from u to

$$\tilde{u} = (D^T D)^{1/2} u + (D^T D)^{-1/2} D^T Mx$$

This gives

$$z = \tilde{M}x + \tilde{D}\tilde{u}, \quad \tilde{M} = (I - D(D^T D)^{-1} D^T)M, \quad \tilde{D} = D(D^T D)^{-1/2}$$

It is now easy to verify that $\tilde{M}$, $\tilde{D}$ satisfy (10.8). These calculations are analogous to the treatment of the cross term in (9.14). Normally

$\lim_{s\to\infty} W_u(s)$ will be invertible, since it means that there is an upper bound on the input power at high frequencies. This implies that also $D^T D$ is invertible. Note also, however, that this transformation may affect the stability of $A - NC$.

Example 10.1:   A DC Motor with Extended Model

Consider a DC motor with transfer function

$$G(s) = \frac{20}{s(s+1)}$$

with state space representation

$$\dot{x} = \begin{bmatrix} 0 & 0 \\ 1 & -1 \end{bmatrix} x + \begin{bmatrix} 20 \\ 0 \end{bmatrix} u$$
$$y = \begin{bmatrix} 0 & 1 \end{bmatrix} x$$

(10.9)

Suppose the requirements on the three transfer functions G_{wu}, T and S are reflected by the weighting functions

$$W_u = 1, \quad W_T = 1, \quad W_S = 1/s$$

Very large penalties are thus associated with the sensitivity function at low frequencies. To realize the transfer function (10.4) for these choices of G, W_u, W_S, W_T, we introduce a third state (see also Figure 10.1)

$$x_3 = \frac{1}{p}(Gu + w)$$

which gives

$$\dot{x} = \begin{bmatrix} 0 & 0 & 0 \\ 1 & -1 & 0 \\ 0 & 1 & 0 \end{bmatrix} x + \begin{bmatrix} 20 \\ 0 \\ 0 \end{bmatrix} u + \begin{bmatrix} 0 \\ 0 \\ 1 \end{bmatrix} w$$

(10.10a)

$$z = \begin{bmatrix} 0 & 0 & 0 \\ 0 & 1 & 0 \\ 0 & 0 & 1 \end{bmatrix} x + \begin{bmatrix} 1 \\ 0 \\ 0 \end{bmatrix} u$$

(10.10b)

$$y = \begin{bmatrix} 0 & 1 & 0 \end{bmatrix} x + w$$

(10.10c)

Here it is easy to verify that (10.8) holds:

$$D^T \begin{bmatrix} M & D \end{bmatrix} = \begin{bmatrix} 1 & 0 & 0 \end{bmatrix} \begin{bmatrix} 0 & 0 & 0 & 1 \\ 0 & 1 & 0 & 0 \\ 0 & 0 & 1 & 0 \end{bmatrix} = \begin{bmatrix} 0 & 0 & 0 & 1 \end{bmatrix}$$

Suppose now that we want to pose more specific requirements: First, that the sensitivity function should decrease to zero "with slope 2" at low frequencies, so that ramp disturbances can be taken care of without any stationary error (the coefficient $s_1 = 0$; see (6.34a)), and, second, that the complementary sensitivity function should be very small around 0.2 rad/s (e.g., due to substantial system variations at this frequency). To capture this, we pick

$$W_S(s) = \frac{0.04}{s^2} \tag{10.11a}$$

$$W_T(s) = \frac{s^2 + s + 0.04}{s^2 + 0.001s + 0.04} = 1 + \frac{0.999s}{s^2 + 0.001s + 0.04} \tag{10.11b}$$

$$W_u(s) = 1 \tag{10.11c}$$

Note that we obtain

$$W_T Gu = x_2 + \frac{0.999p}{p^2 + 0.001p + 0.04} x_2$$

To accommodate this, the states are augmented with

$$x_4 = \frac{1}{p} x_3 = W_S(Gu + w)$$

$$x_5 = \frac{0.999}{p^2 + 0.0001p + 0.04} x_2$$

The last relationship is realized in observer canonical form (see (2.17)) and a sixth state is needed for this. We then obtain

$$\dot{x} = \begin{bmatrix} 0 & 0 & 0 & 0 & 0 & 0 \\ 1 & -1 & 0 & 0 & 0 & 0 \\ 0 & 1 & 0 & 0 & 0 & 0 \\ 0 & 0 & 1 & 0 & 0 & 0 \\ 0 & 0.999 & 0 & 0 & -0.001 & 1 \\ 0 & 0 & 0 & 0 & -0.04 & 0 \end{bmatrix} x + \begin{bmatrix} 20 \\ 0 \\ 0 \\ 0 \\ 0 \\ 0 \end{bmatrix} u + \begin{bmatrix} 0 \\ 0 \\ 1 \\ 0 \\ 0 \\ 0 \end{bmatrix} w \tag{10.12a}$$

$$z = \begin{bmatrix} 0 & 0 & 0 & 0 & 0 & 0 \\ 0 & 1 & 0 & 0 & 1 & 0 \\ 0 & 0 & 0 & 0.04 & 0 & 0 \end{bmatrix} x + \begin{bmatrix} 1 \\ 0 \\ 0 \end{bmatrix} u \tag{10.12b}$$

$$y = \begin{bmatrix} 0 & 1 & 0 & 0 & 0 & 0 \end{bmatrix} x + w \tag{10.12c}$$

Along these lines it is straightforward to build models that reflect various requirements on S, T and G_{wu}. It is important to note that the total order of the resulting model typically is the sum of the orders of the four original transfer functions. This means that it is desirable to work with as simple weighting functions as possible. With normal software support, the construction of the resulting state space model is done automatically from the given transfer functions.

Including a Reference Signal

For the servo problem the reference signal is most easily introduced into the controller according to the recipe (8.3), which is expressed as (8.32) in state space form. The pre-compensator $\tilde{F}_r$ is often adjusted "by hand", even though formal methods exist. The approach (9.11)–(9.12) will normally not work in this case, since z in (10.4a,b,c) does not correspond to any signal that shall track r.

10.3   Optimal $\mathcal{H}_2$ Control

Let us formalize the requirement (10.2) as Design Criterion 6.6, i.e., to minimize

$$V(F_y) = \frac{1}{2\pi} \int \left(|W_S(i\omega)S(i\omega)|_2^2 + |W_T(i\omega)T(i\omega)|_2^2 \right.$$
$$\left. + |W_u(i\omega)G_{wu}(i\omega)|_2^2 \right) d\omega = \frac{1}{2\pi} \int |G_{ec}(i\omega)|_2^2 d\omega = \|G_{ec}\|_2^2$$
$$(10.13)$$

with respect to the controller F_y. The expression (10.13) is also *the $\mathcal{H}_2$-norm for the transfer function matrix G_{ec} in (10.6)*.

From (10.6) we know that $z = G_{ec}w$. If the spectrum of w is $\Phi_w \equiv I$ (that is white noise with intensity I) we have

$$\Phi_z(\omega) = G_{ec}(i\omega)G_{ec}^T(-i\omega)$$

and hence according to (6.35) that $(P = I)$

$$\|z\|^2 = \frac{1}{2\pi}\text{tr}\left(\int \Phi_z(\omega)d\omega \right) = \frac{1}{2\pi} \int \text{tr}(\Phi_z(\omega))d\omega$$
$$= \frac{1}{2\pi} \int |G_{ec}(i\omega)|_2^2 d\omega = V(F_y)$$

(Recall the definition $|A|_2^2 = \text{tr}(AA^*)$.) This means that

$$V(F_y) = \|z\|^2 = \|Mx + Du\|^2 = \|Mx\|^2 + \|u\|^2 \qquad (10.14)$$

where (10.8) is used in the last step.

To find a controller that minimizes this criterion was the topic of the previous chapter. Since (10.7) is in innovations form if $A - NC$ is stable, we have the case that the state $x = \hat{x}$ is known. According to Theorem 9.1 and (9.9) the solution is given by

$$\dot{\hat{x}} = A\hat{x} + Bu + N(y - C\hat{x}) \tag{10.15a}$$

$$u = -L\hat{x} \tag{10.15b}$$

$$L = B^T S \tag{10.15c}$$

$$0 = A^T S + SA + M^T M - SBB^T S \tag{10.15d}$$

that is

$$F_y(s) = L(sI - A + BB^T S + NC)^{-1} N \tag{10.16}$$

If $A - NC$ is not stable, the observer equation (10.15a) must be changed to the corresponding Kalman filter.

Example 10.2: DC Motor

Consider the DC motor (10.9). We seek a controller that gives a sensitivity function that decreases to zero with slope 2 at low frequencies, and that handles considerable measurement disturbances and/or model deviations around 0.2 rad/s. To find such a controller, we extend the model as in (10.12) and solve the $\mathcal{H}_2$-problem (10.15). This gives

$$L = \begin{bmatrix} 0.326 & 1.06 & 0.656 & 0.04 & 0.597 & -3.999 \end{bmatrix}$$

Figure 10.3 shows the corresponding loop functions. The complementary sensitivity function shows that the controller attenuates measurement noise around 0.2 rad/s (or, equivalently, has good robustness against model uncertainties/model deviations at this frequency). Moreover, the sensitivity function decays rapidly to zero when the frequency tends to zero, just as required.

10.4   Optimal $\mathcal{H}_\infty$ Control

The Criterion

G_{ec} in (10.6) is the closed loop that is formed as the feedback $u = -F_y y$ is applied to (10.4). In the previous section, we minimized the $\mathcal{H}_2$-norm of this transfer function with respect to F_y. Let us now instead minimize the $\mathcal{H}_\infty$-norm, that is, the largest singular value of $G_{ec}(i\omega)$ assumed for any value of ω.

$$\|G_{ec}\|_\infty = \max_\omega \bar{\sigma}(G_{ec}(i\omega)) \tag{10.17}$$

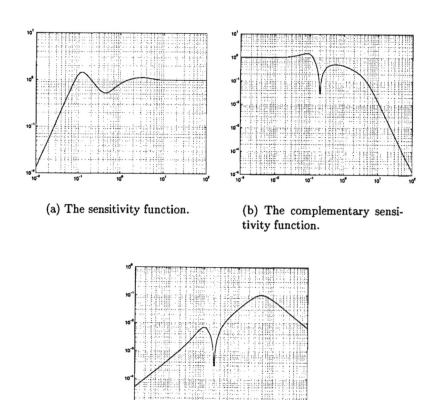

(a) The sensitivity function.

(b) The complementary sensitivity function.

(c) The transfer function G_{wu}.

Figure 10.3: The loop functions when the $\mathcal{H}_2$-norm of the transfer function for the extended DC motor model (10.12) is minimized.

In particular we search for controllers that satisfy

$$\|G_{ec}\|_\infty \leq \gamma \tag{10.18}$$

If this inequality holds, it also follows that

$$|W_S(i\omega)S(i\omega)| \leq \gamma \;\; \forall \omega \tag{10.19a}$$
$$|W_T(i\omega)T(i\omega)| \leq \gamma \;\; \forall \omega \tag{10.19b}$$
$$|W_u(i\omega)G_{wu}(i\omega)| \leq \gamma \;\; \forall \omega \tag{10.19c}$$

This means that we have achieved guaranteed bounds for the most important closed loop frequency functions. Note, though, that (10.18) is a more stringent requirement than (10.19): all inequalities in (10.19) may hold, but if the maxima occur for the same frequency, the right hand side of (10.18) may be up to $\sqrt{3}\gamma$.

The Solution

Starting from the state space description (10.7), suppose that (10.8) holds and that $A - NC$ is stable. Solve the Riccati equation

$$A^T S + SA + M^T M + S(\gamma^{-2}NN^T - BB^T)S = 0 \tag{10.20}$$

Suppose that it has a positive semidefinite solution $S = S_\gamma$ subject to

$$A - BB^T S_\gamma \;\; \text{stable} \tag{10.21}$$

Let

$$L_\infty = B^T S_\gamma \tag{10.22}$$

and introduce the controller

$$\dot{\hat{x}} = A\hat{x} + Bu + N(y - C\hat{x}) \tag{10.23a}$$
$$u = -L_\infty \hat{x} \tag{10.23b}$$

That is

$$F_y(s) = L_\infty(sI - A + BB^T S_\gamma + NC)^{-1}N \tag{10.24}$$

Then it holds that

(a) F_y satisfies (10.18) (and hence all the requirements (10.19)).

(b) If the equation (10.20) has no solution with the properties (10.21), then there is no linear controller that satisfies (10.18).

The solution to the $\mathcal{H}_\infty$ problem thus has exactly the same structure as the $\mathcal{H}_2$ or LQ-solution: feedback from reconstructed states. Another L is used, but it is calculated in a similar fashion. The controller degree equals that of the extended system (10.7). Note that this may be quite high, since it includes the weighting functions W.

If $A-NC$ is not a stable matrix, the observer equation (10.23a) must be replaced by another expression. The optimal observer for this case is somewhat more complex; it involves both a replacement of N and injection of more signals (corresponding to a "worst case" disturbance w; see the proof of Theorem 10.1). See the references for a complete treatment of this case.

User Aspects on $\mathcal{H}_\infty$ Control

It is not easy to realize that (10.24) is the solution to the problem (10.18) – but we shall shortly give a quite simple proof for it. From a user point of view, the technique is still rather easy to apply. Several widely spread software packages are available, that both generate the state space description (10.7) from G, W_S, W_T and W_u, solve for L_∞ and generate the controller. The user steps are then

1. Determine G.

2. Choose W_u, W_S, W_T.

3. Select a γ.

4. Solve (10.20)–(10.22)

   - If no solution exists, go to step 3 and increase γ.
   - If there is a solution, either accept it or go to step 3 and decrease γ to see if a better solution exists.

5. Investigate the properties of the closed loop system. If these are not satisfactory, go back to step 2.

6. Check, if so desired, if the resulting controller can be reduced to lower order (according to Section 3.6) without any negative consequences for the closed loop system.

The important question is of course how to choose the weighting functions in step 2. Primarily, these are consequences of the posed requirements on robustness, sensitivity and input amplitudes. But they must also reflect what is reasonable to achieve, so that impossible and contradictory requirements are avoided. Checking the properties of the $\mathcal{H}_2$-controller for certain choices of W may be a good starting point to judge if the chosen bounds are reasonable. See also Chapter 7 for a discussion of what closed loop properties are possible to achieve.

Usually the W are chosen to be simple functions of first and second order to avoid that the total order of (10.7) becomes too large. In the multivariable case, the weighting matrices are often chosen to be diagonal.

Example 10.3: The DC Motor with Guaranteed Loop Properties

We return to the DC motor of Example 10.1 and now seek a controller that guarantees (10.11), i.e.,

$$|S(i\omega)| < \gamma|W_S^{-1}(i\omega)| = 25\gamma\omega^2$$

$$|T(i\omega)| < \gamma|W_T^{-1}(i\omega)| = \gamma\left|\frac{(i\omega)^2 + 0.001i\omega + 0.04}{(i\omega)^2 + i\omega + 0.04}\right|$$

$$|G_{wu}(i\omega)| < \gamma$$

for all ω and for a γ as small as possible.

We follow the procedure described in this section, and find that the smallest value is $\gamma = 4.107$, which gives

$$L_\infty = \begin{bmatrix} 15.9 & 101.6 & 1283 & 73 & -858 & -7159 \end{bmatrix}$$

The corresponding loop properties are shown in Figure 10.4.

There is extensive software support, especially in MATLAB, to carry out such synthesis. Below follows a MATLAB script that generates Figure 10.4. It does not use any tailor made commands. The function aresolv from the ROBUST CONTROL TOOLBOX solves the Riccati equation with the method of Theorem 9.3.

```
>>% DC motor: 20/(s(s+1))
>>% W_T=(w0s^2+w1s+w2)/(s^2+w3s+w4)
>>% W_S= w/s^2
>>% W_u = 1
>>% The extended model:
>> w0=1;w1=1;w2=0.04;w3=0.001;w4=0.04;
```

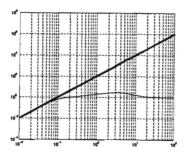

(a) The sensitivity function.

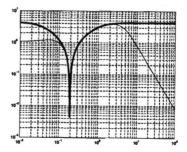

(b) The complementary sensitivity function.

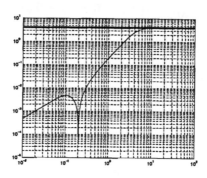

(c) The transfer function G_{wu}.

Figure 10.4: The loop properties when the $\mathcal{H}_\infty$-norm of the closed loop transfer function of the extended DC motor model (10.12) is minimized. The posed upper bounds for the frequency functions are marked by thicker lines.

```
>> w=0.04;
>> A=zeros(6,6);
>> A(2,1)=1;A(2,2)=-1;A(3,2)=1;A(4,3)=1;A(5,5)=-w3;A(5,6)=1;
>> A(6,5)=-w4;A(6,2)=w2-w0*w4;A(5,2)=w1-w0*w3;
>> B=zeros(6,1);B(1,1)=20;
>> M=zeros(3,6);
>> M(2,5)=1;M(2,2)=w0;M(3,4)=w;
>> D=zeros(3,1);D(1,1)=1;
>> N=zeros(6,1);N(3,1)=1;
>> C=[0 1 0 0 0 0];
>> gam=1;
>> while gam>0
>>    gam=input('Gamma '); % The user chooses a gamma
>>    if gam==0,break,end
>>    R=B*B'-gam^(-2)*N*N';
>>    [p1,p2,lamp,perr,wellposed]=aresolv(A,M'*M,R,'Schur');
>>    % p1 and p2 correspond to W2 and W1 i in the theorem on the Riccati equ
>>    P=p2/p1;
>>    L=B'*P; % The feedback
>>    % The closed loop system:
>>    [Ac,Bc,Cc,Dc]=cloop(A,[B,N],[M;C;L;[0 1 0 0 0 0]],...
>>                         [[D;0;0;0],[0;0;0;1;0;0]],5,-1);
>>    [clnum,den]=ss2tf(Ac,Bc,Cc,Dc,2);
>>    lt3=max(real(eig(Ac))); % test that the closed loop system is stable
>>    lt2=min(eig(P)); % test that the solution is positive
>>    if lt1<=0&lt2>=0&strcmp(deblank(wellposed),'TRUE')
>>        disp('Solution accepted'), ok=1;
>>    else
>>        disp('Solution rejected'), ok=0;
>>    end
>>    if ok
>>        figure(1)
>>        ws=logspace(-2,2,1000);
>>        [mags,phase,ws]=bode(clnum(4,:),den,ws);
>>        [magws]=bode(gam*[1 0 0],[w],ws);
>>        loglog(ws,mags,ws,magws,'r'),grid
>>        figure(2)
>>        [magu,phase,ws]=bode(clnum(1,:),den,ws);
>>        loglog(ws,magu,ws,gam*ones(size(magu)));grid
>>        figure(3)
>>        [magt,phase,ws]=bode(clnum(6,:),den,ws);
>>        [magwq,phase]=bode(gam*[1 w3 w4],[w0 w1 w2],ws);
>>        loglog(ws,magt,ws,magwq,'r'),grid,
>>    end
>>end
```

Note in Figure 10.4 that the different requirements determine the solution
in different frequency ranges: at low frequencies, the sensitivity requirement

is just about satisfied, at about 0.1 rad/s it is instead the complementary
sensitivity that determines the solution, and at high frequencies, above 4
rad/s, it is the input amplitude that sets the limit. The posed bound can
thus be achieved quite precisely. This in turn puts the pressure on the user
to find weighting functions W that well capture the real system demands.

Proof of $\mathcal{H}_\infty$ Optimality

In this section we shall prove the (a)-part of the $\mathcal{H}_\infty$-optimality. The
result is formulated as a theorem:

Theorem 10.1 *Given the system (10.7):*

$$\dot{x} = Ax + Bu + Nw \qquad\qquad (10.25a)$$

$$z = Mx + Du \qquad\qquad (10.25b)$$

$$y = Cx + w \qquad\qquad (10.25c)$$

*Assume as in (10.8) that $D^T \begin{bmatrix} M & D \end{bmatrix} = \begin{bmatrix} 0 & I \end{bmatrix}$, and that $A - NC$
is stable (see Remarks 1 and 2 in Section 10.2). Let S be a positive
semidefinite matrix, subject to (10.20):*

$$A^T S + SA + M^T M + S(\gamma^{-2}NN^T - BB^T)S = 0 \qquad (10.26)$$

*and let $L_\infty = B^T S$. If $x(0) = 0$ and u is chosen as $u = -L_\infty \hat{x}$ (with $\hat{x}$
as in (10.15a) we have*

$$\|z\|_2 \leq \gamma\|w\|_2 \qquad\qquad (10.27)$$

for any disturbance signal w.

Remark. The property (10.27) means that the closed loop system
from w to z has a gain – $\mathcal{H}_\infty$-norm – that is less or equal to γ, i.e.,
$\|G_{ec}\|_\infty \leq \gamma$ which is the criterion (10.18). Note also that the theorem
as such does not guarantee that $A - BL_\infty$ is stable. There could be a
non-observable, unstable mode to the closed loop system. The stability
of the closed loop system should therefore be tested separately.

Proof: Let S be an, as yet, undefined, positive semidefinite, constant matrix
and define

$$V(t) = x^T(t)Sx(t) + \int_0^t (z^T(\tau)z(\tau) - \gamma^2 w^T(\tau)w(\tau))d\tau$$

If we can show that $V(t) \leq 0$ for all t, it follows that the integral is
non-positive, which in turn implies that (10.27) holds. Since $V(0) = 0$

it is sufficient to show that $\dot{V}(t) \leq 0$. Note as in (10.14) that $z^T z = x^T M^T M x + u^T u$. This gives

$$
\begin{aligned}
\dot{V} &= \dot{x}^T S x + x^T S \dot{x} + z^T z - \gamma^2 w^T w \\
&= x^T A^T S x + u^T B^T S x + w^T N^T S x + x^T S A x + x^T S B u + x^T S N w \\
&\quad + x^T M^T M x + u^T u - \gamma^2 w^T w \\
&= x^T [A^T S + S A + M^T M - S(BB^T - \gamma^{-2} N N^T) S] x \\
&\quad + (u + B^T S x)^T (u + B^T S x) - \gamma^2 (w - \gamma^{-2} N^T S x)^T (w - \gamma^{-2} N^T S x)
\end{aligned}
$$

The last step above consists of two completions of squares. We see that if S is chosen as the positive semidefinite solution to (10.26) (which clearly requires that such a solution exists) and if u is chosen as $u = -B^T S x$ it follows that $\dot{V} \leq -\gamma^2 (w - \gamma^{-2} N^T S x)^T (w - \gamma^{-2} N^T S x) \leq 0$ and the proof is finished. Note also that the worst case choice of w, i.e., the one that gives equality in (10.27), is given by $w = \gamma^{-2} N^T S x$.

Remark: In this proof we used the infinity norm for signals starting at $t = 0$. For a time invariant linear system the definition of gain will not depend on the starting time. $\square$

10.5 Robust Loop Shaping

A practical and successful method for the design of multivariable controllers was suggested in Glover & McFarlane (1989). In short, the idea is to do a good job with a typically diagonal (decentralized) precompensator W_p to get a good loop gain $G W_p$. See, e.g., Section 8.2. Then this controller is adjusted to achieve a maximum degree of stability margins. This second step is essentially "fully automatic".

Robust Stabilization

First we study the step of modifying a given loop gain to obtain as good stability margins as possible, without destroying bandwidth and other performance measures.

Consider a loop gain $y = Gu$ given in state space form as

$$\dot{x} = Ax + Bu \qquad\qquad\qquad (10.28)$$
$$y = Cx \qquad\qquad\qquad (10.29)$$

It is thus assumed that there is no direct term in G. We seek a feedback law $u = -F_y y$ that stabilizes this system in a "maximally robust way". By this we mean that the closed loop system remains stable under as large variations in G as possible.

This maximally robust stabilizing controller is calculated in the following way

1. Solve the Riccati equations

$$AZ + ZA^T - ZC^TCZ + BB^T = 0 \qquad (10.30a)$$
$$A^TX + XA - XBB^TX + C^TC = 0 \qquad (10.30b)$$

for positive definite matrices X and Z (each equation has a unique such solution if (10.28) is a minimal realization).

2. Let λ_m be the largest eigenvalue of XZ.

3. Introduce

$$\gamma = \alpha(1 + \lambda_m)^{1/2} \qquad (10.31a)$$
$$R = I - \frac{1}{\gamma^2}(I + ZX) \qquad (10.31b)$$
$$L = B^TX \qquad (10.31c)$$
$$K = R^{-1}ZC^T \qquad (10.31d)$$

Here α is a scaling factor, slightly larger than 1, e.g., 1.1. The controller is now given by

$$\dot{\hat{x}} = A\hat{x} + Bu + K(y - C\hat{x}) \qquad (10.32a)$$
$$u = -L\hat{x} \qquad (10.32b)$$

or, in input–output form

$$F_y(s) = L(sI - A + BL + KC)^{-1}K \qquad (10.33)$$

Note again that the controller can be written as feedback from reconstructed states. Apart from the factor R in the observer gain K, this is the controller LQG gives for the criterion $\|u\|^2 + \|y\|^2$ in (10.28) with white measurement noise and white system disturbances directly at the input, that is

$$\dot{x} = Ax + B(u + w)$$
$$y = Cx + n$$

Note also that as $\alpha \to \infty$ we obtain $R = I$ and the controller converges to the LQG solution mentioned above.

In Which Sense Has (10.32) Optimal Stability Margins?

Let us now describe why the controller (10.32) is "good". Represent the system G in factored form as

$$G(s) = M^{-1}(s)N(s) \tag{10.34}$$

$$M^T(-s)M(s) + N^T(-s)N(s) = I \tag{10.35}$$

where M and N are stable transfer functions. Then the controller (10.32) stabilizes all systems that can be written as

$$G_p = (M + \Delta_M)^{-1}(N + \Delta_N) \tag{10.36}$$

where Δ_M and Δ_N are stable transfer function matrices, such that

$$\left\| \begin{bmatrix} \Delta_N & \Delta_M \end{bmatrix} \right\|_\infty < \frac{1}{\gamma} \tag{10.37}$$

This means that γ in (10.31a) should be chosen as small as possible to get the best stability margins. The theoretically best result is obtained for $\alpha = 1$, but practice shows that it is suitable to choose a somewhat larger value.

Since the nominal system can be modified in both "numerator and denominator" according to (10.36) the family G_p contains systems that can have more unstable poles as well as more unstable zeros than G. For a suitably small value of γ this gives a large set of plants stabilized by the single controller. It can also be shown that this implies that crossover frequency and other loop properties cannot be changed very much by the robustification. Moreover, the robust stabilization guarantees reasonable bounds on the loop transfer functions of the type (10.19), (Zhou et al. (1996), Chapter 18).

The robustness measure, given by $1/\gamma$ is a powerful generalization of the traditional gain and phase margins. It has the further advantage that is applicable to multivariable plants.

Example 10.4: DC Motor

For the DC motor we let

$$G(s) = \frac{20}{s(s+1)} = \frac{N(s)}{M(s)}$$

and obtain

$$\begin{aligned}
1 &= N(s)N(-s) + M(s)M(-s) \\
&= N(s)N(-s)(1 + s(s+1)(-s)(-s+1)/400)
\end{aligned}$$

which gives

$$N(s) = \frac{20}{20 + \sqrt{41}s + s^2}, \quad M(s) = \frac{s(s+1)}{20 + \sqrt{41}s + s^2}$$

For the state space representation (10.9) we have from (10.30)

$$Z = \begin{bmatrix} 5.4 & 14.6 \\ 14.6 & 93.5 \end{bmatrix}, \quad X = \begin{bmatrix} 0.32 & 0.05 \\ 0.05 & 0.14 \end{bmatrix}$$

The matrix XZ has eigenvalues 4.32 and 0.12. With $\alpha = 1.1$ in the procedure above we obtain

$$\gamma = 2.53, \quad L = \begin{bmatrix} 1 & 0.27 \end{bmatrix}, \quad K = \begin{bmatrix} 27 \\ 102 \end{bmatrix}$$

The controller given by K and L, thus stabilizes all systems that can be written as

$$\frac{\frac{20}{20+\sqrt{41}s+s^2} + \Delta_N(s)}{\frac{s(s+1)}{20+\sqrt{41}s+s^2} + \Delta_M(s)}$$

for some stable transfer functions $\Delta(s)$ with

$$|\Delta_N(i\omega)|^2 + |\Delta_M(i\omega)|^2 < 1/\gamma^2 = 0.16 \ \forall \omega$$

Loop Shaping Followed by Robustification

The importance of the Glover-McFarlane method lies both in the powerful robustness measure $1/\gamma$ and in its applicability to multivariable systems. Moreover, the trade-off between various closed-loop transfer functions, S, T, G_{wu} is very naturally handled by the design of a *single* precompensator W_p, in contrast to the three possibly conflicting weights in Section 10.4.

The following procedure can be outlined:

1. Start from a dynamical model G, scaled as described in Section 6.1. Renumber the inputs so that G becomes as diagonal as possible. Use, e.g., the RGA method described in Section 8.2.

2. Design a preliminary loop gain $W_2 G(s) W_p(s) W_1$ with the following desired properties

- Choose $W_p(s)$ as a diagonal matrix of compensation links that give the loop gain $G(s)W_p(s)$ a basic desired shape. By this we mean that the singular values shall be confined to areas depicted in Figure 7.2, that the number of integrators is right, that the slope close to the intended crossover frequency is about -1, and preferably steeper at higher frequencies. By and large, the procedure can be based on classical loop shaping thinking for lead-lag compensation of a SISO system.

- Possibly adjust with constant matrices W_1 and W_2 to achieve the correct cross-over frequency (provided the same bandwidth is desired for all channels), better decoupling (diagonal dominance), and better closeness of the singular values around the crossover.

3. Realize the thus obtained loop gain in state space form and form the controller F_y according to (10.28)–(10.33). If it turns out that γ gets a larger value than about 4, the resulting robustness is not so impressive. Then go back and modify $W_p(s)$.

4. Use the controller $u = -W_p(s)W_1F_y(s)W_2y$

Example 10.5: Altitude Control of an Aircraft

A simple, but quite realistic model of the altitude control of an aircraft is given by

$$\dot{x} = Ax + Bu + Nv_1 \tag{10.38a}$$
$$y = Cx + v_2 \tag{10.38b}$$

where the state variables are

x_1: altitude (m)
x_2: speed (m/s)
x_3: pitch angle (degrees)
x_4: pitch rate (degree/s)
x_5: vertical speed (m/s)

and the control inputs

u_1: spoiler angle (tenth of degree)
u_2: aircraft acceleration (m/s^2)
u_3: elevator angle (degrees)

The measured outputs are the three first states. The matrices are

$$A = \begin{bmatrix} 0 & 0 & 1.1320 & 0 & -1.000 \\ 0 & -0.0538 & -0.1712 & 0 & 0.0705 \\ 0 & 0 & 0 & 1.000 & 0 \\ 0 & 0.0485 & 0 & -0.8556 & -1.013 \\ 0 & -0.2909 & 0 & 1.0532 & -0.6859 \end{bmatrix} \tag{10.39a}$$

$$B = \begin{bmatrix} 0 & 0 & 0 \\ -0.120 & 1.000 & 0 \\ 0 & 0 & 0 \\ 4.4190 & 0 & -1.665 \\ 1.5750 & 0 & -0.0732 \end{bmatrix} \tag{10.39b}$$

$$C = \begin{bmatrix} 1 & 0 & 0 & 0 & 0 \\ 0 & 1 & 0 & 0 & 0 \\ 0 & 0 & 1 & 0 & 0 \end{bmatrix} \tag{10.39c}$$

$$G(s) = C(sI - A)^{-1}B \tag{10.39d}$$

Our goal is to design a closed loop system with a bandwidth of about 10 rad/s, integral action in each loop, and well damped step responses. We follow the recipe outlined above, and first study the singular values of the open loop system

>> sigma(A, B, C, D)

These are shown in Figure 10.5a. As a first step in the loop shaping, we cascade a PI-regulator in each loop:

$$W_p(s) = (1 + \frac{5}{s})I$$

The singular values of $G(s)W_p(s)$ are shown in Figure 10.5b. We shall now modify this loop gain using the constant matrices W_1 and W_2. This is a step that involves tricks and some handwaving. See Appendix 10A for a discussion. We choose

$$W_1 = \begin{bmatrix} 6.3186 & 0.4782 & 2.1233 \\ 0.4782 & 3.2274 & 0.2324 \\ 2.1233 & 0.2324 & 11.8837 \end{bmatrix}$$

$$W_2 = \begin{bmatrix} 11.7498 & 0.0575 & 2.6802 \\ 0.0575 & 3.1558 & -0.0031 \\ 2.6802 & -0.0031 & 4.3216 \end{bmatrix}$$

The singular values of $G_m(s) = W_2 G(s) W_p(s) W_1$ are shown in Figure 10.5c. We see that the loop gain has a desired behavior: integrators in all loops, correct crossover, and singular values kept tightly together at the crossover. We realize G_m in state space form and calculate $F_y(s)$ according to (10.28)–

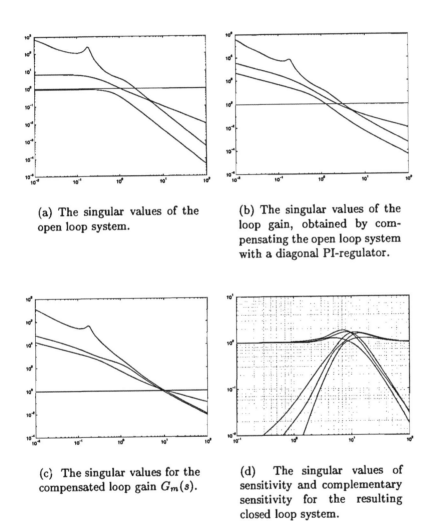

(a) The singular values of the open loop system.

(b) The singular values of the loop gain, obtained by compensating the open loop system with a diagonal PI-regulator.

(c) The singular values for the compensated loop gain $G_m(s)$.

(d) The singular values of sensitivity and complementary sensitivity for the resulting closed loop system.

Figure 10.5: Loop properties for the controller design of the aircraft. In (a)–(c) the line $|G| \equiv 1$ has also been marked.

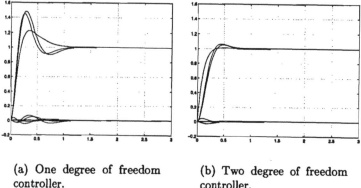

(a) One degree of freedom   (b) Two degree of freedom
controller.   controller.

Figure 10.6: Step responses for the closed loop system. Three different simulations have been carried out, corresponding to a reference signal step for each of the three outputs. The three curves that converge to the value 1 show the response in the corresponding outputs, while the six curves hovering around zero are the responses in the outputs whose setpoints are zero. That these responses are close to zero at all times shows that the closed loop system is well decoupled.

(10.33). The resulting controller is

$$u = -W_p(s)W_1 F_y(s)W_2 y$$

Figure 10.5d shows the singular values of sensitivity and complementary sensitivity for this controller.

Figure 10.6a shows the output responses to steps in the three different reference signals when the controller is

$$u = -W_p(s)W_1 F_y(s)W_2(y - r)$$

that is, a one-degree of freedom controller. Figure 10.6b shows the corresponding curves when the reference signal is introduced via the state feedback according to (9.11).

10.6   Discrete Time Systems

The design process in terms of system norms that was used for continuous time systems can of course equally well be applied to discrete time systems.

The extended discrete time system, analogously to (10.7), takes the form

$$x(t+1) = Ax(t) + Bu(t) + Nw(t) \tag{10.40a}$$
$$z(t) = Mx(t) + Du(t) \tag{10.40b}$$
$$y(t) = Cx(t) + w(t) \tag{10.40c}$$

For the discussion on $\mathcal{H}_2$ control in Section 10.3 the only change is that (10.15) becomes

$$\hat{x}(t+1) = A\hat{x}(t) + Bu(t) + N(y(t) - C\hat{x}(t)) \tag{10.41a}$$
$$u(t) = -L\hat{x}(t) \tag{10.41b}$$
$$L = (B^T SB + I)^{-1} B^T SA \tag{10.41c}$$
$$S = A^T SA + M^T M - A^T SB(B^T SB + I)^{-1} B^T SA \tag{10.41d}$$

This is a direct consequence of Theorem 9.4.

The discrete time $\mathcal{H}_\infty$-theory is more complicated. A common advice is to calculate the controller for a continuous time system. This controller can then be converted to discrete time, e.g., as in (4.1).

10.7 Comments

Main Points of the Chapter

- We have expressed the desired closed loop properties using three weighting functions (matrices), W_S, W_T and W_u, and formalized them as a requirement on the norm of an extended system G_{ec}, (10.6), obtained when the loop around (10.7) is closed.

- Trading off the requirements against each other by minimizing their quadratic norm, is the same as solving a certain LQ-problem. See (10.15).

- Demanding that the requirements shall hold for all frequencies is the same as forcing the $\mathcal{H}_\infty$-norm of G_{ec} to be less than a certain value. A controller that solves this is given by (10.20)–(10.24).

- A third, and useful method for multivariable design is the Glover-McFarlane method to shape the loop gain.

- In all these cases is it recommended to test if the resulting controller F_y can be simplified using the model reduction scheme of Section 3.6.

- Finally, it must be reiterated that any candidate design must be further studied by simulation and examination of all loop transfer functions, before it can be accepted.

Literature

The foundations of loop based design of controllers are described in the classical books James et al. (1947) and Bode (1945). Horowitz developed QFT – Quantitative Feedback Design Theory – during the '60s and onwards, and has summarized the method in Horowitz (1993).

The design methods described in this chapter were developed during the '80s and '90s. A comprehensive treatment with a complete mathematical theory is given in Zhou et al. (1996), while an earlier and more elementary description is given by Maciejowski (1989). The special case of our Theorem 10.1 is called the *disturbance feedforward case* in Zhou et al. (1996). The book Skogestad & Postlethwaite (1996) also gives an insightful treatment with many examples. The Glover-McFarlane method was introduced in Glover & McFarlane (1989) and is thoroughly described in Zhou et al. (1996). The aircraft example 10.5 is from Maciejowski (1989), although solved here by the Glover-McFarlane method rather than by other $\mathcal{H}_\infty$-techniques.

The so called μ-synthesis and structured singular values are closely related to $\mathcal{H}_\infty$-theory. With μ-synthesis more specific knowledge about the system uncertainty can be utilized. It can also be used to guarantee performance under uncertainty, that is, not only stability as in (6.28). Zhou et al. (1996) contains an excellent treatment also of this theory.

Software

There are several MATLAB-toolboxes for control design. The Ro-BUST CONTROL DESIGN TOOLBOX and μ-ANALYSIS AND SYNTHESIS TOOLBOX specifically deal with the methods of this chapter. The key commands concern the solution of Riccati equation (10.20) in a numerically robust way, and support to the user when constructing the extended model (10.7). There is also a QFT-toolbox in MATLAB.

Appendix 10A: To Decrease the Spread of the Loop Gain Singular Values

In Glover-McFarlane's algorithm, an important step is to choose the matrices W_1 and W_2, so that the matrix $W_2 G(i\omega) W_1$ preferably has all singular values equal. Here $G(i\omega)$ is a given transfer function matrix, and ω is a given, fixed frequency, typically the crossover frequency. An important constraint is that the matrices W_i must be real valued, and well conditioned.

There is no known simple and best solution to this problem. We shall here show a possible method that only makes use of the matrix W_1. Start from the singular value decomposition (3.24):

$$A = U\Sigma V^* \qquad (10.42)$$

We need to study (10.42) in more detail for a square matrix A. The matrices are written as

$$U = \begin{bmatrix} u_1 & u_2 & \cdots & u_n \end{bmatrix} \qquad (10.43a)$$

$$V = \begin{bmatrix} v_1 & v_2 & \cdots & v_n \end{bmatrix} \qquad (10.43b)$$

$$\Sigma = \begin{bmatrix} \sigma_1 & 0 & \cdots & 0 \\ 0 & \sigma_2 & \cdots & 0 \\ \vdots & \vdots & \ddots & \vdots \\ 0 & 0 & \cdots & \sigma_n \end{bmatrix} \qquad (10.43c)$$

Then (10.42) can be rewritten as

$$A = \begin{bmatrix} u_1 & u_2 & \cdots & u_n \end{bmatrix} \begin{bmatrix} \sigma_1 & 0 & \cdots & 0 \\ 0 & \sigma_2 & \cdots & 0 \\ \vdots & \vdots & \ddots & \vdots \\ 0 & 0 & \cdots & \sigma_n \end{bmatrix} \begin{bmatrix} v_1 & v_2 & \cdots & v_n \end{bmatrix}^*$$

$$= \begin{bmatrix} u_1 & u_2 & \cdots & u_n \end{bmatrix} \begin{bmatrix} \sigma_1 v_1^* \\ \sigma_2 v_2^* \\ \cdots \\ \sigma_n v_n^* \end{bmatrix} = \sum_{i=1}^{n} u_i \sigma_i v_i^* \qquad (10.44)$$

Based on this expression we ask what are the singular values of $A v_j v_j^*$. We have

$$A v_j v_j^* = \sum_{i=1}^{n} u_i \sigma_i v_i^* v_j v_j^* = u_j \sigma_j v_j^*$$

In this expression we used that V is a unitary matrix so that $v_i v_j^* = 0$ when $i \neq j$ and $v_i v_i^* = 1$. The matrix $A v_j v_j^*$ thus has the singular value σ_j and remaining singular values equal to zero. Analogously it follows that the matrix

$$A(I + \alpha v_j v_j^*)$$

has the same singular values as A, except the jth, which has been changed to $(1 + \alpha)\sigma_j$. This shows how to manipulate a matrix to modify its singular values.

Example 10.6:   Choice of W_1

Consider a system with two inputs and two outputs, such that

$$G(i\omega) = C(i\omega I - A)^{-1}B$$

is a 2×2 matrix. Let its SVD be

$$G(i\omega) = \begin{bmatrix} u_1 & u_2 \end{bmatrix} \begin{bmatrix} \sigma_1 & 0 \\ 0 & \sigma_2 \end{bmatrix} \begin{bmatrix} v_1 & v_2 \end{bmatrix}^*$$

Assume that $\sigma_1 > \sigma_2$. According to the discussion in this appendix we then find that

$$\tilde{G}(i\omega) = G(i\omega)W_1$$

has two equal singular values $(=\sigma_1)$ if W_1 is chosen as

$$W_1 = [I + \frac{\sigma_1 - \sigma_2}{\sigma_2} v_2 v_2^*]$$

A complication is that v_i typically are complex valued vectors, while W_1 should be real. A simple approach is to replace v_i with their real parts and hope for the best. The factor that multiplies $v_i v_i^*$ may then have to be adjusted.

PART III: NONLINEAR CONTROL THEORY

In this part we will consider phenomena occuring in control systems containing nonlinear parts. We will also present design methods based on model predictive control, exact linearization and optimality criteria.

Chapter 11

Describing Nonlinear Systems

All our discussions up to now concerning stability, sensitivity, robustness, etc. are in principle very general. However, in all concrete calculations we have assumed that the system and its controller are linear, i.e., they are described by linear differential equations (or in the frequency domain by transfer functions). The reason for this is mainly that there is a well developed, comparatively simple, mathematical theory for linear systems. All real control systems contain some form of nonlinearity however. This can be seen by considering the consequences of exact linearity. If a linear system reacts with the output y to the input u, then it reacts with the output αy to the input αu for all values of α. For, e.g., a DC motor this implies that the step response should have the same form, regardless of whether the motor current is in the μA range or in the kA range. In practice the former would mean a current, too weak to overcome friction, while the latter would burn up the motor. Similar considerations apply to other physical systems. In many cases the nonlinearity appears as an upper or lower bound of a variable. A valve, for instance, can be positioned in a range between completely open and completely closed. If the control signal from a regulator passes a D/A-converter there is a largest and smallest signal, corresponding to the largest and smallest integer which is accepted.

However, the fact that most systems contain nonlinearities does not imply that linear theory is useless. We will see that one of the most important methods for analyzing nonlinear systems is to approximate them with linear ones.

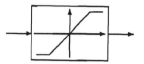

Figure 11.1: Block diagram for a saturation.

11.1   Linear Versus Nonlinear

When studying linear systems we have often used the following properties (without always mentioning them explicitly).

Invariance under scaling. If the input u gives the output y, then αu gives αy.

Additivity. If the inputs u_1 and u_2 give the outputs y_1 and y_2, then $u_1 + u_2$ gives the output $y_1 + y_2$.

Frequency fidelity. A component in the reference or disturbance signal with the frequency ω affects the control error only at the frequency ω.

Sometimes additivity and invariance under scaling are summarized in the principle of *superposition*: If the inputs u_1 and u_2 give the outputs y_1 and y_2, then $\alpha u_1 + \beta u_2$ gives the output $\alpha y_1 + \beta y_2$.

We will now show that even a simple nonlinearity causes a violation of these principles. As mentioned above, one of the most common nonlinearities is a limitation of the amplitude of a physical variable, often referred to as a *saturation*. In the simplest case the mathematical relationship between the input u and the output y is

$$y(t) = \begin{cases} a & u(t) > a \\ -a & u(t) < -a \\ u(t) & |u(t)| \le a \end{cases} \tag{11.1}$$

The block diagram representation is shown in Figure 11.1. We will now investigate the effect of saturation in an otherwise linear control system.

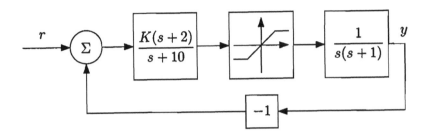

Figure 11.2: Block diagram of DC motor with saturation.

DC Motor with Saturation

Let us consider a simple example of a DC motor servo with a lead compensator and saturation of the control signal as shown in Figure 11.2. Physically the saturation could, e.g., correspond to the limitations of the power amplifier driving the motor. In Figure 11.3 the step responses are shown for the reference signal amplitudes 1 and 5. The step response of $r = 5$ is scaled to the same stationary level as $r = 1$ to make a comparison simpler. We see that the shape of the step response depends heavily on the amplitude, a clear violation of scaling invariance.

Let us now investigate how the system of Figure 11.2 reacts to the addition of reference signals. Figure 11.4a shows the response to a reference signal which is a ramp, $r(t) = 0.9t$. As one would expect, the ramp is followed with a certain stationary error. Figure 11.4b shows the response to a sine signal with the angular frequency $\omega = 10$ and the amplitude 0.5. As expected the response is a sinusoid of the same frequency. Since the frequency is above the bandwidth of the motor, the amplitude is small. Figure 11.4c shows the response when the reference signal is the sum of the signals of the two previous cases, i.e., $r(t) = 0.9t + 0.5 \sin 10t$. Here an interesting phenomenon occurs. The fact that we add a sinusoid to the reference signal leads to an increased steady state error. The sine component itself is not visible in the output. Clearly this contradicts the principle of superposition. It also violates frequency fidelity: the addition of a high frequency component in the reference signal affects the control error at the frequency 0. To understand what happens physically, we can plot the control signal before and after the saturation, see Figure 11.4d. We notice that the addition of a sine component together with the saturation effectively

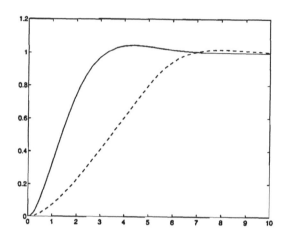

Figure 11.3: The step responses of the DC motor for steps with the amplitude 1 (solid) and 5 (dashed). (The curves are scaled so that the stationary level is the same.)

shrinks the available range for the control signal. The control signal decreases during the negative half periods, but the saturation does not allow a corresponding increase during the positive half periods. The control signal is therefore, on the average, below the saturation level and the ramp is followed with a larger error than would otherwise be the case.

Discussion

In the example above we have seen that neither invariance of scaling nor additivity or frequency fidelity need apply to a nonlinear system. This means that it is much more difficult to verify the functioning of a nonlinear system than a linear one, using simulation and testing. If a linear system has been tested with a step in the reference signal and later with a step in a disturbance signal, the principle of superposition immediately tells us what would happen if both steps are applied simultaneously. For a nonlinear system, on the contrary, all different combinations of inputs, disturbances, initial values, different parameter values, etc. have to be tested or simulated. This difficulty makes it important to get an understanding of the qualitative properties of phenomena that can occur in nonlinear systems. In this chapter and

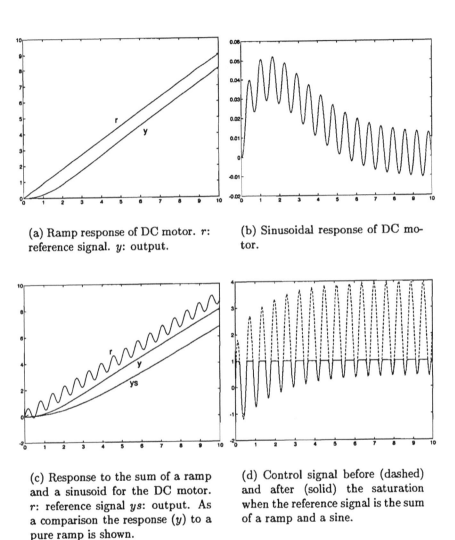

(a) Ramp response of DC motor. r: reference signal. y: output.

(b) Sinusoidal response of DC motor.

(c) Response to the sum of a ramp and a sinusoid for the DC motor. r: reference signal ys: output. As a comparison the response (y) to a pure ramp is shown.

(d) Control signal before (dashed) and after (solid) the saturation when the reference signal is the sum of a ramp and a sine.

Figure 11.4: Superposition is not valid for a DC motor with saturation. The sum of a ramp response, (a), and the response to a sine, (b), is not equal to the response of a ramp plus sine, (c). The reason is the asymmetric saturation, (d).

the following ones, we will show some of the fundamental theory that helps in such an understanding.

11.2   Examples of Nonlinear Systems

Example 11.1:   DC Motor with Saturation

The DC motor described in Figure 11.2 without its controller is given by

$$\dot{x}_1 = x_2 \tag{11.2a}$$
$$\dot{x}_2 = -x_2 + \text{sat } u \tag{11.2b}$$
$$y = x_1 \tag{11.2c}$$

where the function "sat" is defined by

$$\text{sat } u = \begin{cases} u & |u| \leq 1 \\ 1 & u > 1 \\ -1 & u < -1 \end{cases}$$

Example 11.2:   A Generator

Electrical energy is produced by generators that convert mechanical energy into electrical energy. The frequency of the alternating current that is generated is proportional to the speed of rotation of the generator. Let θ denote the angle of rotation of the generator. The laws of rotational mechanics then give the following simplified model.

$$J\ddot{\theta} = M_d + M$$

Here J denotes the moment of inertia of the rotating parts, M_d is the driving torque (e.g., from a steam turbine) and M is the torque generated by the electromagnetic phenomena in the generator. A simplified model for these is

$$M = -f\dot{\theta} + K\sin(\omega_o t - \theta)$$

The first term represents energy losses that give the system some damping. The second term describes the interaction with the rest of the electric power system. If $\theta = \omega_o t$, the generating current is in phase with the rest of the power net, otherwise an accelerating or braking torque is generated. The sign of $\theta - \omega_o t$ shows if the generator delivers or consumes electric power. This is consequently an interesting quantity, which is a natural state variable.

$$x_1 = \theta - \omega_o t$$

Introducing its derivative as a second state

$$x_2 = \dot{\theta} - \omega_o$$

gives the following model.

$$\dot{x}_1 = x_2 \tag{11.3a}$$
$$\dot{x}_2 = u - ax_2 - b\sin x_1 \tag{11.3b}$$

where

$$u = (M_d - f\omega_o)/J, \quad a = f/J, \quad b = K/J$$

and we regard u as an input.

Example 11.3: The Effect of Friction

Consider a system described by the following figure.

An object is moved along a surface by the force F_d. The equation of motion is

$$m\ddot{z} = F_d + F$$

where z is the position, m is the mass, and F is the friction force. Assuming that the maximum static friction is F_0, and that the sliding friction is F_1, gives the friction model

$$F = \begin{cases} -F_1 \operatorname{sgn}\dot{z} & \dot{z} \neq 0 \\ -F_d & \dot{z} = 0, \quad |F_d| \leq F_0 \\ -F_0 \operatorname{sgn}F_d & \dot{z} = 0, \quad |F_d| > F_0 \end{cases} \tag{11.4}$$

(Here sgn is the sign function: $\operatorname{sgn} x = 1$ if $x > 0$, $\operatorname{sgn} x = -1$ if $x < 0$.) Suppose we want to move the object with a constant velocity v_0 along the surface. The reference signal is then $r(t) = v_0 t$ and the control error and its derivatives become

$$e = v_0 t - z, \quad \dot{e} = v_0 - \dot{z}, \quad \ddot{e} = -F_d - F$$

For simplicity we have used $m = 1$. Choosing the state variables

$$x_1 = e, \quad x_2 = \dot{e}$$

the system is described by

$$\dot{x}_1 = x_2 \tag{11.5a}$$
$$\dot{x}_2 = -F_d - F \tag{11.5b}$$

where F_d can be regarded as an input. Using a PD controller gives the relation

$$F_d = Ke + f\dot{e}$$

where K and f are some positive constants. The feedback system is then described by

$$\dot{x}_1 = x_2$$
$$\dot{x}_2 = \begin{cases} -\ell(x) + F_1 \operatorname{sgn}(v_0 - x_2) & x_2 \neq v_0 \\ 0 & x_2 = v_0, \ |\ell(x)| \leq F_0 \\ -\ell(x) + F_0 \operatorname{sgn}\ell(x) & x_2 = v_0, \ |\ell(x)| > F_0 \end{cases} \tag{11.6}$$

where $\ell(x) = Kx_1 + fx_2$.

11.3   Mathematical Description

In Examples 11.1–11.3, the nonlinear system can be described in the following way

$$\dot{x} = f(x, u) \tag{11.7a}$$
$$y = h(x) \tag{11.7b}$$

This is a system of first order nonlinear differential equations, where the right hand side depends on u. We will regard (11.7) as the standard description of nonlinear systems.

By studying the examples we see that there are several interesting special cases. In Example 11.2 the right hand side is an *affine* (linear plus constant) function of u for constant x. Such *control affine* systems can in general be written in the form

$$\dot{x} = f(x) + g(x)u \tag{11.8a}$$
$$y = h(x) \tag{11.8b}$$

In Example 11.3 we get

$$f(x) = \begin{bmatrix} x_2 \\ -ax_2 - b\sin x_1 \end{bmatrix}, \quad g(x) = \begin{bmatrix} 0 \\ 1 \end{bmatrix}$$

In particular we see that all linear systems are control affine.

In Examples 11.1 and 11.3 the systems are *piecewise linear*. This means that the right hand side can be written in the form

$$f(x, u) = Ax + Bu + d \tag{11.9}$$

The matrices A, B, and d are however different in different parts of the state and input spaces. In Example 11.1 the equations are

$$A = \begin{bmatrix} 0 & 1 \\ -1 & 0 \end{bmatrix}, \quad B = \begin{bmatrix} 0 \\ 1 \end{bmatrix}, \quad d = 0, \quad \text{for } |u| < 1$$

$$A = \begin{bmatrix} 0 & 1 \\ -1 & 0 \end{bmatrix}, \quad B = 0, \quad d = \begin{bmatrix} 0 \\ 1 \end{bmatrix}, \quad \text{for } u > 1$$

$$A = \begin{bmatrix} 0 & 1 \\ -1 & 0 \end{bmatrix}, \quad B = 0, \quad d = \begin{bmatrix} 0 \\ -1 \end{bmatrix}, \quad \text{for } u < 1$$

Example 11.3 has the following piecewise linear structure.

$$\dot{x} = \begin{bmatrix} 0 & 1 \\ -K & -f \end{bmatrix} x + \begin{bmatrix} 0 \\ -F_1 \end{bmatrix}, \quad x_2 > v_0 \tag{11.10a}$$

$$\dot{x} = \begin{bmatrix} 0 & 1 \\ -K & -f \end{bmatrix} x + \begin{bmatrix} 0 \\ F_1 \end{bmatrix}, \quad x_2 < v_0 \tag{11.10b}$$

$$\dot{x} = \begin{bmatrix} 0 & 1 \\ 0 & 0 \end{bmatrix} x + \begin{bmatrix} 0 \\ 0 \end{bmatrix}, \quad x_2 = v_0, \ |\ell(x)| \le F_0 \tag{11.10c}$$

$$\dot{x} = \begin{bmatrix} 0 & 1 \\ -K & -f \end{bmatrix} x + \begin{bmatrix} 0 \\ F_0 \end{bmatrix}, \quad x_2 = v_0, \ \ell(x) > F_0 \tag{11.10d}$$

$$\dot{x} = \begin{bmatrix} 0 & 1 \\ -K & -f \end{bmatrix} x + \begin{bmatrix} 0 \\ -F_0 \end{bmatrix}, \quad x_2 = v_0, \ \ell(x) < -F_0 \tag{11.10e}$$

There is another structure which is common among nonlinear systems and is exemplified by Figure 11.2. Figure 11.5 shows how the block diagram can be rewritten in successive steps to give a diagram with a

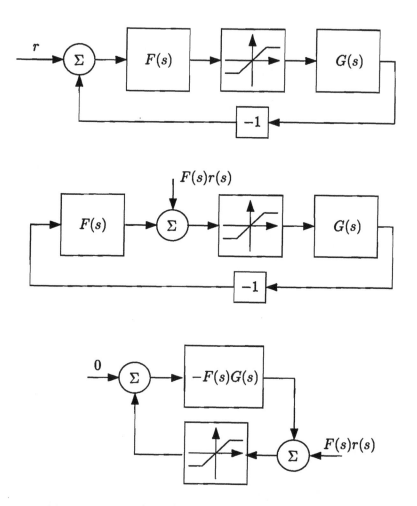

Figure 11.5: Successive reshaping of the block diagram of a DC motor. In the first step the summation point of the reference signal is moved. In the next step the linear blocks are merged.

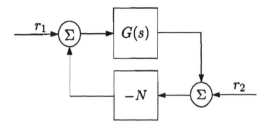

Figure 11.6: Linear system with feedback through a nonlinearity N.

linear block and a static nonlinearity in a feedback loop. The transfer functions involved are

$$F(s) = \frac{K(s+2)}{s+10}, \quad G(s) = \frac{1}{s(s+1)}$$

The resulting block diagram is a special case of the more general diagram of Figure 11.6, where

$$G(s) = \frac{K(s+2)}{s(s+1)(s+10)}$$

and N is the saturation. We have moved the minus sign from the linear to the nonlinear block without any other reason than to be consistent with some of the existing literature. One could of course also include the minus sign in G or N. Note that this is the system structure that is used in the Small Gain Theorem, see Section 1.6.

11.4 Equilibria and Linearization

In many control problems the goal is to keep certain variables constant. If all state variables are constant, the system is said to be in *equilibrium*. For a system described by (11.7a) an equilibrium is given by constant vectors u_0 and x_0 such that

$$f(x_0, u_0) = 0 \tag{11.11}$$

The equilibrium is also sometimes called a *singular point* or a *stationary point*.

Example 11.4:   Equilibria of the Generator

For the generator (11.2), the equilibria for $u = 0$ are given by $x_2 = 0$ and $\sin x_1 = 0$, i.e.,

$$x_{0,1} = 0, \pm\pi, \pm 2\pi, \ldots; \quad x_{0,2} = 0$$

The example shows that a nonlinear system can have several distinct equilibria.

Let us investigate a nonlinear system close to equilibrium. We can then write it in the following form.

Theorem 11.1 *If the function f is differentiable in a neighborhood of the equilibrium point x_0, u_0, then the system (11.7a) can be written as*

$$\dot{z} = Az + Bv + g(z, v) \tag{11.12}$$

where $z = x - x_0$, $v = u - u_0$ and $g(z, v)/(|z| + |v|) \to 0$ when $|z| + |v| \to 0$. The matrices A and B have elements a_{ij} and b_{ij} given by

$$a_{ij} = \frac{\partial f_i}{\partial x_j}, \quad b_{ij} = \frac{\partial f_i}{\partial u_j}$$

where f_i is the i:th row of f and the derivatives are computed at x_0, u_0.

Proof: This follows directly from a Taylor series expansion of f.   □

Example 11.5:   A Generator

For the generator (11.3) we get the linearization

$$\dot{z} = \begin{bmatrix} 0 & 1 \\ -b & -a \end{bmatrix} z + \begin{bmatrix} 0 \\ 1 \end{bmatrix} v + g(z, v)$$

at $x_2 = 0$, $x_1 = 0, \pm 2\pi, \pm 4\pi, \ldots$, and

$$\dot{z} = \begin{bmatrix} 0 & 1 \\ b & -a \end{bmatrix} z + \begin{bmatrix} 0 \\ 1 \end{bmatrix} v + g(z, v)$$

at $x_2 = 0$, $x_1 = \pm\pi, \pm 3\pi, \ldots$.

Intuitively, it seems reasonable to approximate the nonlinear system (11.12) with the linear system

$$\dot{z} = Az + Bv \qquad\qquad (11.13)$$

This linear system is called the *linearization* of the nonlinear system at the equilibrium x_0, u_0. The methods we discussed in Chapters 2–10 are in fact usually used with linear models which have been created in this way. It is of great interest to know to what extent properties of (11.13) are also properties of the original nonlinear system.

In the coming chapters we will present some methods for analyzing the properties of nonlinear systems and in that context we will also discuss the validity of linearized models.

11.5   Comments

Main Points of the Chapter

- The superposition principle is not valid for nonlinear systems.

- Most nonlinear systems can be written in the form

$$\dot{x} = f(x, u)$$
$$y = h(x)$$

An important special case is

$$\dot{x} = f(x) + g(x)u$$
$$y = h(x)$$

An equilibrium (singular point, stationary point) corresponding to the constant value u_0 is a point x_0 such that

$$f(x_0, u_0) = 0$$

- A linearization of

$$\dot{x} = f(x, u)$$

around the equilibrium x_0, u_0 is the system

$$\dot{z} = Az + Bv$$

where $z = x - x_0$, $v = u - u_0$. The matrices A and B have elements a_{ij} and b_{ij} given by

$$a_{ij} = \frac{\partial f_i}{\partial x_j}, \quad b_{ij} = \frac{\partial f_i}{\partial u_j}$$

where the derivatives are evaluated at x_0, u_0.

Literature

Nonlinear control theory is based on methods used in physics, in particular mechanics, to describe nonlinear phenomena, since the days of Newton. The basic mathematical results are presented in many textbooks on differential equations. A nice exposition with emphasis on results that are useful in applications is given in Verhulst (1990). Many books discuss nonlinear systems from a control point of view. Among books which discuss the more general aspects we mention Slotine & Li (1991), Leigh (1983), Khalil (1996) and Cook (1994).

Software

In MATLAB there is SIMULINK which can be used for modeling and simulation of general nonlinear systems. Of interest for this chapter are in particular the commands:

trim Determination of equilibria.

linmod Linearization.

Chapter 12

Stability of Nonlinear Systems

Stability, instability and asymptotic stability were defined in Definition 1.3 for a solution of the system equations. This definition is easily applied to equilibria.

12.1 Stability of Equilibria

Let the system (11.7a) have the equilibrium (x_0, u_0) and consider the system obtained by fixing u to u_0.

$$\dot{x}(t) = f(x(t), u_0) \tag{12.1a}$$
$$\text{where } f(x_0, u_0) = 0 \tag{12.1b}$$

Let $x^*(t)$ be the solution of (12.1a) with $x^*(0) = x_0$. It follows that $x^*(t) = x_0$ for all t.

Definition 12.1 *The equilibrium x_0 of the system (12.1a) is said to be* **stable,** **unstable** *or* **asymptotically stable** *if the solution $x^*(t)$ has the corresponding property according to Definition 1.3.*

An equilibrium is thus asymptotically stable, if all solutions starting close to it, remain close and eventually converge towards it. This is obviously a desirable property of a control system in most cases.

Even better is if all solutions converge to the equilibrium. This leads to the definition

321

Definition 12.2 *An equilibrium x_0 of the system (12.1a) is* **globally asymptotically stable** *if it is stable and*

$$x(t) \rightarrow x_0, \quad t \rightarrow \infty$$

for every initial value $x(0)$.

Intuitively it seems likely that an equilibrium of a nonlinear system is asymptotically stable if its linear approximation (11.13) is. We have the following theorem.

Theorem 12.1 *Let the system (12.1a) be described by (11.12) close to an equilibrium x_0. If all eigenvalues of A have strictly negative real parts, then x_0 is an asymptotically stable equilibrium.*

Proof: This is a special case of a more general result (Theorem 12.6) which will be shown below.   □

It is possible to show a similar result concerning instability.

Theorem 12.2 *Let the system (12.1a) be described by (11.12) close to the equilibrium x_0. If A has an eigenvalue with strictly positive real part, then x_0 is an unstable equilibrium.*

The two theorems 12.1 and 12.2 do not tell us what happens when the linearization has eigenvalues on the imaginary axis, but not in the right half plane. Such equilibria can be both stable and unstable as shown by the following examples.

Example 12.1: Imaginary Eigenvalues, Asymptotic Stability

Let the system be

$$\begin{aligned}
\dot{x}_1 &= x_2 - x_1^3 \\
\dot{x}_2 &= -x_1 - x_2^3
\end{aligned}$$
(12.2)

with the linear approximation

$$\begin{aligned}
\dot{x}_1 &= x_2 \\
\dot{x}_2 &= -x_1
\end{aligned}$$

near the origin. The eigenvalues are $\pm i$. The linearized system is thus stable but not asymptotically stable. (A solution is $x_1 = C \sin t$, $x_2 = C \cos t$.) We can investigate the properties of the system by checking the distance to the origin along a solution. We then have to calculate the change of $x_1^2(t) + x_2^2(t)$ as a function of time, along solutions of (12.2). We get

$$\frac{d}{dt}(x_1^2(t) + x_2^2(t)) = 2x_1(t)\dot{x}_1(t) + 2x_2(t)\dot{x}_2(t)$$

Substituting $\dot{x}_1$ and $\dot{x}_2$ from (12.2) gives

$$\frac{d}{dt}(x_1^2 + x_2^2) = 2x_1(x_2 - x_1^3) + 2x_2(-x_1 - x_2^3) = -2x_1^4 - 2x_2^4 \leq 0$$

suppressing the argument t. We see that the time derivative of the (squared) distance to the origin is negative along solutions that satisfy (12.2). The distance to the origin is consequently decreasing. Since the time derivative of the distance is strictly negative as long as x_1 and x_2 are not both zero, the distance will decrease to zero (this will be proved in Theorem 12.3 below) so that the solution approaches the origin.

Example 12.2: Imaginary Eigenvalues, Instability

For the system

$$\begin{aligned} \dot{x}_1 &= x_2 + x_1^3 \\ \dot{x}_2 &= -x_1 + x_2^3 \end{aligned} \qquad (12.3)$$

the linearization is the same as in the previous example. Analogous calculations give

$$\frac{d}{dt}(x_1^2 + x_2^2) = 2x_1\dot{x}_1 + 2x_2\dot{x}_2$$

$$= 2x_1(x_2 + x_1^3) + 2x_2(-x_1 + x_2^3) = 2x_1^4 + 2x_2^4$$

The distance to the origin will increase without bound and the origin is thus an unstable equilibrium.

12.2 Stability and Lyapunov Functions

In Examples 12.1 and 12.2 we investigated the stability by checking whether the squared distance $x_1^2(t) + x_2^2(t)$ increased or decreased with time. This idea can be generalized. Suppose we study the system $\dot{x} = f(x)$ with the equilibrium x_0. Further, assume that there is a function V with the properties

$$V(x_0) = 0; \quad V(x) > 0, \ x \neq x_0; \quad V_x(x)f(x) \leq 0 \qquad (12.4)$$

(Here V_x is the row vector $(\partial V/\partial x_1,\ldots,\partial V/\partial x_n).$) V can be interpreted as a generalized distance from x to the point x_0. This generalized distance decreases for all solutions of $\dot{x} = f(x)$, since

$$\frac{d}{dt}V(x(t)) = V_x(x(t))\dot{x}(t) = V_x(x(t))f(x(t)) \leq 0$$

A function V satisfying (12.4) in some neighborhood of the equilibrium x_0 is called a (local) *Lyapunov function*.

By adding some requirements to the Lyapunov function properties we get the following stability test.

Theorem 12.3 *An equilibrium x_0 of the system $\dot{x} = f(x)$ is globally asymptotically stable if there exists a function V, satisfying (12.4) for all values of x, and in addition satisfying*

$$V_x(x)f(x) < 0, \; x \neq x_0 \tag{12.5a}$$
$$V(x) \to \infty, |x| \to \infty \tag{12.5b}$$

Proof: If $x(t)$ is a solution of the differential equation, then

$$\frac{d}{dt}V(x(t)) = V_x(x(t))f(x(t)) < 0, \quad \text{if } x(t) \neq x_0$$

The value of V is thus strictly decreasing. In particular, $V(x(t)) \leq V(x(t_0))$, for all t greater than the initial time t_0. From (12.5b) it then follows that the solution is bounded. Since V is decreasing along a trajectory, and V is bounded below by zero, it follows that $V(x(t))$ converges to some value $c \geq 0$, as t tends to infinity. Assume that $c > 0$. Then we have $c \leq V(x(t)) \leq V(x(t_0))$. The trajectory thus remains in a compact set in which $V_x(x)f(x) < 0$. It follows that $\dot{V} = V_x f \leq -\epsilon$ for some positive constant ϵ. This implies that V goes to minus infinity, a contradiction. It follows that we must have $c = 0$. This means that $x(t)$ has to converge to x_0. $\square$

Example 12.3:   Imaginary Eigenvalues, Continued

In Example 12.1 we used the function $V = x_1^2 + x_2^2$ to show asymptotic stability.

Example 12.4:   Fourth Order V-function

Let the system be

$$\dot{x}_1 = x_2, \quad \dot{x}_2 = -x_2 - x_1^3$$

With $V = \alpha x_1^4 + x_2^2$ we get

$$V_x f = (4\alpha - 2)x_1^3 x_2 - 2x_2^2$$

The choice $\alpha = 1/2$ gives $V_x f = -2x_2^2 \leq 0$. We cannot use Theorem 12.3 directly, since we do not have strict inequality in (12.5a).

The difficulty occurring in the example above is common and has resulted in an extension of Theorem 12.3.

Theorem 12.4 *An equilibrium x_0 of the system $\dot{x} = f(x)$ is globally asymptotically stable if it is possible to find a function V, which*

1. *satisfies (12.4) for all values of x*

2. *satisfies (12.5b)*

3. *has the property that no solution of the differential equation (except $x(t) = x_0$) lies entirely in the set $V_x(x)f(x) = 0$*

Proof: Using reasoning analogous to the proof of Theorem 12.3, one can see that the solution must converge to a set where $V_x f = 0$. It can be shown that the limit set must itself be a solution, which means that it has to be x_0. □

Example 12.5: Fourth Order V-function, Continued

In Example 12.4 we obtained $V_x f = -2x_2^2 \leq 0$. We use Theorem 12.4. If a solution lies entirely in the set where $V_x f = 0$, then x_2 has to be identically zero. This can only happen when x_1 is zero. It follows that the origin is a globally asymptotically stable equilibrium.

In the theorems and examples we have looked at so far, the properties (12.4) have been valid in the whole state space and we have been able to prove global asymptotic stability. The situation is not often this favorable. Instead, V satisfies (12.4) only for x-values belonging to some subset N of the state space.

Let us consider a subset M_d defined in the following way:

$$M_d = \{x : V(x) \leq d\} \tag{12.6}$$

for some positive number d. Let M_d lie entirely in N, see Figure 12.1. For a solution starting in M_d one has

$$\frac{d}{dt}V(x(t)) = V_x(x(t))f(x(t)) \leq 0$$

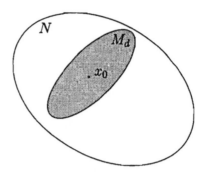

Figure 12.1: Subset where $V_x f \leq 0$ (N) and subset where $V(x) \leq d$ (M_d).

The last inequality is valid as long as the solution remains in M_d, since M_d is a part of N and (12.4) is valid in N. This means that V is decreasing along a solution as long as it remains in M_d. It follows that a solution starting in the interior of M_d, satisfying $V(x(t_0)) < d$, cannot leave M_d. This is because it cannot leave M_d without passing the boundary at some time t_1, which would imply $V(x(t_1)) = d > V(x(t_0))$, contradicting the fact that V decreases. Once it is known that solutions of (12.1a) remain in M_d, it is possible to use the same reasoning as in Theorems 12.3 and 12.4. This gives the following result.

Theorem 12.5 *Assume that there exists a function V and a positive number d such that (12.4) is satisfied for the system (12.1a) in the set*

$$M_d = \{x : V(x) \leq d\}$$

Then a solution starting in the interior of M_d remains there. If, in addition, no solutions (except the equilibrium x_0) remain in the subset of M_d where $V_x(x)f(x) = 0$, then all solutions starting in the interior of M_d will converge to x_0.

Example 12.6:   Stability of the Generator

Consider again the model of a generator in Example 11.2, equation (11.3). Let us investigate the stability of the equilibrium $x = 0$ corresponding to $u = 0$. The state equations are

$$\dot{x}_1 = x_2$$
$$\dot{x}_2 = -ax_2 - b\sin x_1$$

For physical reasons we have $a > 0$, $b > 0$. To find a Lyapunov function we can use the following reasoning. In deriving the model we said that the term

$$-K \sin(\omega_o t - \theta) = bJ \sin x_1$$

represents a torque on the axis. The corresponding potential energy is obtained by integrating with respect to x_1, which gives

$$\text{potential energy} = bJ(1 - \cos x_1)$$

where we have defined the potential energy to be zero at the origin. The rotational energy is $Jx_2^2/2$. We can thus interpret

$$\frac{J}{2}x_2^2 + bJ(1 - \cos x_1)$$

as the total energy. Since there is an energy loss represented by the term $-ax_2$, we can expect the total energy to be decreasing, and use it as a candidate for a Lyapunov function. We define

$$V = \frac{1}{2}x_2^2 + b(1 - \cos x_1)$$

(J is just a scaling which can be removed.) This choice gives

$$V_x f = \begin{bmatrix} b \sin x_1 & x_2 \end{bmatrix} \begin{bmatrix} x_2 \\ -ax_2 - b \sin x_1 \end{bmatrix} = -ax_2^2 \leq 0$$

For $b = 1$ the set $M_{1.5}$ corresponding to $V(x) \leq 1.5$ is shown in Figure 12.2. We note that this set is not connected — there is a dotted region around every equilibrium where x_1 is an even multiple of π. Since solutions are continuous, they cannot jump between the disconnected subsets. It follows that one only has to consider that subset of $M_{1.5}$ where the solution starts. Theorem 12.5 can then be applied, for instance, to the connected part of $M_{1.5}$ which contains the origin. Since there is no solution except the origin contained entirely in $V_x f = 0$, we conclude that the origin is an asymptotically stable equilibrium, and that all solutions starting in the connected dotted area around the origin converge to the origin. The same reasoning goes through for every set $V(x) \leq d$ for $d < 2$. For $d > 2$ we note that there is a connected set containing all the equilibria. It is then not possible to conclude from Theorem 12.5 which equilibrium (if any) a solution converges to.

Lyapunov Functions for Almost Linear Systems

In Section 11.4 we discussed how a nonlinear system could be approximated by a linear system close to an equilibrium. It is then natural to consider Lyapunov functions for the linear approximation.

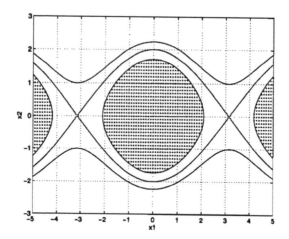

Figure 12.2: Lyapunov function for the generator with $b = 1$. The set $V \leq 1.5$ is dotted. The curves $V = 2$ and $V = 2.5$ are also shown.

Hopefully they will also give valuable information about the stability of the nonlinear system. Consider the linear system

$$\dot{x} = Ax \tag{12.7}$$

We will try a quadratic Lyapunov function

$$V(x) = x^T P x$$

where P is a symmetric positive definite matrix. We get

$$\frac{d}{dt}V(x(t)) = \dot{x}^T(t)Px(t) + x^T(t)P\dot{x}(t) = x^T(t)\left(A^T P + PA\right)x(t)$$

If we demand that

$$\frac{d}{dt}V(x(t)) = -x^T(t)Qx(t)$$

where Q is positive semidefinite, we get the equation

$$A^T P + PA = -Q$$

This equation is called a *Lyapunov equation*. If we succeed in finding positive definite matrices P and Q satisfying it, we have succeeded in constructing a Lyapunov function which can be used in Theorem 12.3. We have the following result.

Lemma 12.1 *Let all the eigenvalues of the matrix A have strictly negative real parts. Then, for every symmetric positive semidefinite matrix Q, there is a symmetric positive semidefinite matrix P such that*

$$A^T P + PA = -Q \qquad (12.8)$$

If Q is positive definite, then so is P.

Conversely, if there are positive semidefinite matrices P and Q such that (12.8) is satisfied and (A, Q) is detectable (see Definition 3.3), all eigenvalues of A have strictly negative real parts.

Proof: See Appendix 12A.   □

Given an A-matrix whose eigenvalues have strictly negative real parts, the lemma shows that we can choose an arbitrary positive definite Q and solve the Lyapunov equation to get a positive definite P. Using $V = x^T P x$ in Theorem 12.3 we can then show that the system is globally asymptotically stable. However, this fact is already known from Theorem 3.10. It is more interesting to consider a nonlinear system

$$\dot{x} = Ax + g(x) \qquad (12.9)$$

where g is sufficiently small near the origin.

Theorem 12.6 *Consider the system (12.9). Assume that all eigenvalues of A have strictly negative real parts and that $g(x)/|x| \to 0$ when $x \to 0$. Let P and Q be positive definite matrices satisfying (12.8). Then there exists a $d > 0$ such that $V(x) = x^T P x$ satisfies (12.4) for the system (12.9) in the set $V(x) \leq d$.*

Proof: See Appendix 12A.   □

Remark The function V of this theorem can be used in Theorem 12.5 to calculate the set M_d. The technique used in the proof can in principle give an explicit value of d. This value will however often be quite conservative.

12.3   The Circle Criterion

In Section 11.3 we saw that a system with one single nonlinear block could be represented in the form shown in Figure 11.6. A common special case is a nonlinearity which is static and unique, described by some function f, see Figure 12.3. The output of the nonlinearity is given by $y_2(t) = -f(e_2(t))$. We assume the function f to be continuous

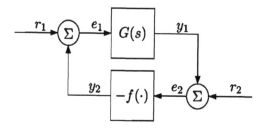

Figure 12.3: A linear system with nonlinear feedback given by f.

and satisfying the condition

$$f(0) = 0, \quad k_1 \leq \frac{f(y)}{y} \leq k_2 \text{ for } y \neq 0 \qquad (12.10)$$

for some positive numbers k_1 and k_2. This means that the graph of f is confined to a cone shaped region, as in Figure 12.4.

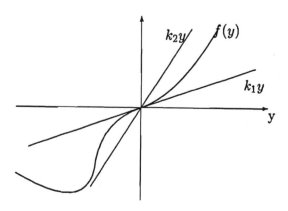

Figure 12.4: The nonlinearity and its bounds.

It is possible to use the Small Gain Theorem (Theorem 1.1) directly to show stability. This however requires

$$k_2 \sup_\omega |G(i\omega)| < 1$$

which normally is too restrictive. As shown in the chapters on control design, the loop gain is usually large, at least at some frequencies, in order to reduce the effects of disturbances and model uncertainty.

It is possible to reformulate the problem before using the Small Gain Theorem to get a more useful result. Define the average gain k by the relation

$$k = \frac{k_1 + k_2}{2} \tag{12.11}$$

and the deviation from the average by

$$\tilde{f}(y) = f(y) - ky \tag{12.12}$$

A simple calculation shows that, for $y \neq 0$,

$$\left| \frac{\tilde{f}(y)}{y} \right| \leq r, \quad \text{with } r = \frac{k_2 - k_1}{2} \tag{12.13}$$

We can now write

$$e_1 = r_1 - f(r_2 + y_1) = r_1 - \{k(r_2 + y_1) + \tilde{f}(r_2 + y_1)\}$$
$$= (r_1 - kr_2) - ky_1 - \tilde{f}(r_2 + y_1) = \tilde{r}_1 - ky_1 - \tilde{f}(r_2 + y_1) \tag{12.14}$$

where $\tilde{r}_1 = r_1 - kr_2$. The system is thus equivalent to the system of Figure 12.5. The blocks within the dashed lines are equivalent to the

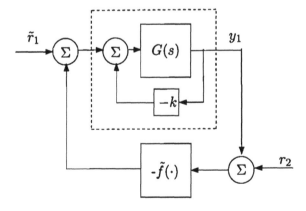

Figure 12.5: Modified block diagram.

linear system

$$\tilde{G} = \frac{G}{1 + kG} \tag{12.15}$$

so that the block diagram becomes that of Figure 12.6. Applying the

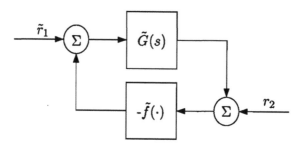

Figure 12.6:   Modified block diagram for application of the Small Gain Theorem.

Small Gain Theorem to the block diagram of Figure 12.6 gives the condition

$$r|\tilde{G}(i\omega)| < 1 \quad \text{or} \quad \frac{1}{|\tilde{G}(i\omega)|} > r$$

Here we have used the results of Examples 1.6 and 1.8 to calculate the gains of $\tilde{G}$ and $\tilde{f}$. By substituting $\tilde{G}$ from (12.15) one gets

$$\left| \frac{1}{G(i\omega)} + k \right| > r \tag{12.16}$$

This condition has to hold for all ω, and its geometric interpretation is shown in Figure 12.7. The condition (12.16) can be interpreted as a requirement that the distance from $1/G(i\omega)$ to the point $-k$ is greater than r. This is equivalent to saying that the curve $1/G(i\omega)$ (the "inverse Nyquist curve") lies outside the circle marked in Figure 12.7. However, it is more convenient to have a criterion involving the ordinary Nyquist curve. Such a criterion can be obtained by transforming the diagram of Figure 12.7 so that every point z in the complex plane is mapped to $1/z$. The curve $1/G(i\omega)$ is then mapped onto the curve $G(i\omega)$, i.e., the ordinary Nyquist curve. The circle of Figure 12.7 is mapped onto a new circle. The points of intersection with the real axis are mapped

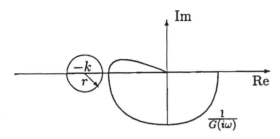

Figure 12.7: Conditions on $1/G(i\omega)$.

as follows.

$$-k + r \quad \rightarrow \quad \frac{1}{-k+r} = -\frac{1}{k_1}$$

$$-k - r \quad \rightarrow \quad \frac{1}{-k-r} = -\frac{1}{k_2}$$

The circle of Figure 12.7 is thus transformed into a circle according to Figure 12.8.   For the Small Gain Theorem to be applicable we have

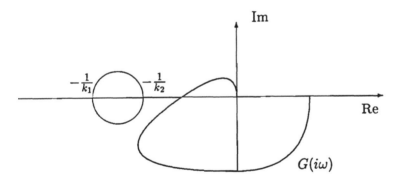

Figure 12.8:   Conditions on the Nyquist curve according to the Circle Criterion.

to require that the Nyquist curve lies outside this circle. However, the Small Gain Theorem also requires the linear subsystem ($\tilde{G}$ in our case) not to have any poles in the right half plane. Since $\tilde{G}$ was formed using feedback around G with the gain $-k$, this is equivalent to a requirement that the Nyquist curve of G does not encircle the point $-1/k$, which lies inside the circle of Figure 12.8. The result of this chain of reasoning is summarized in the following theorem.

Theorem 12.7 The Circle Criterion. Version 1. *Assume that the transfer function G has no poles in the right half plane and that the nonlinearity f satisfies (12.10). A sufficient condition for input output (from r_1, r_2 to e_1, e_2, y_1, y_2) stability of the system in Figure 12.3 is then that the Nyquist curve (i.e., $G(i\omega)$, $0 \leq \omega < \infty$) does not enter or enclose the circle of Figure 12.8.*

A variation of the circle criterion is the following one.

Theorem 12.8 The Circle Criterion. Version 2. *Let the assumptions of Theorem 12.7 be satisfied and assume that the system G has a controllable and observable state space description*

$$\dot{x} = Ax + Bu, \quad y = Cx$$

The origin, $x = 0$, is a globally asymptotically stable equilibrium of the system if $r_1 = 0$, $r_2 = 0$.

Proof: The proof involves the construction of a quadratic Lyapunov function $V(x) = x^T P x$. We refrain from giving the details here.   □

Comments. There are some observations that can be made concerning the Circle Criterion.

- It is a *sufficient* condition for stability. The fact that it is not satisfied does *not* imply instability.

- If f is linear, the circle becomes a point, and we get the ordinary Nyquist criterion for linear systems.

- The first version of the Circle Criterion admits more general linear systems, e.g., systems with delays. (A system with delay has no finite dimensional state space description.)

Example 12.7: The Circle Criterion

Consider the system

$$G(s) = \frac{1}{s(s+1)(s+2)}$$

Its Nyquist curve is shown in Figure 12.9. A vertical line with real part -0.8 lies entirely to the left of the Nyquist curve. Since such a line can be interpreted as a circle with infinite radius, the Circle Criterion is satisfied with $k_1 = 0$ and $k_2 = 1/0.8 = 1.25$. This fact guarantees stability if the

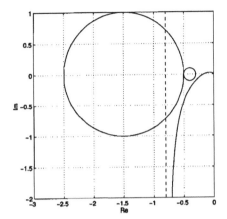

Figure 12.9: Nyquist curve for G (solid) and some circles that satisfy the circle criterion.

nonlinearity is a dead zone or a saturation, see Figure 12.10, provided the slope of the linear part does not exceed 1.25. There are, however, many other possibilities to draw a circle satisfying the Circle Criterion. In the figure there is a circle through -0.5 and -2.5, corresponding to $k_1 = 0.4$ and $k_2 = 2$, and a circle through -0.5, -0.3 corresponding to $k_1 = 2$, $k_2 = 3.33$. It is possible to prove stability for different nonlinearities by choosing different circles.

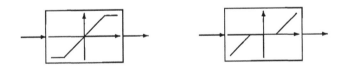

Figure 12.10: Saturation (to the left) and dead zone (to the right).

12.4   Comments

Main Points of the Chapter

- An equilibrium is said to be asymptotically stable if all solutions starting close enough also stay close and eventually converge to the equilibrium.

- An equilibrium, for which all eigenvalues of the linearization have strictly negative real parts, is asymptotically stable.

- An equilibrium, for which some eigenvalue of the linearization has a strictly positive real part, is unstable.

- A Lyapunov function is a function V satisfying

$$V(x_0) = 0; \quad V(x) > 0, \ x \neq x_0; \quad V_x(x)f(x) \leq 0 \quad (12.17)$$

for x in some neighborhood of an equilibrium x_0.

- If there exists a function V satisfying (12.17) in a set $M_d = \{x : V(x) \leq d\}$ for some positive d, then all solutions starting in the interior of M_d will remain there. If, in addition, no solution (except x_0) lies entirely in the set $V_x f = 0$, then all solutions starting in the interior of M_d converge to x_0.

- Let a system with transfer function $G(s)$, having no poles in the right half plane, be part of a feedback loop:

$$u = r_1 - f(y + r_2), \quad f(0) = 0, \quad k_1 \leq \frac{f(y)}{y} \leq k_2$$

If the Nyquist curve of G does not encircle or enter the circle whose diameter is defined by the points $-1/k_1$ and $-1/k_2$, then the system is input output stable (the Circle Criterion).

Literature

The Russian scientist A. M. Lyapunov published a comprehensive theory of stability in 1892. His articles have been reprinted and published as a book in connection with the centennial, Lyapunov (1992). Lyapunov functions have become a standard tool and are treated in many text books, e.g., Leigh (1983), Khalil (1996), Cook (1994), Vidyasagar (1993). All these books also treat the Circle Criterion which is a result from about 1960. There are many variants of the Circle Criterion in the literature, e.g., the "Popov criterion" and the "off-axis circle criterion".

Software

In MATLAB there is the command
   lyap which solves the Lyapunov equation.

Appendix 12A. Proofs

Proof of Lemma 12.1

Proof: The elements of e^{At} are of the form

$$p(t)e^{\lambda_i t}$$

where λ_i is an eigenvalue of A and p is a polynomial, see the proof of Theorem 3.7. We can form the matrix

$$P = \int_0^\infty e^{A^T t} Q e^{At} \, dt$$

The right hand side is a convergent integral since all eigenvalues have strictly negative real parts. Substituting this value of P into (12.8) gives

$$A^T P + PA = \int_0^\infty \left(A^T e^{A^T t} Q e^{At} + e^{A^T t} Q e^{At} A \right) dt =$$
$$\int_0^\infty \frac{d}{dt} \left(e^{A^T t} Q e^{At} \right) dt = -Q$$

We have thus found a solution of (12.8). If Q is positive definite, then

$$z^T P z = \int_0^\infty z^T e^{A^T t} Q e^{At} z \, dt > 0$$

for $z \neq 0$, since the integrand is always strictly positive. Consequently, P is positive definite, whenever Q is.

To show the opposite implication, consider the matrix

$$M(t) = e^{A^T t} P e^{At} + \int_0^t e^{A^T s} Q e^{As} ds$$

Differentiation with respect to time gives

$$\dot{M}(t) = e^{A^T t} (A^T P + PA + Q) e^{At} = 0$$

We conclude that M is constant so that $M(t) = M(0) = P$. This gives the equality

$$P = e^{A^T t} P e^{At} + \int_0^t e^{A^T s} Q e^{As} ds \tag{12.18}$$

Now let A have the eigenvalue λ (for simplicity assumed to be real) with corresponding eigenvector v. From $Av = \lambda v$ it follows that $e^{At} v = e^{\lambda t} v$. From (12.18) we then get

$$v^T P v = v^T P v e^{2\lambda t} + v^T Q v \int_0^t e^{2\lambda s} ds$$

If $\lambda \geq 0$ then the integral diverges and it follows that $v^T Q v = 0$. This implies that $(A - KQ)v = Av = \lambda v$. The non-negative eigenvalue λ is an eigenvalue of $A - KQ$ for all K, contradicting the detectability. It follows that we must have $\lambda < 0$. $\qquad\square$

Proof of Theorem 12.6

Proof: With the notation $V = x^T P x$, $f(x) = Ax + g(x)$ we get

$$V_x(x)f(x) = 2x^T P(Ax + g(x)) = x^T(PA + A^T P)x + 2x^T P g(x)$$
$$= -x^T Q x + 2x^T P g(x)$$

Let μ be the smallest eigenvalue of the matrix Q, and let m and M be the smallest and largest eigenvalues respectively of P. We then have the relation

$$m|z|^2 \leq z^T P z \leq M|z|^2, \quad \mu|z|^2 \leq z^T Q z, \quad \text{all } z \qquad (12.19)$$

The assumed properties of g imply the existence of a function $\rho(x)$ such that $\lim_{x \to 0} \rho(x) = 0$ and $|g(x)| \leq \rho(x)|x|$. We can then get the estimate

$$V_x(x)f(x) = -x^T Q x + 2x^T P g(x) \leq -|x|^2(\mu - 2M\rho(x))$$

The fact that ρ tends to zero as x tends to zero implies the existence of a constant $\nu > 0$ such that $\rho(x) < \mu/(2M)$ whenever $|x| \leq \nu$. It follows that $V_x(x)f(x) \leq 0$ for $|x| \leq \nu$. If d is chosen such that $d \leq \nu^2 m$, then

$$V(x) \leq d \Rightarrow m|x|^2 \leq x^T P x \leq \nu^2 m \Rightarrow |x| \leq \nu \Rightarrow V_x(x)f(x) \leq 0$$

$\qquad\square$

Chapter 13

Phase Plane Analysis

We have seen that the linear approximation of a system gives information about the stability of an equilibrium. Insight into other properties can also be expected. We will discuss this matter in detail for the case of two state variables only. An advantage of this special case is that the results can easily be presented graphically. A diagram where one state variable is plotted as a function of the other is called a *phase plane*. The curves in the phase plane are called *trajectories* or *orbits*. We will begin with a discussion of linear systems.

13.1   Phase Planes for Linear Systems

Consider the system

$$\dot{x} = Ax \tag{13.1}$$

We know that the eigenvalues of A play an important role for the dynamic properties of (13.1). We will see that also the eigenvectors are important.

Lemma 13.1 *Consider the system (13.1) and let λ be an eigenvalue of A with corresponding eigenvector v. Then*

$$x(t) = \alpha\, e^{\lambda t}\, v \tag{13.2}$$

(where α is an arbitrary scalar) is a solution of (13.1).

Proof: Substitution of (13.2) into (13.1) gives

$$\alpha\lambda e^{\lambda t} v$$

for the left hand side and

$$\alpha e^{\lambda t} A v = \alpha \lambda e^{\lambda t} v$$

for the right hand side.   □

The lemma is illustrated in Figure 13.1 for the case of a real eigenvalue and a two-dimensional state space. It follows that rays from the

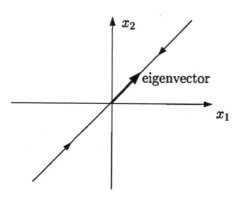

Figure 13.1: Eigenvector and solutions of a linear differential equation.

origin which are tangent to the eigenvector are also trajectories. If the eigenvalue is negative, the solution tends towards the origin otherwise away from it.

It is possible to classify singular points depending on whether the eigenvalues are complex or real and whether the real parts are positive or negative. We will concentrate on the case where A is a two by two matrix.

Two Real, Different, Negative Eigenvalues

Assume that the system (13.1) has two real eigenvalues, λ_1 and λ_2, satisfying

$$\lambda_1 < \lambda_2 < 0$$

Let v_1 and v_2 be corresponding eigenvectors. According to Lemma 13.1 there are then trajectories approaching the origin along straight lines tangent to v_1 and v_2. Since solutions along v_1 approach the origin faster than solutions along v_2, it is customary to call v_1 and v_2 the fast and slow eigenvector, respectively. Solutions which do not start on an eigenvector will be linear combinations

$$x(t) = c_1 e^{\lambda_1 t} v_1 + c_2 e^{\lambda_2 t} v_2 \tag{13.3}$$

where c_1 and c_2 are constants. When t goes to infinity the second term of (13.3) will dominate, so trajectories will approach the origin along the slow eigenvector. For t tending to minus infinity the first term will dominate, so initially trajectories will be parallel to the fast eigenvectors. The phase plane will look like Figure 13.2. An equilibrium of this type is called a *stable node*.

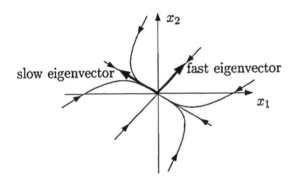

Figure 13.2: Stable node.

Two Real Distinct Positive Eigenvalues

When both eigenvalues are real, distinct and positive the trajectories look the same as in the previous case. The difference is that the exponential functions in the right hand side of (13.3) grow with time. This gives the same figure as in 13.2 but with the trajectories traveling in the opposite direction. Such an equilibrium is called an *unstable node*, see Figure 13.3.

Two Real Eigenvalues of Opposite Sign

Let the two eigenvalues have opposite signs, $\lambda_1 < 0 < \lambda_2$ and call the corresponding eigenvectors v_1 and v_2. Since λ_1 is negative, trajectories

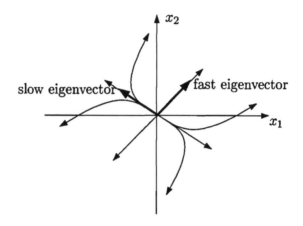

Figure 13.3: Unstable node.

along v_1 will approach the origin and it is natural to call v_1 the "stable eigenvector". Analogously, v_2 is called the "unstable eigenvector" since trajectories are directed away from the origin. A general trajectory is a linear combination as in (13.3), where one component decreases towards zero and the other one increases towards infinity. Trajectories will therefore tend to the stable eigenvector when moving backwards in time and towards the unstable one when moving forwards in time. The phase plane will then have the general appearance of Figure 13.4. This type of equilibrium is called a *saddle point*.

Multiple Eigenvalues

In the situation where the two eigenvalues of a second order system coincide, $\lambda_1 = \lambda_2 = \lambda$, there are two possibilities, depending on the number of linearly independent eigenvectors. When there is just one independent eigenvector we have again a node, but with the slow and fast eigenvector coinciding. The node can be either stable or unstable depending on the sign of λ. Figure 13.5 shows such a node. If there are two linearly independent eigenvectors, then all linear combinations of them are also eigenvectors. This means that all straight lines through the origin correspond to trajectories, see Figure 13.6. This type of equilibrium is called a *star node*.

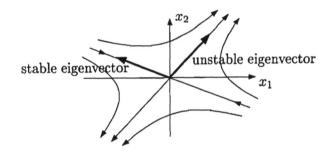

Figure 13.4: Saddle point.

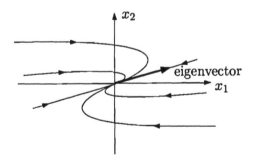

Figure 13.5: Stable node.

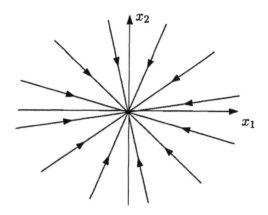

Figure 13.6: Stable star node.

Complex Eigenvalues

An example of a system with complex eigenvalues is

$$\dot{x} = \begin{bmatrix} \sigma & -\omega \\ \omega & \sigma \end{bmatrix} x \tag{13.4}$$

which has the eigenvalues $\sigma \pm i\omega$. With polar coordinates

$$x_1 = r\cos\phi, \quad x_2 = r\sin\phi$$

the system can be written

$$\dot{r} = \sigma r \tag{13.5}$$
$$\dot{\phi} = \omega \tag{13.6}$$

The trajectories become spirals since ϕ is monotone. If $\sigma < 0$ then the distance to the origin decreases all the time and the phase plane of Figure 13.7 is obtained. An equilibrium of this type is called a *stable focus*. If $\sigma > 0$, r increases and the trajectories spiral outwards from the origin. This is known as an *unstable focus*. Finally, if $\sigma = 0$ then the distance to the origin remains constant and the trajectories are circles as shown in Figure 13.8. The phase plane is called a *center*. Now consider an arbitrary second order system

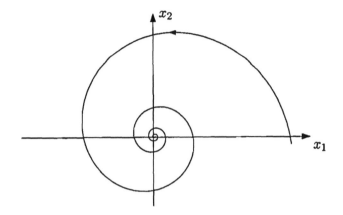

Figure 13.7: Stable focus.

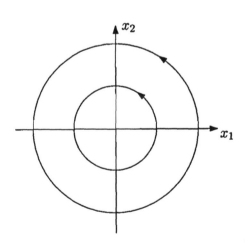

Figure 13.8: Center.

$$\dot{x} = Ax$$

By making the transformation $x = T\tilde{x}$, where T is a nonsingular matrix, one obtains the system

$$\dot{\tilde{x}} = T^{-1}AT\tilde{x} \tag{13.7}$$

If A has the complex eigenvalues $\sigma \pm i\omega$ with complex eigenvectors $v \pm iw$, then the choice

$$T = \begin{bmatrix} v & -w \end{bmatrix}$$

gives, after some calculations

$$T^{-1}AT = \begin{bmatrix} \sigma & -\omega \\ \omega & \sigma \end{bmatrix}$$

Any second order system with complex eigenvalues can thus be transformed into the form (13.4). Complex eigenvalues with negative real parts give a stable focus, while positive real parts give an unstable focus, and purely imaginary eigenvalues give rise to a center. The phase planes will usually be deformed versions of Figures 13.7 and 13.8 due to the influence of the transformation T. It is often possible to sketch approximately how the figure is deformed by checking the angles at which the trajectories cross the coordinate axes. One simply puts $x_1 = 0$ or $x_2 = 0$ into the system equations and forms the ratio $\dot{x}_2/\dot{x}_1$. See also Section 13.2 below.

13.2   Phase Planes for Nonlinear Systems

The Phase Plane Close to an Equilibrium

Let us now return to the nonlinear system. We assume that the right hand side is continuously differentiable. Disregarding the input and assuming the equilibrium to be at the origin for simplicity, we can write

$$\dot{x} = Ax + g(x), \quad |g(x)|/|x| \to 0, \quad |x| \to 0 \tag{13.8}$$

according to Theorem 11.1. It is natural to apply the linear analysis of the preceding section to the matrix A. We can expect the nonlinear system to have almost the same phase plane as the linear part close to the equilibrium, since g is very small compared to Ax there.

Let us consider the case of a system with two states having a linearization with complex eigenvalues $\sigma \pm i\omega$ where $\omega \neq 0$. The same variable transformations as in Section 13.1, will replace equations (13.5) and (13.6) by

$$\dot{r} = \sigma r + \rho_1(r, \phi), \quad |\rho_1(r, \phi)|/r \to 0, \ r \to 0$$
$$\dot{\phi} = \omega + \rho_2(r, \phi), \quad |\rho_2(r, \phi)| \to 0, \ r \to 0$$

where ρ_1 and ρ_2 are functions coming from the nonlinear part g of (11.12). If σ is nonzero then the first equation will be dominated by the linear part close to the equilibrium. The second equation shows that ϕ will be monotonically increasing or decreasing if r is small enough so that $|\rho_2| < |\omega|$. If the linear part is a focus, it follows that the nonlinear system has the same qualitative behavior close to the equilibrium. If the linear part is a center, the situation is different. The r-equation is then

$$\dot{r} = \rho_1(r, \phi)$$

Depending on the nonlinearity r, can either increase, decrease or remain constant. The nonlinear behavior can thus be similar to a stable or unstable focus or a center (or possibly something entirely different). This fact was illustrated in Examples 12.1 and 12.2.

A similar analysis can be made for other types of equilibria and leads to the following conclusions.

- If the linearized system has a *node, focus* or *saddle point*, then the nonlinear system has a phase plane with the same qualitative properties close to the equilibrium. If the linearized system has a *star node*, then the qualitative behavior remains the same for the nonlinear system if in (13.8) we have $|g(x)| \leq C|x|^{1+\delta}$ close to the origin, for some positive constants C and δ.

- If the linearized system has a *center*, then the nonlinear system can behave qualitatively like a center or focus (stable or unstable) close to the origin.

The Phase Plane Far from Equilibria

It is possible to form an idea of phase plane properties of a second order system

$$\dot{x}_1 = f_1(x_1, x_2)$$
$$\dot{x}_2 = f_2(x_1, x_2)$$

(13.9)

by forming the derivative

$$\frac{dx_2}{dx_1} = \frac{\dot{x}_2}{\dot{x}_1}$$

which gives

$$\frac{dx_2}{dx_1} = \frac{f_2(x_1, x_2)}{f_1(x_1, x_2)}$$

(13.10)

This is a differential equation where the time is eliminated. The right hand side of (13.10) gives the slope of the trajectories as each point. In particular the slope is horizontal at points where

$$f_2(x_1, x_2) = 0$$

and vertical where

$$f_1(x_1, x_2) = 0$$

By studying the slopes at interesting points, e.g., the coordinate axes, it is often possible to get a good qualitative feeling for the system behavior. From the limits

$$\lim_{x_1 \to \pm\infty} \frac{f_2(x_1, x_2)}{f_1(x_1, x_2)}, \quad \lim_{x_2 \to \pm\infty} \frac{f_2(x_1, x_2)}{f_1(x_1, x_2)}$$

it is possible to get some idea of system properties for large amplitudes. In some special cases it is also possible to solve (13.10) explicitly.

Some Examples

We finish our discussion of phase planes by showing the complete phase planes of our two examples.

Example 13.1: Phase Plane of the Generator

Consider the generator of Example 11.2, with $a = 1$, $b = 2$, and $u = 0$. The linearizations around the equilibria were computed in Example 11.5. For $x_2 = 0$, $x_1 = \nu 2\pi$, we get

$$A = \begin{bmatrix} 0 & 1 \\ -2 & -1 \end{bmatrix}$$

with the characteristic equation

$$\lambda^2 + \lambda + 2 = 0$$

The eigenvalues are complex with negative real parts, so these equilibria are stable foci. For $x_2 = 0$, $x_1 = \nu\pi$, we get instead

$$A = \begin{bmatrix} 0 & 1 \\ 2 & -1 \end{bmatrix}$$

with the eigenvalues 1 and -2. This is a saddle point with the unstable eigenvector $[1 \quad 1]^T$ ($\lambda = 1$) and the stable eigenvector $[1 \quad -2]^T$ ($\lambda = -2$). The phase plane will thus be periodic in the x_1 direction with a focus at every even multiple of π and a saddle point at every odd multiple. To get more information on the phase plane we could check the slope of trajectories. Equation (13.10) gives

$$\frac{dx_2}{dx_1} = -1 - \frac{2 \sin x_1}{x_2}$$

We see that trajectories have vertical slope when they pass the x_1 axis and that the slope tends to -1 as $|x_2| \to \infty$. The phase plane is shown in Figure 13.9.

Example 13.2: Phase Plane of the System with Friction

Let us consider the mechanical system with friction given by (11.6). We use the parameter values $K = 1$, $v_0 = 1$, $F_0 = 1.5$, and $F_1 = 1$. As we noted when discussing the equations (11.10), the system is piecewise linear. With the parameter values above we have

$$\begin{aligned} \dot{x}_1 &= x_2 \\ \dot{x}_2 &= -x_1 - fx_2 - 1 \end{aligned} \quad \text{if } x_2 > 1 \qquad (13.11a)$$

$$\begin{aligned} \dot{x}_1 &= x_2 \\ \dot{x}_2 &= -x_1 - fx_2 + 1 \end{aligned} \quad \text{if } x_2 < 1 \qquad (13.11b)$$

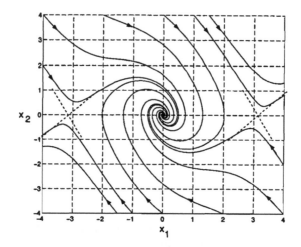

Figure 13.9: Phase plane for the generator. The directions of the stable and unstable eigenvectors are shown with dotted lines at $x_2 = 0$, $x_1 = \pm\pi$.

$$\begin{aligned}\dot{x}_1 &= 1 \\ \dot{x}_2 &= 0\end{aligned} \quad \text{if } x_2 = 1 \text{ and } |x_1 + fx_2| \leq 1.5 \qquad (13.11c)$$

(The set described by $x_2 = 1$, $x_1 + fx_2 > 1.5$ is not important. In this set we have $\dot{x}_2 < 0$, and trajectories starting there immediately enter (13.11b). Analogously the set $x_2 = 1$, $x_1 + fx_2 < 1.5$ is unimportant.)

For (13.11a) there is an equilibrium at $x_1 = -1$, $x_2 = 0$, which is outside the region where the equations are valid. For (13.11b) there is an equilibrium at $x_1 = 1$, $x_2 = 0$. The equations (13.11c) have no equilibrium. Since x_1 is increasing, solutions starting there will eventually leave the region and enter (13.11b). Both for (13.11a) and (13.11b) the linear dynamics is described by the matrix

$$\begin{bmatrix} 0 & 1 \\ -1 & -f \end{bmatrix}$$

with the characteristic equation

$$\lambda^2 + f\lambda + 1 = 0$$

For $f = 0$ the eigenvalues are imaginary and the phase plane has a center. For $0 < f < 4$ the eigenvalues are complex with negative real parts and the dynamics is described by a stable focus. In particular, we consider the case of f being a small positive number. The trajectories are then almost circular with a slow spiraling motion towards the equilibrium. For $0 < f < 4$ we get the following piecewise linear system:

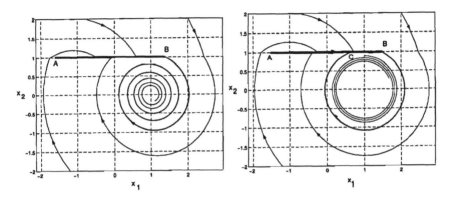

(a) Phase plane of a mechanical system with friction. The damping is $f = 0.1$.

(b) Phase plane of a mechanical system with friction. The damping is $f = 0.02$.

Figure 13.10: Phase plane of a mechanical system with friction for different values of the damping.

1. $x_2 > 1$
   The upper part of a stable focus with equilibrium ($x_1 = -1$, $x_2 = 0$) outside the region. For small values of f, the trajectories are approximately circles centered at $x_1 = -1$, $x_2 = 0$.

2. $x_2 < 1$
   A stable focus around an equilibrium at $x_1 = 1$, $x_2 = 0$.

3. $x_2 = 1$, $|x_1 + f| \leq 1.5$.
   The region consists of one single trajectory between $x_1 = -1.5 - f$ and $x_1 = 1.5 - f$. At the latter point it joins a trajectory of region 2.

For $f = 0.1$ we get the phase plane of Figure 13.10a. Region 3 is the solid line between points A and B in the figure.

If the damping term is reduced to $f = 0.02$ we get the interesting phenomenon of Figure 13.10b. We see that trajectories that leave region 3 at B in the phase plane return to region 3 at C. There is thus a closed curve in the phase plane corresponding to a periodic solution. It is clear that all trajectories arriving at the segment AB will end in this periodic motion. A solution which is a limit of trajectories in this way is called a *limit cycle*. From the geometry of the phase plane we see that the limit cycle is only formed if B is to the right of $x_1 = 1$. This situation occurs when the static friction is greater than the sliding friction.

13.3   Comments

Main Points of the Chapter

- It is possible to get a good idea of the properties of a nonlinear system close to an equilibrium by studying the eigenvalues and eigenvectors of the linear approximation. This is especially true in two dimensions where the result can be plotted (the phase plane).

- Two real distinct eigenvalues of the same sign give a tangent node, stable for negative real values, unstable for positive ones.

- Complex eigenvalues with nonzero real parts give a focus, stable for negative real parts, unstable for positive ones.

- Real eigenvalues of opposite signs give a saddle point.

- In all these cases a nonlinear system described by

$$\dot{x} = Ax + g(x), \qquad g(x)/|x| \to 0, \ x \to 0$$

  has a phase plane which essentially agrees with that of the linear part close to the origin.

- In two dimensions a good qualitative picture of the phase plane is obtained by plotting the slope of the trajectories at different points.

Literature

The classification of equilibria according to their linearizations was done by 19th century mathematicians, in particular Poincaré. An overview with many examples is given in Verhulst (1990). Solutions which do not approach any equilibrium or limit cycle but remain bounded can show so called chaotic behavior, described in e.g., Thompson & Stewart (1986). The use of mathematical results of phase plane analysis in control theory became common at the middle of the 20th century. It is treated in e.g., Slotine & Li (1991), Leigh (1983) and Cook (1994).

Software

In MATLAB trajectories can be simulated using SIMULINK. There are also the commands

trim Determination of equilibria.

linmod Linearization.

eig Determination of eigenvalues and eigenvectors.

to analyze the character of the equilibria.

Chapter 14

Oscillations and Describing Functions

We will now look at some simple nonlinear systems, in cases where the variables do not approach an equilibrium. Consider the linear feedback system of Figure 14.1. Its Nyquist curve for $K = 4$ is shown in Figure 14.2. The Nyquist criterion tells us that the system is unstable.

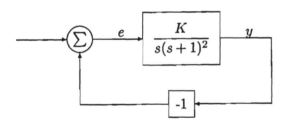

Figure 14.1: Feedback system.

We can thus expect signals that grow without limit. However, in almost all real systems the signals saturate. Figure 14.3 shows a saturation at the input. Simulating this system with $K = 4$ and the saturation level 1 gives the result of Figure 14.4. We see that there is an oscillation which grows to a certain level and then stabilizes. This is a typical behavior of nonlinear systems. Stability analysis in the nonlinear case is therefore often connected to the problem of showing existence or absence of oscillations. We will discuss methods for doing that.

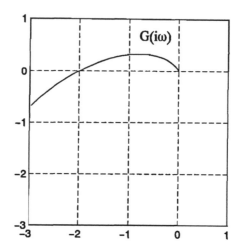

Figure 14.2: Nyquist curve for $K = 4$.

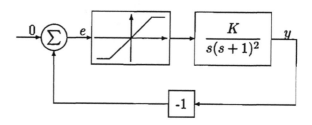

Figure 14.3: Feedback system with saturation.

14.1   The Describing Function Method

The describing function method addresses problems characterized by the block diagram of Figure 14.5. This is a special case of the block diagram shown in Figure 11.6, where r_1 and r_2 are zero. We thus assume that no external signals activate the system. The system is a feedback loop consisting of a linear part with transfer function $G(s)$ and a static nonlinearity described by the function f. Now assume that we apply the signal

$$e(t) = C \sin \omega t$$

at the input of f. At the output we then get the periodic function $w(t) = f(C \sin \omega t)$. Expanding this function into a Fourier series we

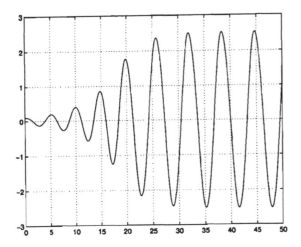

Figure 14.4: Output of a system with saturation, $K = 4$.

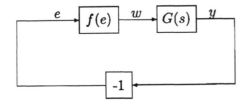

Figure 14.5: Block diagram for the describing function method.

can write w in the form

$$w(t) = f(C \sin \omega t) = f_0(C) + A(C) \sin(\omega t + \phi(C))$$
$$+ A_2(C) \sin(2\omega t + \phi_2(C)) + A_3(C) \sin(3\omega t + \phi_3(C)) + \cdots$$
$$(14.1)$$

The coefficients $f_0(C)$, $A(C)$, $\phi(C)$, etc. in this expression depend on C but not on ω. This is because we consider a static nonlinearity f. The shape of the output signal within one period is then independent of the frequency, and it is possible to scale the time so that the period is normalized to 2π. This is used in the Fourier coefficient formulas (14.4,14.5) presented below.

It is now easy to calculate how each term is affected by the linear system. For the fundamental component (with the angular frequency ω) the gain is given by $|G(i\omega)|$ and the change in phase by $\psi(\omega) = \arg G(i\omega)$. We now make the following assumptions.

1. $f_0(C) = 0$. This is the case e.g., if f has an odd symmetry around the origin ($f(-x) = -f(x)$).

2. The linear system acts as a low pass filter so that the components with angular frequencies 2ω, 3ω and so on are attenuated much more that the fundamental component with angular frequency ω.

We then get

$$y(t) \approx A(C)\,|G(i\omega)|\,\sin(\omega t + \phi(C) + \psi(\omega))$$

We see that there (approximately) will be a self-sustained oscillation if this signal, after passing the -1-block, agrees with the signal we assumed at the input, i.e., $C\sin\omega t$. We then get the following system of equations

$$A(C)\,|G(i\omega)| = C \tag{14.2a}$$
$$\phi(C) + \psi(\omega) = \pi + \nu\,2\pi, \tag{14.2b}$$

where ν is an integer. This is a system of nonlinear equations in the two unknowns C and ω. Solving it gives the amplitude and frequency of an oscillation. If there is no solution, a likely reason is that there is no oscillation in the system.

We define the complex number

$$Y_f(C) = \frac{A(C)e^{i\phi(C)}}{C} \tag{14.3}$$

which we call the *describing function* of the nonlinearity. Introducing the Fourier coefficients

$$a(C) = \frac{1}{\pi}\int_0^{2\pi} f(C\sin\alpha)\cos\alpha d\alpha = A(C)\sin\phi(C) \tag{14.4}$$

$$b(C) = \frac{1}{\pi}\int_0^{2\pi} f(C\sin\alpha)\sin\alpha d\alpha = A(C)\cos\phi(C) \tag{14.5}$$

we can also write

$$Y_f(C) = \frac{b(C) + ia(C)}{C} \tag{14.6}$$

The absolute value of Y_f is the ratio of the amplitudes of the fundamental component after and before the nonlinearity. We can view Y_f as an *amplitude dependent gain*. Writing G in polar form

$$G(i\omega) = |G(i\omega)|\, e^{i\psi(\omega)}$$

we can restate the equations (14.2a) and (14.2b) in the form

$$Y_f(C)G(i\omega) = -1 \qquad (14.7)$$

We see that the calculation of an oscillation consists of two steps.

1. Calculate $Y_f(C)$ from (14.6).

2. Solve (14.7).

Example 14.1: Ideal Relay

Consider the nonlinearity

$$f(e) = \begin{cases} 1 & \text{if } e > 0 \\ -1 & \text{if } e < 0 \end{cases} \qquad (14.8)$$

i.e., an ideal relay. Using (14.5) gives

$$b(C) = \frac{1}{\pi} \int_0^\pi \sin\alpha\, d\alpha + \frac{1}{\pi} \int_\pi^{2\pi} (-1)\sin\alpha\, d\alpha = \frac{4}{\pi}$$

while (14.4) gives

$$a(C) = \frac{1}{\pi} \int_0^\pi \cos\alpha\, d\alpha + \frac{1}{\pi} \int_\pi^{2\pi} (-1)\cos\alpha\, d\alpha = 0$$

The describing function is thus real and given by

$$Y_f(C) = \frac{4}{\pi C} \qquad (14.9)$$

The describing function calculated in this example is real. From (14.4) it follows that this is always the case for single-valued nonlinearities. Nonlinearities with complex valued describing functions are, e.g., backlash and hysteresis, see Examples 14.7 and 14.8 in Appendix 14A.

14.2   Computing Amplitude and Frequency of Oscillations

Direct Algebraic Solution

In simple cases it is possible to solve (14.7) directly. We illustrate this fact in an example. Consider the system of Figure 14.6 where the nonlinearity f is given by (14.8). From Example 14.1 we have

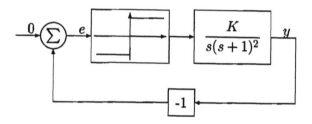

Figure 14.6: Relay system.

$$Y_f(C) = \frac{4}{\pi C}$$

Furthermore

$$G(i\omega) = \frac{K}{i\omega(1+i\omega)^2} = \frac{-2K\omega - iK(1-\omega^2)}{\omega(1+\omega^2)^2}$$

Since Y_f is real and positive, (14.7) can only be satisfied if G is real and negative. This implies $\omega = 1$. Since

$$G(i1) = -\frac{K}{2}$$

we get

$$-\frac{K}{2}\frac{4}{\pi C} = -1$$

showing that $C = 2K/\pi$. For $K = 1$ we predict an oscillation with amplitude 0.64 and angular frequency 1. Figure 14.7 shows the result of a simulation that agrees well with these values.

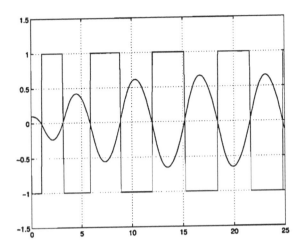

Figure 14.7: Oscillation in a relay system. The square wave is the output of the relay while the sine-like curve is y.

Graphical Interpretation of the Solution

If Equation (14.7) is rewritten as

$$G(i\omega) = -\frac{1}{Y_f(C)}$$

we see that the solution can be interpreted as the intersection of two curves in the complex plane, namely the Nyquist curve $G(i\omega)$, plotted as a function of ω and $-1/Y_f(C)$, plotted as a function of C.

Consider the example solved algebraically in the previous section. In Figure 14.8 the Nyquist curve is shown together with $-1/Y_f(C)$. We see that an intersection takes place at $\omega = 1$ and $C \approx 0.6$.

Stability of Oscillations

We have seen that an intersection between $-1/Y_f(C)$ and the Nyquist curve indicates the presence of an oscillation in the system. It is of interest to know if the oscillation will continue after a small disturbance. To get a feeling for this, one can investigate the effect of trying to increase or decrease the amplitude of the oscillation. Consider the situation shown in Figure 14.9a. Let C_o be the C-value corresponding to the intersection of the curves. Now consider an amplitude with $C < C_o$, i.e., a point on $-1/Y_f(C)$ to the right of

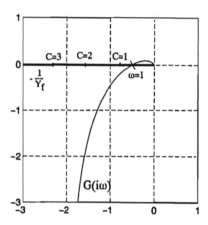

Figure 14.8: Nyquist curve and describing function.

the intersection. If we replace the nonlinearity with a constant gain equal to $Y_f(C)$, the system will be unstable according to the Nyquist criterion. It is then reasonable to assume that an amplitude $C < C_o$ in the nonlinear system will tend to grow. If instead $C > C_o$, then the corresponding point on the $-1/Y_f(C)$-curve is to the left of the intersection. Replacing the nonlinearity with a constant gain would give a stable system according to the Nyquist criterion. The amplitude should thus decrease. Intuitively we conclude that the amplitude of an oscillation would tend to return to the value C_o when perturbed. The situation depicted in Figure 14.9a therefore indicates an oscillation with stable amplitude. Note that the relay example in the previous section was of this type.

If the curves are instead related as shown in Figure 14.9b the result of the reasoning will be reversed. A value $C > C_o$ should give an increasing amplitude while $C < C_o$ should give a decreasing one. The amplitude would then move away from the value C_o and we expect an unstable oscillation. It also seems likely that disturbances below a certain threshold give rise to oscillations that vanish, while large disturbances could start oscillations that grow without bound.

This reasoning about amplitudes of oscillation can also be extended to the case when the curves G and $1/Y_f$ do not intersect, and we consequently expect no oscillation. There are two possibilities. Either $-1/Y_f(C)$ lies completely outside the Nyquist curve, as in

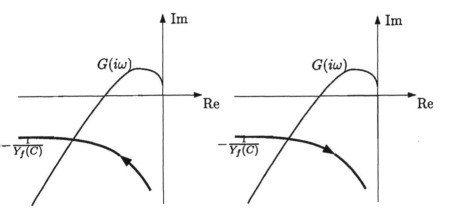

(a) Indication of a stable oscillation. The arrow is in the direction of increasing C-values.

(b) Indication of an unstable oscillation. The arrow is in the direction of increasing C-values.

Figure 14.9: Stability of oscillations.

Figure 14.10a or completely inside as in Figure 14.10b. In the former case we expect oscillations that vanish, in analogy with linear stability theory, in the latter case signal amplitudes that increase without bound.

Example 14.2: Control System with Saturation

In the introduction to the chapter we studied the system in Figure 14.1, which has a saturation nonlinearity. The describing function of a saturation is given in Appendix 14A. The Nyquist curve of the system was shown in Figure 14.2. Introducing $-1/Y_f(C)$ into this diagram gives Figure 14.11. The diagram predicts an oscillation with amplitude approximately 2.3 and with angular frequency $\omega = 1$. This agrees quite well with the time response of Figure 14.4. We also note that Figure 14.11 predicts an oscillation with stable amplitude. This is confirmed by the time response, where the amplitude of the oscillation grows and then stabilizes.

Approximations

It is important to remember that the describing function approach as described here is approximative. The results must therefore be interpreted with caution. In cases where the linear part gives

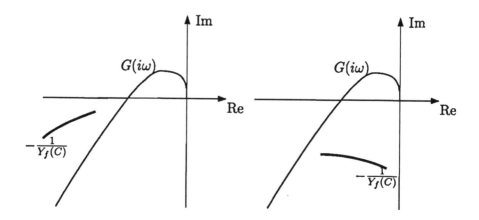

(a) Indication of a vanishing oscillation.

(b) Indication of an oscillation whose amplitude grows without bound.

Figure 14.10: Two situations where $-1/Y_f$ does not intersect the Nyquist curve.

insufficient low pass filtering there might be a large deviation between predicted and actual values of amplitude and frequency. In some cases there will also be indication of an oscillation which actually does not take place. Conversely there are cases where an oscillation occurs although it is not predicted by describing function theory.

It is also possible to estimate the errors made by the approximations of describing function analysis. In some cases one can then make mathematically precise statements about oscillations. This approach is described in Khalil (1996).

14.3   Comments

Main Points of the Chapter

- The describing function of a static nonlinearity can be interpreted as an amplitude dependent gain.

- If a linear system $G(s)$ and a static nonlinearity are connected in

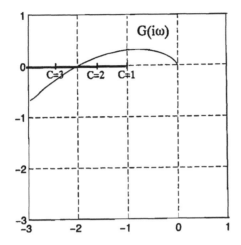

Figure 14.11: Nyquist curve and describing function for saturation

a negative feedback loop, then the oscillation condition is

$$Y_f(C)G(i\omega) = -1$$

where Y_f is the describing function of the nonlinearity, C is the amplitude and ω the angular frequency of the oscillation.

- By plotting $-1/Y_f(C)$ and $G(i\omega)$ in the complex plane, amplitude and frequency of an oscillation can be determined. It is also possible to investigate the stability of oscillations.

Literature

The idea of approximating a periodic solution by a sinusoid and then performing Fourier analysis has old roots in mathematics and mathematical physics. In control theory, describing functions have found widespread use in applications. The methodology is described in detail in Atherton (1982), which also has many extensions, i.e., asymmetric oscillations, external inputs, harmonics, and several frequencies. The book Khalil (1996) also contains a thorough discussion about describing functions and the mathematical conditions for actually proving the existence of an oscillation.

Software

In MATLAB the command nyquist, which draws a Nyquist diagram, is also useful in describing function calculations.

Appendix 14A: Some Describing Functions

Example 14.3: Cubic Gain

The cubic nonlinearity

$$f(u) = u^3$$

has the describing function

$$Y_f(C) = \frac{3C^2}{4}$$

Example 14.4: Relay with Dead Zone

A relay with dead zone is described by the relation

$$f(u) = \begin{cases} 1 & u > D \\ 0 & |u| \leq D \\ -1 & u < -D \end{cases}$$

and has the graph

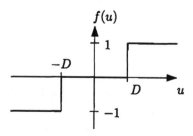

The describing function is

$$Y_f(C) = \frac{4}{\pi C}\sqrt{1 - D^2/C^2}, \quad C > D$$

Example 14.5: Saturation

A saturation is described by the relation

$$f(u) = \begin{cases} 1 & u > 1 \\ u & |u| \le 1 \\ -1 & u < -1 \end{cases}$$

and has the graph

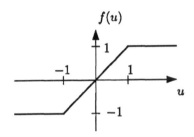

The describing function is

$$Y_f(C) = \begin{cases} \frac{2}{\pi}(\arcsin\frac{1}{C} + \frac{1}{C}\sqrt{1 - C^{-2}}) & C > 1 \\ 1 & C \le 1 \end{cases}$$

Example 14.6: Dead Zone

A dead zone is described by the graph

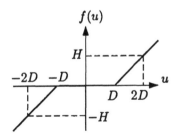

Its describing function is

$$Y_f(C) = \begin{cases} \frac{H}{D} - \frac{2H}{\pi D}\left(\arcsin(D/C) + \frac{D}{C}\sqrt{1 - (\frac{D}{C})^2}\right) & C \ge D \\ 0 & C < D \end{cases}$$

Example 14.7:   Relay with Hysteresis

A relay with hysteresis is described by the graph

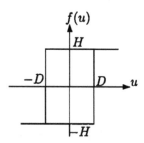

For u-values between $-D$ and D the value of $f(u)$ is not uniquely defined but is given by the following rule. If u has been greater than D then the value of $f(u)$ remains at $f(u) = H$ until u becomes less than $-D$, when it is changed to $-H$. If u has been less than $-D$ the value $-H$ remains until u becomes larger than D. The describing function is well defined if $C \geq D$, and is then given by

$$\text{Re}Y_f(C) = \frac{4H}{\pi C}\sqrt{1 - D^2/C^2}$$
$$\text{Im}Y_f(C) = -\frac{4D}{\pi C^2}$$

Example 14.8:   Backlash

A backlash is described by the graph

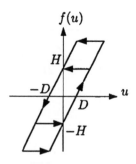

If u is increasing as a function of time, then $f(u)$ is given by the sloping line on the right hand side, if u is decreasing by left hand one. If u changes from being increasing to being decreasing or vice versa, then $f(u)$ is constant during the transition. This nonlinearity is an idealized description of what happens in, e.g., a pair of gears. If $C \geq D$ then the describing function is well defined and given by

$$\mathrm{Re}Y_f(C) = \frac{H}{\pi D}\left(\frac{\pi}{2} + \arcsin(1 - \frac{2D}{C}) + 2(1 - \frac{2D}{C})\sqrt{\frac{D}{C}(1 - \frac{D}{C})}\right)$$

$$\mathrm{Im}Y_f(C) = -\frac{4H}{\pi C}(1 - \frac{D}{C})$$

Chapter 15

Controller Synthesis for Nonlinear Systems

We have discussed how to describe nonlinear systems, how to investigate their stability, how they behave near equilibria and how oscillations can be detected. Now we will discuss controller structures and design methods.

15.1 Linear Design and Nonlinear Verification

In Theorem 12.1 we saw that stability of the linear approximation normally results in local stability for a nonlinear system. In Chapter 13 we have also seen that nonlinear systems often are well approximated by their linearization close to an equilibrium. Since controllers often have the task of keeping the system variables close to an equilibrium, it is reasonable to base the design on a model which is a linear approximation. This model is typically computed according to Theorem 11.1. The controller can then be synthesized using one of the methods described in Chapters 8–10. It is natural to pay special attention to sensitivity and robustness aspects, since we know that the model neglects nonlinear effects.

After designing a controller it should be evaluated using a nonlinear model. Often this is done using simulation. It is then important to remember what was said at the introduction of Chapter 11: nonlinear systems do not satisfy the superposition principle. It is thus necessary to perform the simulations with different combinations of initial states, reference signals, and disturbance signals and different magnitudes of the variables. One might also try using methods from Chapter 12 to investigate stability or use the describing function method of Chapter 14 to detect oscillations. Of course, the controller must eventually be evaluated using the real physical system.

15.2   Nonlinear Internal Model Control

The principles that were discussed in Section 8.3 can also be used for nonlinear systems. As an example, consider Figure 15.1, where a saturation is followed by a first order system. The IMC principle then

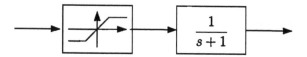

Figure 15.1: First order system with input saturation.

gives the structure of Figure 15.2. When discussing IMC for linear

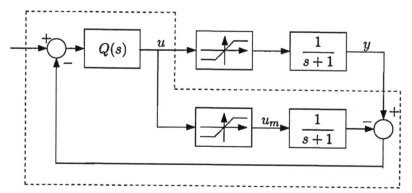

Figure 15.2: Controller structure for IMC of first order system with saturation. The controller is the subsystem within the dashed boundary.

systems in Section 8.3 we said that a natural choice of Q often is

$$Q = \frac{1}{(\lambda s + 1)^n} \cdot S^{-1}$$

where S is the system to be controlled. In our case there is no exact inverse since the saturation is not invertible. We can then, e.g., take the inverse of only the linear system. Choosing $n = 1$ we get

$$Q = \frac{s+1}{\lambda s + 1}$$

Assuming the reference to be zero, the controller is described by

$$u = -\frac{s+1}{\lambda s + 1}y + \frac{1}{\lambda s + 1}u_m$$

If $|u| < 1$, then $u = u_m$, and we get

$$\left(1 - \frac{1}{\lambda s + 1}\right)u = -\frac{s+1}{\lambda s + 1}y$$

After simplification this gives

$$u = -\frac{s+1}{\lambda s}y$$

i.e.,, a PI-controller with gain $1/\lambda$. This is precisely the result obtained using linear IMC. If $|u| > 1$ then we get instead

$$u = -\frac{s+1}{\lambda s + 1}y \pm \frac{1}{\lambda s + 1}$$

In this expression there is no longer any integration. We have arrived at a PI-controller whose integral part is not active during those time intervals, when the input to the controlled system is saturated. This is normally an advantage. A weakness of linear PI-controllers is precisely the fact that the integral part can reach very high levels during control signal saturation. As an example, we can consider the problem of controlling the system with an output disturbance v. We choose $\lambda = 0.2$ and compare our IMC-controller to a PI-controller having the same parameters, i.e., the controller

$$5\frac{s+1}{s}$$

Figure 15.3a shows the disturbance v and the outputs of the linear PI-controller and the IMC control. We see that the PI-controller produces a fairly large undershoot, which is avoided by the IMC controller. The explanation, with the control signals, is shown in Figure 15.3b. The control signal of the PI-controller has large negative values because the integral part is increased to large values during saturation. It then takes a long time for the controller variables to return to reasonable values.

The phenomenon of an integrator getting large output values during control signal saturation is often called *integrator windup* or *reset windup*. The IMC principle thus gives a PI-controller having "anti-windup", i.e., a built-in protection against integrator windup. There are also many other ways of protecting PI-controllers from windup.

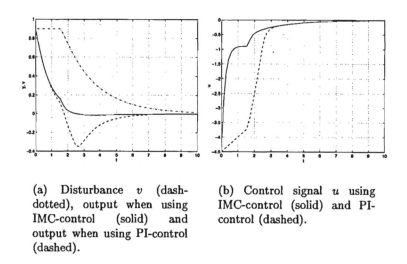

(a) Disturbance v (dash-dotted), output when using IMC-control (solid) and output when using PI-control (dashed).

(b) Control signal u using IMC-control (solid) and PI-control (dashed).

Figure 15.3: Comparison of PI-control and IMC when the control signal saturates.

15.3 Parametric Optimization

Parametric optimization is a very useful method to determine parameters in controllers, for both linear and nonlinear systems. In the nonlinear case the typical situation is as follows. A system

$$\dot{x} = f(x, u), \quad y = h(x)$$

is governed by a controller, which is also a dynamic system

$$\dot{z} = m(z, y, r; \eta), \quad u = k(z, y, r; \eta)$$

with state z, reference signal r and an adjustable parameter vector η. The system performance is described by a criterion of the form

$$V = \int_0^{t_f} L(x, z, r; \eta)\, dt$$

that we want to minimize. Normally it is not possible to calculate explicitly how V depends on η. It is, however, possible to calculate V numerically for a given value of η by simulating the system and its controller. If the functions f, h, m, k and L are regular enough it is also possible to calculate derivatives of V. The minimization of V can thus be performed by a numerical algorithm for nonlinear optimization. There is commercial software available to do this in, e.g., MATLAB. A difficulty is the absence of scaling invariance for nonlinear systems. If the optimization is done, e.g., for a step response, then the optimal controller will, in general, be different for different step amplitudes. To take this fact into account it might be necessary to calculate V as a weighting of the responses for several different step amplitudes.

Example 15.1:   DC Motor with Saturation

Consider once again the DC motor of Example 11.1, Figure 11.2. The control bound is assumed to be $|u| \leq 1$. Let us write the regulator in the form

$$K \frac{s+b}{s+5b}$$

(i.e., a lead compensator) and try to find suitable values of K and b. We use the criterion

$$V = \int_0^1 (|e(t)| + |u(t)|)\, dt$$

where e is the control error and u is the signal after the saturation. We consider the case of a step with amplitude 2. In Figure 15.4a the criterion function V is shown as a function of b and K. The figure has been generated by simulating the system for a large number of values of b and K. The figure shows that the criterion is minimized by b-values close to 1 (a more detailed plot indicates $b = 1.1$). As a function of K the criterion is rather flat close to the optimum. In Figure 15.4b the criterion is shown as a function of K, for $b = 1.1$. The optimum is obtained for a $K \approx 8$. Figure 15.4c shows the step response for the optimal controller setting $b = 1.1$, $K = 8$.

Example 15.2:   The Linear-Quadratic Case

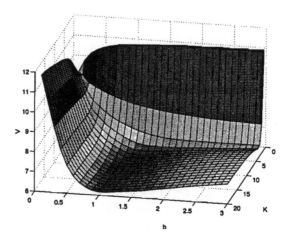

(a) The criterion as a function of b and K.

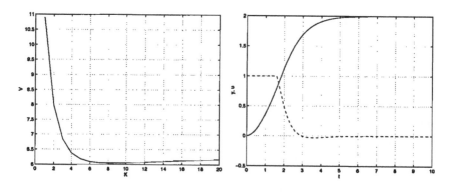

(b) The criterion for $b = 1.1$, as a function of K.

(c) Step response using the optimal controller ($b = 1.1$, $K = 8$). Solid curve: y, dashed curve: u.

Figure 15.4: Criterion and step response using parametric optimization of lead compensator for DC motor with saturation.

If the system is linear and L is a quadratic function, then the criterion V will consist of second moments of various signals, depending on η, i.e., integrals of the form

$$I_n = \frac{1}{2\pi} \int_{-\infty}^{\infty} \left| \frac{B_n(i\omega, \eta)}{A_n(i\omega, \eta)} \right|^2 d\omega$$

with

$$A_n(s, \eta) = s^n + a_{n-1}s^{n-1} + \ldots + a_0$$
$$B_n(s, \eta) = b_{n-1}s^{n-1} + \ldots + b_0 \tag{15.1}$$

where the coefficients may depend on η in an arbitrary manner. These integrals can be efficiently calculated using a state space representation of B/A and Theorem 5.3. If the covariance matrix of the state Π_x is known, then the norm of Cx equals $C\Pi_x C^T$ according to (5.9). In MATLAB the computation can be done as follows: We first make a state space realization:

```
>> num = [bn1 bn2 ... b1 b0];
>> den = [1 an1 an2 ... a1 a0];
>> [A,B,C] = tf2ss(num,den);
```

We then solve (5.62) using

```
PI = lyap(A,B*B');
```

Finally we get the integral by

```
>> J = C*PI*C';
```

The discrete time case is completely analogous. The only difference is that lyap is replaced by dlyap.

15.4 Other Approaches

Model Predictive Control

A natural framework for controller design is the predictive power of a model of the controlled object. Using such a model, the influence of controller action on the interesting variables can be predicted. The control signal can be chosen so that a given criterion is minimized, subject to physically motivated bounds on the signals. There are several methods based on this philosophy, such as MPC (model predictive control), GPC (generalized predictive control) and DMC (dynamic matrix control). We will present them in Chapter 16.

Exact Linearization

A useful method of handling nonlinearities is exact linearization. The basic idea is to use an inner loop which exactly compensates the nonlinearities. In the outer loop, the system then becomes linear and a linear design is possible. The difficulty of compensating the nonlinearity varies strongly with the application. In some cases it is trivial, in other cases extensive calculations might be necessary to decide if it is even possible to compensate exactly. We will discuss methods of exact linearization in Chapter 17.

Optimal Control

In Chapter 9 we saw controllers for linear systems designed by optimization of a quadratic criterion. For nonlinear systems it is natural to allow more general criteria, as well as explicit bounds on the control signals. The difference from parametric optimization, as described in Section 15.3, is that no parameterized structure is assumed. Instead, the optimum among all possible controls is sought. The control can be represented either as an open loop signal or as a feedback from the state. This methodology is described in Chapter 18.

15.5   State Feedback and Observer

Several of the methods we have mentioned lead to a state feedback solution, e.g., exact linearization and optimal control. If only an output signal is measured it is natural to use the same method as in the linear case, i.e., feedback from an observer. The structure is, in principle, the same as in the linear case, see Section 8.4, but the feedback and the observer contain nonlinear elements. Let us assume that the system to be controlled is

$$\dot{x} = f(x, u), \quad y = h(x) \tag{15.2}$$

and that we have calculated a control law

$$u = \ell(x) \tag{15.3}$$

such that the closed loop system

$$\dot{x} = f(x, \ell(x))$$

has the properties we want. In analogy with the linear case, we would like to construct an observer, i.e., a dynamical system with u and y as inputs and a state $\hat{x}$, with the property that the *observer error*

$$\tilde{x} = x - \hat{x}$$

tends to zero (in the absence of disturbances and model error). There are several approaches to the construction of such an observer.

Nonlinear Filter

If the disturbances can be described as stochastic processes it is natural to view the calculation of $\hat{x}$ as a filtering problem. The filter is constructed to minimize some criterion, like the expectation of the squared prediction error. For linear systems with Gaussian disturbances this approach leads to the Kalman filter. For nonlinear systems state variables and outputs usually are not Gaussian even if the disturbances are. Then it is not enough to calculate only means and covariances, as in the Kalman filter. Instead, it is necessary to compute the conditional expectation of the state, given the measurements. Mathematically, the problem can be solved, but the solution leads to a partial differential equation and has a complexity which is too high to be realistic in on line applications. An additional complication is that the combination of an optimal filter and an optimal state feedback usually does not give the overall optimal solution. This is in contrast to the linear case where the overall optimal solution can be separated into an optimal state feedback and a Kalman filter (the Separation Theorem, see Theorem 9.1).

Some Different Approaches to Observers

The simplest observer is the model itself:

$$\dot{\hat{x}} = f(\hat{x}, u) \tag{15.4}$$

If $x(0) = \hat{x}(0)$ we have, in the absence of modeling errors and disturbances, that $x(t) = \hat{x}(t)$ for all positive t. In favorable situations f can be such that the estimation error goes to zero even if $x(0) \neq \hat{x}(0)$. Normally, however, one has to make some correction based on the measured output. A natural possibility is to make an additive correction in analogy with the linear case.

$$\dot{\hat{x}} = f(\hat{x}, u) + K(y - h(\hat{x})) \tag{15.5}$$

There are different strategies for choosing K. One possibility is to compute the linearization of the system around an equilibrium and then take a K which gives good dynamics for an observer based on the linearization. This will give a constant K which guarantees the observer error to tend to zero close to the equilibrium. Another possibility is to use the present value of $\hat{x}$ to make a linearization and then calculate a Kalman filter based on that linearization. If this is done at every time instant (in practice at every sampling time) the result is an *extended Kalman filter*. A complication is that it is normally not possible to use the formulas for the Kalman filter which were presented in Chapter 5. This is because these formulas assume time invariance and stationarity. Since the linearization point given by $\hat{x}$ varies, it is necessary to use the Kalman filter equations for the time-varying case. They are found, e.g., in Anderson & Moore (1979).

For systems with the special structure

$$\dot{x} = f(x_1, u) + Ax, \quad y = x_1 \tag{15.6}$$

where x_1 is the first component of x, the following observer structure can be used

$$\dot{\hat{x}} = f(y, u) + A\hat{x} + K(y - C\hat{x}) \tag{15.7}$$

where $C = [1 \; 0 \; \dots \; 0]$. The observer error $\tilde{x} = x - \hat{x}$ then satisfies the equation

$$\dot{\tilde{x}} = (A - KC)\tilde{x} \tag{15.8}$$

and if the pair A, C is observable, K can be chosen so that the observer error tends to zero exponentially. The observer is thus globally convergent (in the absence of modeling errors and disturbances). The structure (15.6) is of course very special. However, for systems not having this structure it is sometimes possible to find a nonlinear transformation of the state and the output which leads to a description of the form (15.6). The ideas are related to exact linearization which will be described in Chapter 17. Mathematically, it is however fairly complicated and we refrain from a detailed discussion here.

Some Properties and Problems

One problem using nonlinear observers is to guarantee stability. Locally, this can often be done using Theorem 12.1 for a linearization of the observer. With the exception of the methodology of (15.6)–(15.8) above, it is usually not possible to guarantee convergence from arbitrary initial conditions. In some cases it is possible to show such stability with the help of Lyapunov functions.

It is also noteworthy that there is no guarantee that a good state feedback combined with a good observer will give a good overall result (essentially independently of the interpretation of "good"). We noted above that the principle of separation is not valid for nonlinear systems. As far as stability is concerned, it is possible to construct examples where a state feedback stabilizes globally, but where the global stability is destroyed by an observer, even if this observer has arbitrarily fast exponential convergence. However, the following local result holds. Let the controlled system be given by (15.2) with a linearization around the origin according to Theorem 11.1, given by

$$
\begin{aligned}
&\dot{x} = Ax + Bu, \quad y = Cx \\
&A = f_x(0,0), \quad B = f_u(0,0), \quad C = h_x(0)
\end{aligned}
\tag{15.9}
$$

The closed loop system with an observer according to (15.5) is then given by

$$
\dot{x} = f(x, u)
\tag{15.10a}
$$
$$
y = h(x)
\tag{15.10b}
$$
$$
\dot{\hat{x}} = f(\hat{x}, u) + K(y - h(\hat{x}))
\tag{15.10c}
$$
$$
u = -\ell(\hat{x})
\tag{15.10d}
$$

Setting $L = \ell_x(0)$, the linearization of (15.10) is

$$
\frac{d}{dt}\begin{bmatrix} x \\ \hat{x} \end{bmatrix} = \begin{bmatrix} A & -BL \\ KC & A - BL - KC \end{bmatrix}\begin{bmatrix} x \\ \hat{x} \end{bmatrix}
\tag{15.11}
$$

This is the same result as we would obtain if we made a linear observer design directly for (15.9). In the same way as we did in the linear case, we can conclude that the poles of (15.11) are the poles of $A - BL$ and $A - KC$. If these poles lie in the left half plane we can use Theorem 12.1 to draw the conclusion that the combination of state feedback and observer (15.10) has the origin as an asymptotically stable equilibrium. Furthermore, using the reasoning of Chapter 13 we can draw the conclusion that the properties of (15.10) close to the origin are well described by (15.11).

15.6   Comments

Main Points of the Chapter

- A common methodology is to use linear methods for controller design, but to verify the resulting design with a nonlinear model.

- There are a number of nonlinear design methods

  - Nonlinear IMC.
  - Parametric optimization.
  - Model predictive control.
  - Exact linearization.
  - Optimal control.

- Several of the nonlinear design methods lead to a configuration with a nonlinear feedback and a nonlinear observer.

Literature

The literature on MPC, exact linearization and optimal control is described in the relevant chapters. General overviews of nonlinear control design are given in Leigh (1983), Slotine & Li (1991), Atherton (1982) and Cook (1994).

Software

For parametric optimization MATLAB has the NONLINEAR CONTROL DESIGN BLOCKSET. Software for MPC, exact linearization and optimal control is described in the relevant chapters.

Chapter 16

Model Predictive Control

16.1 Basic Idea: Predict the Output

An important property of a model is its *predictive power*, i.e., it gives us the possibility to predict future values of interesting variables. A natural use of a model for control design is to calculate expected future values of the controlled variables as a function of possible control actions. With this knowledge, it is possible to choose a control action which is the best one according to some criterion. More formally we can proceed as follows

1. At time t compute or predict a number of future outputs $\hat{y}(t+k|t), k = 1, \ldots, M$. They will in general depend on future inputs $u(t+j), j = 0, 1, \ldots, N$ and on measurements known at time t.

2. Choose a criterion based on these variables and optimize it with respect to $u(t+j), j = 0, 1, \ldots, N$.

3. Apply $u(t)$ to the physical plant.

4. Wait for the next sampling instant $t+1$ and go to 1.

This is a very general and flexible method. It is of great value that it is easy to include in the criterion realistic constraints on the magnitude and rate of change of the control variables. The method is called *model predictive control, (MPC)* and we will discuss various aspects of it in this chapter.

16.2 k-step Prediction for Linear Systems

Usually, discrete time or sampled data models are used in MPC. There are several ways of calculating the predicted future outputs, depending on approximations and noise models used. We first consider linear system dynamics.

The most comprehensive approach is to use a state space model with a realistic noise model according to Chapter 5:

$$x(t+1) = Ax(t) + Bu(t) + Nv_1(t) \tag{16.1a}$$
$$y(t) = Cx(t) + v_2(t) \tag{16.1b}$$

The state is estimated using an observer or a Kalman filter according to Section 5.8 and we use $\hat{x}(t) = \hat{x}(t|t)$ to denote the estimate of the state at t, based on $y(k), k \leq t$ and $u(k), k < t$. Since v_1 and v_2 are unpredictable, we get the estimated future states from

$$\hat{x}(t+k+1|t) = A\hat{x}(t+k|t) + Bu(t+k), \quad k = 0, 1, \ldots \tag{16.2}$$

and the predicted outputs are

$$\hat{y}(t+k|t) = C\hat{x}(t+k|t) \tag{16.3}$$

These equations show that $\hat{y}(t+k|t)$ is a linear function of $\hat{x}(t|t)$ and $u(t+j), j = 0, \ldots, k-1$. In principle, we can write

$$Y_t = D_m U_t + D_x \hat{x}(t|t) \tag{16.4a}$$

where D_m and D_x are known matrices and where

$$Y_t = \begin{bmatrix} \hat{y}(t+M|t) \\ \vdots \\ \hat{y}(t+1|t) \end{bmatrix} \tag{16.4b}$$

$$U_t = \begin{bmatrix} u(t+N) \\ \vdots \\ u(t) \end{bmatrix} \tag{16.4c}$$

Note that

- If $N < M-1$ (which is common) we must have a rule for choosing $u(N+1), \ldots, u(M-1)$. The most common one is to take the control signal equal to the last computed value $u(N)$.

- Often (16.4) is written in terms of a U_t based on the differences $\Delta u(t) = u(t) - u(t-1)$.

16.3   The Criterion and the Controller

Usually a quadratic criterion is chosen

$$V(U_t) = Y_t^T Q_y Y_t + U_t^T Q_u U_t \tag{16.5}$$

Since Y_t is a linear (affine) function of U_t, this criterion can easily be minimized analytically, and we get a solution of the form

$$u(t) = L\hat{x}(t|t) \tag{16.6}$$

for the last component of U_t. This is a linear controller.

Optimization with control constraints

The great advantage of the MPC approach does not lie in the solution of (16.6) – for large values of M and N this is more easily done using the techniques of Chapter 9. The point is instead that it is easy to handle constraints of the form

$$|u(t)| \leq C_u \quad \text{and} \quad |y(t)| \leq C_y$$

which are almost always present in applications. We have seen control constraints in Examples 1.2, 11.1, 14.2 and 15.1. The problem of minimizing the quadratic criterion (16.5), $V(U_t)$ under a linear constraint

$$R_u U_t \leq C_t \tag{16.7}$$

has no explicit solution, but it is a convex quadratic programming (QP) problem, for which very efficient numerical methods exist. This problem has to be solved on-line at each sample, but for normal problem sizes and sampling speeds this is no problem. In chemical process control, where this approach is common, the sampling interval can be on the order of minutes. Note that the result is a *nonlinear controller* despite the fact that the model (16.1a) is linear, since we take the constraints (16.7) into account.

Choosing the Horizons M and N

The output horizon M is usually chosen to cover a typical settling time for the system. We want to look past the transient of the system response to the input. The control signal horizon N is usually chosen to be significantly shorter – it determines the size of the optimization problem.

Example 16.1: A DC Motor

Consider the DC motor of Example 10.1. With the sampling interval $T = 0.15$ the sampled state space description is according to (4.16) (Note the factor 20 in the numerator in Example 10.1.)

$$x(t + T) = Ax(t) + Bu(t)$$
$$y(t) = Cx(t)$$

$$A = \begin{bmatrix} 1 & 0.1393 \\ 0 & 0.8607 \end{bmatrix}, \quad B = \begin{bmatrix} 0.2142 \\ 2.7858 \end{bmatrix}, \quad C = \begin{bmatrix} 1 & 0 \end{bmatrix}$$

(16.8)

We want the output to equal the reference value r and we choose a criterion of the form (16.5), but without a penalty on u:

$$V(U_t) = \|Y_t - R\|^2$$

where R is a vector whose elements are all equal to r. We take $M = 8$, which approximately corresponds to the system time constant $(8 \cdot 0.15 = 1.2)$ and $N = 2$. We assume that the control values after $t + N$ are chosen equal to zero, since there is an integrator in the system. This gives

$$Y_t = \begin{bmatrix} \hat{y}(t+8|t) \\ \hat{y}(t+7|t) \\ \vdots \\ \hat{y}(t+1|t) \end{bmatrix}$$

$$= \begin{bmatrix} CA^8 \\ CA^7 \\ \vdots \\ CA \end{bmatrix} x(t) + \begin{bmatrix} CA^6B & CA^7B \\ CA^5B & CA^6B \\ \vdots & \vdots \\ 0 & CB \end{bmatrix} \begin{bmatrix} u(t+1) \\ u(t) \end{bmatrix} = Hx(t) + SU_t$$

The criterion can be written

$$V(U_t) = \|Hx(t) + SU_t - R\|^2$$

For a given value of $x(t)$ and r this is a quadratic criterion in U_t with a minimum in

$$U_t^* = -(S^T S)^{-1} S^T Hx(t) + (S^T S)^{-1} S^T R$$

$$= - \begin{bmatrix} -2.4962 & -0.1825 \\ 2.7702 & 0.5115 \end{bmatrix} x(t) + \begin{bmatrix} 2.7702 \\ -2.4962 \end{bmatrix} r =$$

$$- \begin{bmatrix} -2.4962 & -0.1825 \\ 2.7702 & 0.5115 \end{bmatrix} \begin{bmatrix} y(t) - r \\ x_2(t) \end{bmatrix}$$

Using the MPC principle, $u(t)$ is now calculated from this formula giving the state feedback

$$u(t) = -\begin{bmatrix} 2.7702 & 0.5115 \end{bmatrix} \begin{bmatrix} y(t) - r & x_2(t) \end{bmatrix}^T$$

This control law gives a closed loop system with poles in 0 and -0.1576. In Figure 16.1a it is shown how it works in the presence of the control variable constraint $|u(t)| \le 1$.

A major point of MPC is, however, that it is possible to take the input constraint into account by minimizing $V(U_t)$ under the constraints $|u(t)| \le 1$, $|u(t+1)| \le 1$. Note that this is not the same thing as bounding the value that minimizes $V(U_t)$. This is shown in Figure 16.1c. It illustrates the level sets of the criterion $V(U_t)$ which are ellipses around U_t^* with the matrix

$$S^T S = \begin{bmatrix} 15.3707 & 13.0423 \\ 13.0423 & 11.2696 \end{bmatrix}$$

defining the major and minor axes. The minimizing value of $u(t)$ is attained when the square defining the constraints just touches the lowest level set, i.e., approximately 0.4. The minimizing value without constraints is $u(t) = -2.6490$.

Figure 16.1b shows the performance of the controller which minimizes $V(U_t)$ under control signal constraints. We note that it performs considerably better by "knowing" that it cannot use large negative values to brake, but instead changes sign of the control signal earlier.

The computation of the minimizing u under constraints can be done using the following straightforward MATLAB-code. (S and H are the matrices defined above and $r = 0$)

```
>> L=S\H;
>> u=-L*x;
>> if max(abs(u))>1
>>    S1=S(:,1);S2=S(:,2);kk=0;
>>    for u1=[-1,1]
>>       kk=kk+1;
>>       u2=-(S2\H)*x-S2\S1)*u1;
>>       if u2>1,u2=1;end,if u2<-1,u2=-1;end
>>       V(kk)=norm(H*x+S*[u1;u2])^2;
>>       U1(kk)=u1;U2(kk)=u2;
>>    end
>>    for u2=[-1,1]
>>       kk=kk+1;
>>       u1=-(S1\H)*x-(S1\S2)*u2;
>>       if u1>1,u1=1;end,if u1<-1,u1=-1;end
```

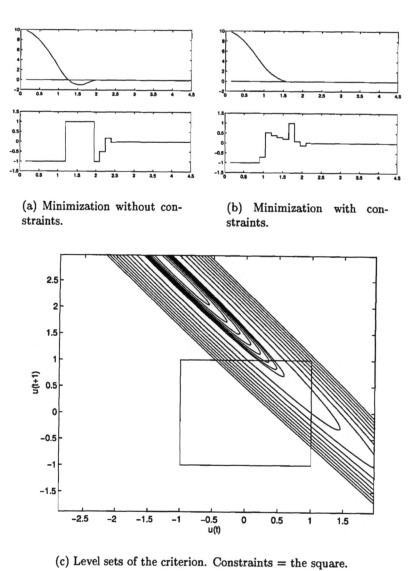

(a) Minimization without constraints.

(b) Minimization with constraints.

(c) Level sets of the criterion. Constraints = the square.

Figure 16.1: MPC of the DC motor. The upper curves of (a) and (b) are the output, and the lower ones the input. Initial value $x(0) = [10, 0]^T$ and $r = 0$. The constraint is $|u(t)| \leq 1$. The level sets are for $x(t) = [1.5 \ -6]^T$.

```
>>        V(kk)=norm(H*x+S*[u1;u2])^2;
>>        U1(kk)=u1;U2(kk)=u2;
>>    end
>>    [dum,ks]=min(V);u=U1(ks);
>> else
>>    u=u(1);
>> end
```

Example 16.2: Distillation Column

We return to the distillation column of Examples 8.2 and 8.3. The same control signals are used, and in this case all three outputs (including the temperature of the lowest bottom) are used in the feedback. The sampling interval is chosen to be 4 minutes.

The criterion (16.5) is minimized with $M = 100$ and $N = 20$, and a matrix Q giving the penalty 1 to $\hat{y}_1(t + k|t)$ and $\hat{y}_2(t + k|t)$ for all k, and the penalty 0 to $\hat{y}_3(t + k|t)$. The penalty matrix Q_u is chosen to be the unit matrix. The result is shown in Figure 16.2.

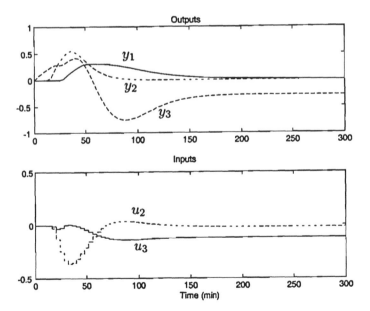

Figure 16.2: MPC-control of a distillation column. No constraints in the criterion. At time 0 a disturbance with amplitude 0.5 occurs in v_1. All reference values are zero.

We now demand that y_3, the bottom reflux temperature, does not go below -0.5. Furthermore, we add constraints requiring the control signal to be between ± 0.5 and the control rate of change to be between ± 0.2 per sample. The same criterion is used as before, but the control signal horizon is reduced to $N = 5$ (to get faster computations). The resulting MPC control is shown in Figure 16.3. This controller achieves the constraint on y_3.

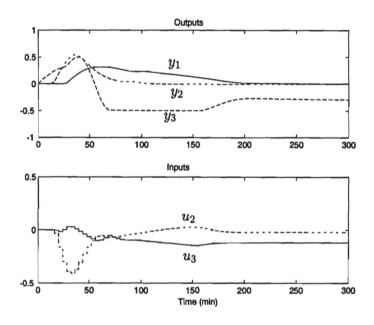

Figure 16.3: Same MPC problem as in Figure 16.2, but with constraints on y_3 and the control signals.

Nonlinear System Dynamics

Now suppose that there is a noise free sampled system dynamics given by

$$x(t+1) = f(x(t), u(t)), \qquad y(t) = h(x(t))$$

Let $\hat{x}(t) = \hat{x}(t|t)$ be a good estimate of the state at time t, for instance generated by a nonlinear observer according to Section 15.5. Reasonable k-step predictors can then be calculated in analogy with

(16.2)–(16.3) using

$$\hat{x}(t + k + 1|t) = f(\hat{x}(t + k|t), u(t + k)),$$
$$\hat{y}(t + k|t) = h(\hat{x}(t + k|t)) \quad k = 0, 1, \dots$$

We can now proceed as we did in Sections (16.4)–(16.5) – possibly with a more general, non-quadratic criterion V – and minimize the criterion with respect to U_t. This minimization is typically more demanding than in the linear case and must normally be done using a numerical search. If N is not too large it is however often a tractable problem. MPC of this generality can also be seen as an approximate method of solving *optimal control* problems. The theory of optimal control will be treated in Chapter 18.

16.4   Comments

Main Points of the Chapter

The basic idea of MPC is straightforward: let the model calculate future outputs as a function of future inputs. Choose the inputs that optimize a criterion based on future values of the signals. A special advantage is that it is easy to include constraints on the variables.

Literature

The method is presented under various names in the literature. MPC is probably the most common one. If the model is estimated simultaneously from observed data the term *Generalized Predictive Control, (GPC)*, is used. An early version of the method was introduced in the 1970's at the Shell company by Cutler and Ramaker who called it *Dynamic Matrix Control, (DMC)*. (The "Dynamic matrix" is D_m in (16.4).) MPC is discussed thoroughly in Chapter 27 of Seborg, Edgar & Mellichamp (1989) and also in the book Prett & Garcia (1988). GPC is treated in e.g. Camacho & Bordons (1995).

The methods play an important role in contemporary chemical and petrochemical industry.

Software

Commercial software is available e.g. in a MATLAB Toolbox, THE MODEL PREDICTIVE CONTROL TOOLBOX.

Chapter 17

To Compensate Exactly for Nonlinearities

When nonlinearities come up in different applications, it is common to use an inner feedback loop to remove much of the nonlinear effects. The system can be controlled as a linear system in an outer control loop. In some cases it turns out to be possible to remove the nonlinear effects completely in the inner control loop. Recently, a systematic methodology has been developed to do this. Often the term *exact linearization*, is used to stress the fact that the effect of the nonlinearity is completely removed. To show the basic ideas we will look at some examples.

17.1   Examples of Exact Linearization

We begin with an example from mechanics.

Example 17.1:   Exact Linearization of a Mechanical System

Consider the system of Figure 17.1. Let x_1 be the position of the cart that moves with the velocity x_2. We assume that the spring force is described by a nonlinear function $\phi(x_1)$ of the position and that the damper force is a nonlinear function $\psi(x_2)$ of the velocity. The system is then described by the equations.

$$\dot{x}_1 = x_2$$
$$\dot{x}_2 = -\phi(x_1) - \psi(x_2) + u \tag{17.1}$$

This is a nonlinear system. The special structure means, however, that it is simple to find a feedback that compensates for the nonlinearities. If r is a

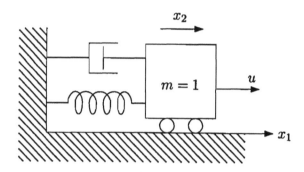

Figure 17.1: Mechanical system. Unit mass with damper and spring.

reference signal, we can use

$$u = r + \phi(x_1) + \psi(x_2) - a_1 x_1 - a_2 x_2 \tag{17.2}$$

giving the closed loop system

$$\dot{x}_1 = x_2$$
$$\dot{x}_2 = -a_1 x_1 - a_2 x_2 + r \tag{17.3}$$

This is a linear relation between reference signal and position. By choosing the coefficients a_1 and a_2 we can get any desired pole placement. Since the equation (17.2) can be interpreted as a calculation showing the force required to get a certain acceleration, the method is sometimes called "computed force". In industrial robot applications where the input signal often is a torque, the method is called "computed torque".

The example we gave is very special. We shall look at a case where it is not as obvious how to compensate for the nonlinearity.

Example 17.2:   Speed Control of Aircraft

We will consider a highly simplified model of speed control for an aircraft. Let the speed be x_1 and the thrust of the engine mx_2, where m is the mass of the aircraft. Let the aerodynamic drag be given by the nonlinear function $mD(x_1)$. The relationship between force and acceleration for the aircraft then gives

$$\dot{x}_1 = -D(x_1) + x_2$$

Let the desired thrust be the input u to the system. We assume that the engine responds with a certain time constant, giving

$$\dot{x}_2 = -x_2 + u$$

A state space description of the system is thus

$$\dot{x}_1 = -D(x_1) + x_2 \tag{17.4}$$
$$\dot{x}_2 = -x_2 + u$$

Now, it is not possible to compensate directly for the nonlinearity $D(x_1)$, by choosing u, since those quantities are not in the same equation.

To treat the more complicated cases which are suggested by our second example, we must develop a background theory. This is done in the following sections.

17.2   Relative Degree

We must first decide what nonlinearities to consider. It turns out to be useful to assume the system to be in control affine form (11.8)

$$\dot{x} = f(x) + ug(x) \tag{17.5a}$$
$$y = h(x) \tag{17.5b}$$

where f, g and h are infinitely differentiable functions. The state x is assumed to be a vector with n components, while u and y are scalars. We allow essentially arbitrary nonlinearities in the state but assume a simple affine dependence in u. Examples 17.1 and 17.2 were both of this type.

Obviously y does not depend explicitly on u, since u is not explicitly present in h. If u is changed instantaneously there will be no immediate change in y; the change comes gradually via x. To check the corresponding behavior of $\dot{y}$ we can differentiate the output equation (17.5b). We then get

$$\dot{y} = h_x(x)\dot{x} = h_x(x)\left(f(x) + g(x)u\right) \tag{17.6}$$

where

$$h_x = \left(\frac{\partial h}{\partial x_1}, \ldots, \frac{\partial h}{\partial x_n}\right)$$

which can be written

$$\dot{y} = h_x f + u h_x g$$

Here we have suppressed the x-dependence in the notation. We see that $\dot{y}$ depends directly on u, at least for some values of x, if and only if $h_x g \not\equiv 0$. (Here $\equiv$ means ("identically equal to".) We will say that the system has *relative degree 1* if

$$h_x g \not\equiv 0$$

Now assume that $h_x g \equiv 0$. Differentiating once more we get

$$\ddot{y} = (h_x f)_x f + u(h_x f)_x g$$

The system is said to have relative degree 2 if

$$h_x g \equiv 0, \quad (h_x f)_x g \not\equiv 0$$

In that case $\ddot{y}$ depends directly on u. Otherwise we can continue differentiating. It is useful to introduce some notation for the repeated differentiations. Introduce the partial differential operators

$$L_f = f_1 \frac{\partial}{\partial x_1} + \ldots + f_n \frac{\partial}{\partial x_n}, \quad L_g = g_1 \frac{\partial}{\partial x_1} + \ldots + g_n \frac{\partial}{\partial x_n}$$

L_f is called the *Lie-derivative* in the direction of f. Letting ν denote the relative degree, we can rewrite our calculations as follows

$$
\begin{aligned}
\nu = 1: \quad & L_g h \not\equiv 0 \\
\nu = 2: \quad & L_g h \equiv 0, \quad L_g L_f h \not\equiv 0 \\
& \text{etc.}
\end{aligned}
\tag{17.7}
$$

It is thus natural to make the following definition

Definition 17.1 *The relative degree ν is the least positive integer such that*

$$L_g L_f^{\nu-1} h \not\equiv 0$$

Definition 17.2 *A system has strong relative degree ν if the relative degree is ν and furthermore*

$$L_g L_f^{\nu-1} h \neq 0, \quad \text{all } x$$

For linear systems the relative degree is related to well-known quantities.

Theorem 17.1 *For a SISO linear system*

$$\dot{x} = Ax + Bu \qquad (17.8a)$$

$$y = Cx \qquad (17.8b)$$

the relative degree ν can be calculated as
(a) the difference in degree between denominator and numerator of the transfer function
(b) the least positive number such that

$$CA^{\nu-1}B \neq 0$$

Proof: For a step response we have the initial value theorem

$$0 \neq \lim_{t \to 0} y^{(\nu)} = \lim_{s \to \infty} s \cdot s^{\nu}G(s) \cdot 1/s$$

which gives (a). Differentiating (17.8a), (17.8b) gives

$$L_g L_f^j h = CA^j B$$

which proves (b). □

17.3   Input-Output Linearization

Consider a system with strong relative degree ν. From the definition and computations of the previous section we have

$$y^{(\nu)} = L_f^{\nu} h + u L_g L_f^{\nu-1} h, \quad L_g L_f^{\nu-1} h \neq 0$$

We see an interesting possibility: introduce the feedback

$$u = \frac{1}{L_g L_f^{\nu-1} h}(r - L_f^{\nu} h) \qquad (17.9)$$

This is possible since $L_g L_f^{\nu-1} h \neq 0$. The resulting relationship is

$$y^{(\nu)} = r$$

from reference signal to output. By using (17.9), which is a nonlinear state feedback ($L_g L_f^{\nu-1} h$ and $L_f^{\nu} h$ are functions of x), we have obtained a linear system from reference signal to output. To stress the fact that it is not a linear approximation, it is often called *exact input-output linearization*

Example 17.3:   A Simple Input-Output Linearization

Consider the system

$$\dot{x}_1 = x_2^3$$
$$\dot{x}_2 = u \tag{17.10}$$
$$y = x_1 + x_2$$

Since

$$\dot{y} = x_2^3 + u$$

the system has strong relative degree 1. The feedback

$$u = -x_2^3 + r$$

gives the linear reference-to-output relation

$$\dot{y} = r$$

There is unfortunately a disadvantage in this methodology, which becomes clear with a change of variables. Let us introduce the new variable

$$z = \phi(x) \tag{17.11}$$

with

$$
\begin{pmatrix}
\phi_1 \\
\phi_2 \\
\vdots \\
\phi_\nu \\
\phi_{\nu+1} \\
\vdots \\
\phi_n
\end{pmatrix}
=
\begin{pmatrix}
h \\
L_f h \\
\vdots \\
L_f^{\nu-1} h \\
\text{``arbitrary''} \\
\vdots \\
\text{``arbitrary''}
\end{pmatrix}
\tag{17.12}
$$

Remark. Even if the choice of the last components of z is essentially arbitrary, it must of course be done so that ϕ becomes invertible, i.e., it must be possible to compute x for a given z. It is possible to show that this can be done, at least locally in the state space.

The coordinate change means that

$$z_1 = y, \quad z_2 = \dot{y}, \dots, z_\nu = y^{(\nu-1)}$$

and the system description becomes

$$\dot{z}_1 = z_2$$
$$\dot{z}_2 = z_3$$
$$\vdots$$
$$\dot{z}_\nu = \xi(z) + u\eta(z) \qquad\qquad (17.13)$$
$$\dot{z}_{\nu+1} = \psi_1(z, u)$$
$$\vdots$$
$$\dot{z}_n = \psi_{n-\nu}(z, u)$$
$$y = z_1$$

for some functions ξ, η and ψ. The linearizing feedback is

$$u = (r - \xi)/\eta$$

which gives the following closed loop system

$$\dot{z}_1 = z_2$$
$$\dot{z}_2 = z_3$$
$$\vdots$$
$$\dot{z}_\nu = r \qquad\qquad (17.14)$$
$$\dot{z}_{\nu+1} = \psi_1(z, (r - \xi)/\eta)$$
$$\vdots$$
$$\dot{z}_n = \psi_{n-\nu}(z, (r - \xi)/\eta)$$
$$y = z_1$$

We see that the whole system has not been linearized. There is possibly still some nonlinear dynamics affecting the state variables $z_{\nu+1}, \dots, z_n$. This dynamics is not visible in the output and is called the *zero dynamics* of the system. If one performs the calculations above for a linear system, it is possible to show that the poles of the zero dynamics equals the zeros of the transfer function from u to y. It is natural to view instability of the zero dynamics as a generalization of the concept of non-minimum phase to nonlinear systems.

We will illustrate the importance of the zero dynamics by looking at two examples.

Example 17.4:   Nice Zero Dynamics

Consider again the system

$$\dot{x}_1 = x_2^3$$
$$\dot{x}_2 = u \qquad\qquad (17.15)$$
$$y = x_1 + x_2$$

Introducing new states according to (17.11), (17.12), gives

$$z_1 = x_1 + x_2, \quad z_2 = x_2$$

In the new variables we then get

$$\dot{z}_1 = z_2^3 + u$$
$$\dot{z}_2 = u \qquad\qquad (17.16)$$
$$y = z_1$$

The feedback is

$$u = r - z_2^3$$

where r is the reference signal. The result is

$$\dot{z}_1 = r$$
$$\dot{z}_2 = -z_2^3 + r \qquad\qquad (17.17)$$
$$y = z_1$$

We see in particular that

$$\dot{y} = r$$

The dynamics which is not visible in the output signal is

$$\dot{z}_2 = -z_2^3 + r$$

It is easy to see that this system is globally asymptotically stable for any constant r.

Example 17.5:   Bad Zero Dynamics

Consider instead the system

$$\dot{x}_1 = -x_2^2 + u$$
$$\dot{x}_2 = u \qquad\qquad (17.18)$$
$$y = x_1$$

n this case it is unnecessary to make a coordinate change since we already
have $x_1 = y$ and the system has relative degree 1. The feedback is

$$u = x_2^2 + r$$

giving

$$\dot{x}_1 = r$$
$$\dot{x}_2 = x_2^2 + r \qquad\qquad\qquad (17.19)$$
$$y = x_1$$

We see that there are problems already for $r = 0$. A small initial value of x_2
gives a solution that rapidly approaches infinity. This means that the control
signal will also tend to infinity.

The examples show that the proposed input output linearization only
works if the zero dynamics is well-behaved.

17.4   Exact State Linearization

We note that the potential problems with the zero dynamics disappear
f $\nu = n$. This is such an important special case that it deserves a
theorem.

Theorem 17.2 *If a system (17.5) has strong relative degree n (= the
number of state variables), there is a state feedback making the system
exactly linear in the state variables $z_1 = y,...,z_n = y^{(n-1)}$.*

Proof: In the z-variables the system is

$$\dot{z}_1 = z_2$$
$$\dot{z}_2 = z_3$$
$$\vdots \qquad\qquad\qquad (17.20)$$
$$\dot{z}_n = \xi(z) + u\eta(z)$$
$$y = z_1$$

for some functions ξ and η. The feedback

$$u = (r - \xi)/\eta$$

makes the system linear, with r as input.   □

This form of feedback is called *exact state linearization* or just *exact linearization*.

Remark 1. A natural variation is to use the feedback

$$u = (r - \xi(z) - a_1 z_1 - \cdots - a_n z_n)/\eta(z)$$

in (17.20), where the a_i are chosen to give a good pole placement.

Remark 2. It is not obvious that one should aim at a linear dynamics. A more general feedback would be

$$u = (-\xi(z) + \tilde{\xi}(z))/\eta(z)$$

replacing the nonlinear function ξ by another arbitrary function $\tilde{\xi}$. One could thus try to replace "bad" nonlinearities by "good" nonlinearities.

Remark 3. The description (17.20) is often regarded as a *controller canonical form* for nonlinear systems.

Example 17.6:   Exact Linearization of a Sine Function

For the system

$$\begin{aligned}
\dot{x}_1 &= x_2 \\
\dot{x}_2 &= \sin x_3 \\
\dot{x}_3 &= u \\
y &= x_1
\end{aligned} \tag{17.21}$$

we get

$$\dot{y} = x_2, \quad \ddot{y} = \sin x_3, \quad y^{(3)} = u \cos x_3$$

The system has thus relative degree 3, as long as $|x_3| < \pi/2$. The following change of variables is one-to-one if $|x_3| < \pi/2$

$$z_1 = x_1, \quad z_2 = x_2, \quad z_3 = \sin x_3$$

In the new variables the system description is

$$\begin{aligned}
\dot{z}_1 &= z_2 \\
\dot{z}_2 &= z_3 \\
\dot{z}_3 &= u \cos(\arcsin z_3) \\
y &= z_1
\end{aligned} \tag{17.22}$$

The feedback

$$u = (r - z_1 - 3z_2 - 3z_3)/\cos(\arcsin(z_3))$$

gives a linear system with all its poles in -1. In the original variables the feedback is

$$u = (r - x_1 - 3x_2 - 3\sin x_3)/\cos x_3$$

Now suppose we have a system with relative degree $\nu < n$. We can argue as follows: since we have to use a feedback from all state variables, we can regard some other function of the state as output. This alternative output can then be chosen so that the relative degree is n. Denote the new output

$$\tilde{y} = c(x)$$

We see that the function c has to satisfy

$$L_g L_f^j c = 0, \quad j = 0, \dots, n-2; \quad L_g L_f^{n-1} c \neq 0 \qquad (17.23)$$

This is a system of partial differential equations for the function c. One can show that (17.23) sometimes lacks a solution. Even when a solution exists, it is not certain that it can be calculated explicitly. Here we find a definite limit for the use of exact linearization. There are, however, many cases when (17.23) is satisfied by simple functions c, which can be guessed using some physical insight. We show this by continuing Example 17.2.

Example 17.7: Exact Linearization of Aircraft Speed Control

In Example 17.2 we showed that the speed control dynamics is

$$\begin{aligned}
\dot{x}_1 &= -D(x_1) + x_2 \\
\dot{x}_2 &= -x_2 + u
\end{aligned} \qquad (17.24)$$

We did not define any output, and we are thus free to choose one satisfying (17.23). Since u is absent from the first differential equation, we see directly that we get relative degree 2 by using x_1 as the output. The coordinate change is

$$\begin{array}{ll}
z_1 = x_1 & x_1 = z_1 \\
z_2 = -D(x_1) + x_2 & x_2 = D(z_1) + z_2
\end{array}$$

and in the new variables the dynamics becomes

$$\begin{aligned}
\dot{z}_1 &= z_2 \\
\dot{z}_2 &= -D(z_1) - D'(z_1)z_2 - z_2 + u
\end{aligned}$$

The dynamics is linearized by the control law

$$u = r + D(z_1) + D'(z_1)z_2 + z_2 - a_1 z_1 - a_2 z_2 \qquad (17.25)$$

Note that we can perform the exact linearization for any differentiable function $D(x_1)$ describing the drag. However, the function must, of course, be known in order to calculate u from (17.25).

We have now seen several examples where exact linearization can be used. Among the limitations we can mention:

- The method requires feedback from the state. If some state variables are not measured, they have to be reconstructed. This can be done by a *nonlinear observer*, see Section 15.5.

- Since the linearization is achieved using feedback, the system usually remains linear only as long as the control signal is not saturated.

- In some cases the linearization can only be done in part of the state space. We saw such a case in Example 17.6.

- If the physical system differs from the model, the closed loop system is no longer exactly linear. The sensitivity and robustness therefore usually cannot be analyzed using methods from linear theory.

- It is not obvious that nonlinearities should be removed. They might be good for the system performance, and then one should not waste control signal energy to remove them through feedback.

Nevertheless, exact linearization has been used with success in a number of applications including

1. Industrial robots. The "computed torque" methods is in principle a generalization of the solution we saw in Example 17.1.

2. Aircraft, helicopters. We saw a very simplified case in Examples 17.2 and 17.7.

3. Electrical generators.

4. Chemical reactors.

17.5   Comments

Main Points of the Chapter

- The relative degree of a system states how many times the output has to be differentiated to get an explicit dependence on the input.

- If the input output dynamics is made linear, the hidden zero dynamics might cause problems.

- If the relative degree equals the number of state variables, the problem of zero dynamics disappears, and an exact linearization of the dynamics from reference to transformed state variables becomes possible.

- For many systems one can obtain a relative degree which is equal to the number of state variables by redefining the output.

- Exact linearization should be done with care so that useful nonlinearities are not removed.

Literature

The mathematical theory of exact linearization and zero dynamics was developed during the 1980s mainly by Jacubczyk, Respondek, Hunt, Su, Meyer, Byrnes and Isidori. The theory is explained in detail in Isidori (1995), where the multivariable case is also considered.

Software

The computations using the operators L_f and L_g that come up in exact linearization are easily handled in computer algebra tools like MAPLE and MATHEMATICA.

Chapter 18

Optimal Control

In very general terms, control problems can be formulated as follows: "Choose the control signal so that the system behaves as well as possible". The difficulty often lies in the formulation of what "as well as possible" means. If it is possible to give a mathematical criterion, there is a completely systematic method for control design: solve the mathematical optimization problem. We have used this point of view in Chapter 9. There the system is assumed to be linear, and the criterion is a quadratic function. This approach can, however, be extended to much more general situations. We will first look at a classic example.

18.1 The Goddard Rocket Problem

Around 1910 one of the pioneers of space research, R. H. Goddard, investigated the possibility of space research using rockets. With the technology available at the time, it was not self-evident that interesting altitudes could be reached (and even less that other planets could be visited). One of Goddard's inventions was a liquid propellant rocket engine whose thrust could be controlled. A natural question was then: how should the thrust vary with time in order to reach the highest altitude? Mathematically the problem can be formulated as follows.

Let the rocket have the mass m and the velocity v. The Newtonian force relation then gives the equation

$$\dot{v} = \frac{1}{m}(u - D(v, h)) - g \tag{18.1a}$$

where g is the gravitational acceleration, u is the engine thrust, which is the control input, and $D(v, h)$ is the aerodynamic drag, depending on the velocity v and the altitude h. (Since m is not constant, $\dot{m}$ appears in the equation. This effect is, however, incorporated in u, see (18.1c) below.) If the rocket ascends vertically we have by definition

$$\dot{h} = v \tag{18.1b}$$

The rocket mass decreases as the propellant is consumed. Since it is the ejected propellant that gives the thrust we have

$$\dot{m} = -\gamma u \tag{18.1c}$$

for some constant γ. The thrust is limited by the engine construction so that

$$0 \le u \le u_{max} \tag{18.1d}$$

Suppose the altitude is measured from the starting point and the rocket has weight m_0 with full tanks, and m_1 with empty tanks. Then

$$v(0) = 0, \quad h(0) = 0, \quad m(0) = m_0, \quad m(t_f) \ge m_1 \tag{18.1e}$$

where t_f is the final time. The optimization problem is thus

$$\text{maximize } h(t_f) \tag{18.1f}$$

under the restrictions given by (18.1a)–(18.1e). If there is no aerodynamic drag, i.e., $D(v, h) = 0$, then the solution is intuitively clear. The control signal ought to be u_{max} until all propellant is ejected, and then zero. The reason is that propellant which is used at a certain altitude has been lifted there, consuming energy. It is therefore advantageous to use all propellant at the lowest possible altitude. In the presence of aerodynamic forces the situation is more complicated. The previously mentioned strategy might lead to a high velocity while the rocket is still in the dense part of the atmosphere where the drag is high. Much energy will then be wasted in overcoming the drag. It might be desirable to keep down the speed until a higher altitude is reached. Exactly how this tradeoff should be made cannot be decided by intuition, and this is where a theory of optimal control is needed.

The Goddard optimization problem is an example of an *optimal control problem*. We note some properties of the problem.

- The control signal, $u(t)$, is defined on a time interval $0 \le t \le t_f$ and the value at every time instant is to be chosen. Since there are infinitely many such values, this is mathematically an *infinite dimensional* optimization problem.

- The quantity to be optimized, $h(t_f)$, is related to the control signal via the system dynamics, i.e., the differential equations (18.1a)–(18.1c).

- The control signal is constrained, (18.1d).

- There are state constraints at the beginning and at the end, (18.1e).

- The final time t_f is not determined beforehand, but is part of the optimization problem.

We will see that several of these points are typical for optimal control problems, even if not all of them come up in all problems.

Generating Reference Signals

We note that the Goddard's problem formulation leads to an *open loop control*, since we seek the control signal as a function of time. Now suppose that we have found an optimal control $u^*(t)$ with corresponding state trajectories $v^*(t)$, $h^*(t)$, $m^*(t)$. If this control signal is used to control a real rocket, then the actual values of velocity, altitude and mass will of course not correspond exactly to those computed, due to differences between model and reality. But this is exactly the type of problem that was treated in Chapter 9. We could for instance let $v^*(t)$ be a reference signal for the velocity, and construct a regulator that controls the thrust in order to keep the actual velocity close to the reference value. If the deviations from the reference value are small, the system dynamics can probably be approximated by a linear model. It is then possible to use the linear design techniques discussed earlier.

This example illustrates a common use of optimal control theory, namely the *generation of reference trajectories*.

Optimal Feedback

One way of handling differences between model and reality would be to describe the solution to the optimal control problem not as a function of time but as a feedback from the state to the control variable. Since we know that feedback often reduces sensitivity, this seems to be a reasonable approach. It is also a natural extension of the LQ-methodology treated in Chapter 9. Instead of linear systems with quadratic criteria we allow general nonlinear systems with arbitrary criteria. This is potentially a very powerful way of attacking control problems. We will see that there are at least two major problems. The first one is the computation of the optimal feedback in a reasonably explicit form. The second one is the analysis of sensitivity and robustness. We will, however, be able to show examples of optimal feedback which are not of LQ type.

18.2　The Maximum Principle

We will now start the discussion of necessary conditions that a control signal has to satisfy in order to be optimal. They are usually called "the maximum principle" for reasons that will become clear.

Formulation of the Optimal Control Problem

Guided by the discussion above we can formulate a general problem in optimal control.

$$\text{minimize} \int_0^{t_f} L(x(t), u(t))\, dt + \phi(x(t_f)) \tag{18.2a}$$

$$\dot{x}(t) = f(x(t), u(t)) \tag{18.2b}$$

$$u(t) \in U, \quad 0 \le t \le t_f \tag{18.2c}$$

$$x(0) = x_0, \quad \psi(x(t_f)) = 0 \tag{18.2d}$$

Here $x(t)$ and $u(t)$ are vectors with n and m elements, respectively, while ψ is a vector with r elements. The control signal is *constrained* by the condition $u(t) \in U$, where U is an arbitrary set in R^m. If u is a scalar, then U is usually an interval, so that the constraint takes the form $u_{min} \le u \le u_{max}$. The *criterion* to be minimized is given by L and ϕ. There are also *terminal constraints* given by ψ.

The Maximum Principle in a Special Case

To begin with we will look at a special case of optimal control, namely the situation where t_f is fixed, there are no terminal constraints and

$L = 0$ in the criterion. The problem is then

$$\text{minimize } \phi(x(t_f)) \tag{18.3a}$$
$$\dot{x}(t) = f(x(t), u(t)) \tag{18.3b}$$
$$u(t) \in U, \quad 0 \le t \le t_f \tag{18.3c}$$
$$x(0) = x_0 \tag{18.3d}$$

Now assume that we have a control signal $u^*(t)$ with corresponding state trajectory $x^*(t)$, satisfying (18.3b), (18.3c) and (18.3d). To check optimality of this control input, we can vary it in such a way that it is identical with u^* except during a short interval of length ϵ at time $t = t_1$:

$$u(t) = \begin{cases} \bar{u}, & t_1 - \epsilon \le t \le t_1 \\ u^*(t), & \text{otherwise} \end{cases} \tag{18.4}$$

We only consider the case $\bar{u} \in U$ so that the new control signal satisfies (18.3c). The solution generated by (18.3b), with the new control signal is denoted $x(t, \epsilon)$ to stress the dependence on ϵ. We will investigate what happens as ϵ tends to zero. Obviously we have

$$x(t, \epsilon) = x^*(t), \quad 0 \le t \le t_1 - \epsilon$$

since the old and new control signals are the same in this time interval. For the solution of the differential equation (18.3b) we have that

$$x^*(t_1) = x^*(t_1 - \epsilon) + \epsilon f(x^*(t_1), u^*(t_1)) + \rho_1(\epsilon)$$
$$x(t_1, \epsilon) = x^*(t_1 - \epsilon) + \epsilon f(x(t_1, \epsilon), \bar{u}) + \rho_2(\epsilon)$$

so that

$$x(t_1, \epsilon) - x^*(t_1) = \epsilon(f(x(t_1, \epsilon), \bar{u}) - f(x^*(t_1), u^*(t_1))) + \rho_3(\epsilon)$$

Here ρ_1, ρ_2 and ρ_3 are functions such that

$$\rho_j(\epsilon)/\epsilon \to 0, \quad \text{as } \epsilon \to 0 \tag{18.5}$$

Since $x(t_1, \epsilon) - x^*(t_1)$ is of the same order as ϵ, we can write

$$x(t_1, \epsilon) - x^*(t_1) = \epsilon(f(x^*(t_1), \bar{u}) - f(x^*(t_1), u^*(t_1))) + \rho_4(\epsilon) \tag{18.6}$$

where ρ_4 also satisfies (18.5). We will now investigate the behavior of $x(t, \epsilon)$ for $t > t_1$. To do that, we calculate

$$\eta(t) = \left.\frac{\partial x(t, \epsilon)}{\partial \epsilon}\right|_{\epsilon=0}$$

From (18.6) we get

$$\eta(t_1) = f(x^*(t_1), \bar{u}) - f(x^*(t_1), u^*(t_1)) \tag{18.7}$$

Since $x(t, \epsilon)$ satisfies (18.3b) we have

$$\frac{\partial}{\partial t} x(t, \epsilon) = f(x(t, \epsilon), u(t))$$

Differentiating this equation with respect to ϵ gives

$$\frac{\partial}{\partial t} \frac{\partial x(t, \epsilon)}{\partial \epsilon} = f_x(x(t, \epsilon), u(t)) \frac{\partial x(t, \epsilon)}{\partial \epsilon}$$

where the Jacobian f_x is an $n \times n$ matrix whose element in position (i, j) is

$$(f_x)_{i,j} = \frac{\partial f_i}{\partial x_j} \tag{18.8}$$

Setting $\epsilon = 0$ gives

$$\dot{\eta}(t) = f_x(x^*(t), u^*(t))\, \eta(t) \tag{18.9}$$

This equation is called the *variational equation* and can be used to investigate how the changed control signal (18.4) affects the criterion. A Taylor expansion of ϕ together with the definition of η give

$$\phi(x(t_f, \epsilon)) = \phi(x^*(t_f)) + \epsilon\phi_x(x^*(t_f))\eta(t_f) + \rho_5(\epsilon) \tag{18.10}$$

where ρ_5 satisfies (18.5) and ϕ_x is a row vector whose i:th component is

$$(\phi_x)_i = \frac{\partial \phi}{\partial x_i} \tag{18.11}$$

If u^* is optimal, then no other choice of control signal satisfying (18.3c) can give the criterion a smaller value. This means that

$$\phi(x(t_f, \epsilon)) \geq \phi(x^*(t_f)) \tag{18.12}$$

Equation (18.10) shows that this inequality can only be satisfied when ϵ tends to zero if

$$\phi_x(x^*(t_f))\eta(t_f) \geq 0 \tag{18.13}$$

This inequality together with (18.9) and (18.7) give necessary conditions for optimality. They are, however, difficult to interpret.

A nicer formulation of the necessary conditions can be obtained by using a property of the variational equation. Suppose we introduce a vector $\lambda(t)$ satisfying the differential equation

$$\dot{\lambda}(t) = -f_x(x^*(t), u^*(t))^T \lambda(t) \tag{18.14}$$

called the *adjoint equation*. Then (18.9) and (18.14) directly give

$$\frac{d}{dt}\lambda(t)^T\eta(t) = \dot{\lambda}(t)^T\eta(t) + \lambda(t)^T\dot{\eta}(t) = 0$$

which shows that

$$\lambda(t)^T\eta(t) = \text{constant} \tag{18.15}$$

Let us now choose λ as the solution to (18.14) satisfying the boundary condition

$$\lambda(t_f) = \phi_x(x^*(t_f))^T$$

Then (18.13) can be rewritten

$$\lambda(t_f)^T\eta(t_f) \geq 0$$

Since the product $\lambda^T\eta$ is constant we get

$$\lambda(t_1)^T\eta(t_1) \geq 0$$

and finally, using (18.7)

$$\lambda(t_1)^T\left(f(x^*(t_1), \bar{u}) - f(x^*(t_1), u^*(t_1))\right) \geq 0 \tag{18.16}$$

Since $\bar{u}$ and t_1 are arbitrary in our derivation, this relation must be satisfied by all $\bar{u} \in U$ and all $0 \leq t_1 \leq t_f$. We have thus shown the following basic result.

Theorem 18.1 *Let the optimization problem (18.3) have a solution* $u^*(t)$, $x^*(t)$. *Then the following relations have to be satisfied.*

$$\min_{u \in U} \lambda(t)^T f(x^*(t), u) = \lambda(t)^T f(x^*(t), u^*(t)), \quad 0 \le t \le t_f \quad (18.17)$$

where

$$\dot{\lambda}(t) = -f_x(x^*(t), u^*(t))^T \lambda(t), \quad \lambda(t_f) = \phi_x(x^*(t_f))^T \quad (18.18)$$

with f_x *and* ϕ_x *defined by (18.8), (18.11).*

Proof: The statement follows directly from (18.16) if we note that $\bar{u} = u^*$ gives equality.   □

Remark. The theorem states *necessary conditions* for optimality. They may not be sufficient. A pair u^*, x^* satisfying the conditions of the theorem is called an *extremal*. An extremal is thus not necessarily an optimum.

Example 18.1: Steering a Boat

Consider a boat moving in an area where the current varies. The position of the boat is given by the coordinates x_1 and x_2. The control variables u_1 and u_2 are the velocities in the x_1 and x_2 directions. The magnitude of the velocity is assumed to be constant and equal to 1. The velocity of the water is assumed to be $v(x_2)$ in the x_1 direction and 0 in the x_2 direction. Consider the problem of starting at the origin, and finding a control so that the boat moves as far as possible in the x_1 direction in T units of time. The control problem is then

$$\begin{aligned}
&\min(-x_1(T)) && (18.19a) \\
&\dot{x}_1 = v(x_2) + u_1 && (18.19b) \\
&\dot{x}_2 = u_2 && (18.19c) \\
&x_1(0) = 0, \quad x_2(0) = 0 && (18.19d) \\
&u_1^2 + u_2^2 = 1 && (18.19e)
\end{aligned}$$

Letting $v'(x_2)$ denote the derivative of $v(x_2)$ we get

$$f_x = \begin{bmatrix} 0 & v'(x_2) \\ 0 & 0 \end{bmatrix}$$

The adjoint equation is then

$$\dot{\lambda}_1 = 0, \quad \dot{\lambda}_2 = -v'(x_2)\lambda_1$$

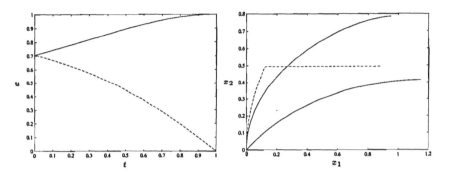

(a) The control signals u_1 (solid) and u_2 (dashed).

(b) An extremal (the lowest curve) and two suboptimal trajectories.

Figure 18.1: An extremal of the boat steering problem: control signals (a) and state trajectory (b).

with the boundary conditions $\lambda_1(T) = -1$ and $\lambda_2(T) = 0$. The control signal is the solution of

$$\min_{u_1^2+u_2^2=1} \; (\lambda_1(v(x_2) + u_1) + \lambda_2 u_2)$$

This can be interpreted as the minimization of the scalar product between the vectors $[\lambda_1 \; \lambda_2]$ and $[u_1 \; u_2]$ under the condition that the length of $[u_1 \; u_2]$ is 1. The minimum is attained when $[u_1 \; u_2]$ and $[\lambda_1 \; \lambda_2]$ have opposite directions. This gives

$$u_1 = -\frac{\lambda_1}{\sqrt{\lambda_1^2 + \lambda_2^2}}, \quad u_2 = -\frac{\lambda_2}{\sqrt{\lambda_1^2 + \lambda_2^2}}$$

The simplest situation is when $v(x_2)$ is a linear function, i.e., $v'(x_2) = 1$. Then

$$\lambda_1 = -1, \quad \lambda_2 = t - T \tag{18.20}$$

and the optimal control is

$$u_1 = \frac{1}{\sqrt{1 + (t - T)^2}}, \quad u_2 = \frac{T - t}{\sqrt{1 + (t - T)^2}}$$

In Figure 18.1a the control signals are shown and in Figure 18.1b the extremal in the x_1-x_2 plane is shown together with some non optimal trajectories.

Integral Term in the Criterion

To extend Theorem 18.1 to problems where the criterion contains an integral is fairly simple. We thus consider the problem

$$\text{minimize } \int_0^{t_f} L(x(t), u(t)) \, dt + \phi(x(t_f)) \tag{18.21a}$$

$$\dot{x}(t) = f(x(t), u(t)) \tag{18.21b}$$

$$u(t) \in U, \quad 0 \le t \le t_f \tag{18.21c}$$

$$x(0) = x_0 \tag{18.21d}$$

It turns out to be useful to introduce the function

$$H(x, u, \lambda) = L(x, u) + \lambda^T f(x, u) \tag{18.22}$$

called the *Hamiltonian* and we have the following result..

Theorem 18.2 *Let the optimal control problem (18.21) have a solution $u^*(t)$, $x^*(t)$. Then the following relations must be satisfied.*

$$\min_{u \in U} H(x^*(t), u, \lambda(t)) = H(x^*(t), u^*(t), \lambda(t)), \quad 0 \le t \le t_f \tag{18.23}$$

where λ satisfies

$$\dot{\lambda}(t) = -H_x(x^*(t), u^*(t), \lambda(t))^T, \quad \lambda(t_f) = \phi_x(x^*(t_f))^T \tag{18.24}$$

Proof: Introduce an extra state x_0 satisfying

$$\dot{x}_0 = L(x, u), \quad x_0(0) = 0$$

The criterion then becomes

$$\min_{u \in U} x_0(t_f) + \phi(x(t_f))$$

and the adjoint equation is

$$\frac{d}{dt} \begin{bmatrix} \lambda_0(t) \\ \lambda(t) \end{bmatrix} = - \begin{bmatrix} 0 & 0 \\ L_x^T & f_x^T \end{bmatrix} \begin{bmatrix} \lambda_0(t) \\ \lambda(t) \end{bmatrix}, \quad \begin{bmatrix} \lambda_0 \\ \lambda \end{bmatrix}(t_f) = \begin{bmatrix} 1 \\ \phi_x^T \end{bmatrix}$$

We see that $\lambda_0(t) \equiv 1$ and that

$$\dot{\lambda} = -L_x^T - f_x^T \lambda$$

which together with (18.22) gives (18.24). Since $\lambda_0 = 1$, equation (18.17) of Theorem 18.1 directly gives (18.23). $\square$

Terminal Constraints

We will now consider a more general version of the optimal control problem, where the terminal time t_f is not necessarily specified, but has to be chosen as part of the optimization problem. We allow terminal constraints of the form

$$\psi(t_f, x(t_f)) = 0$$
$$\psi(t_f, x(t_f)) = \begin{bmatrix} \psi_1(t_f, x(t_f)) & \cdots & \psi_r(t_f, x(t_f)) \end{bmatrix}^T$$

The optimal control problem is then

$$\text{minimize} \int_0^{t_f} L(x(t), u(t)) \, dt + \phi(t_f, x(t_f)) \tag{18.25a}$$

$$\dot{x}(t) = f(x(t), u(t)) \tag{18.25b}$$

$$u(t) \in U, \quad 0 \le t \le t_f \tag{18.25c}$$

$$x(0) = x_0, \quad \psi(t_f, x(t_f)) = 0 \tag{18.25d}$$

When handling constraints in optimal control problems it is useful to consider how constraints are treated in ordinary minimization problems. When minimizing an ordinary function $f(x)$, where x is an n-vector, a necessary condition for an optimum at x^* is

$$f_x(x^*) = 0 \tag{18.26}$$

If we add a constraint so that the optimization problem becomes

$$\min f(x), \text{ under the constraint } g(x) = 0 \tag{18.27}$$

with g an m-vector, then (under certain regularity conditions) a necessary condition for optimality is

$$f_x(x^*) + \mu^T g_x(x^*) = 0 \tag{18.28}$$

for some choice of the m-vector μ. The constraint is thus handled by considering the function

$$f(x) + \mu^T g(x) \tag{18.29}$$

depending on the *Lagrange multiplier* μ. In analogy with this approach the terminal constraint $\psi = 0$ can be handled by replacing that part of the criterion which affects the end point, i.e., $\phi(t_f, x(t_f))$ with

$$\phi(t_f, x(t_f)) + \mu^T \psi(t_f, x(t_f)) \tag{18.30}$$

in Theorem 18.1. Problems where this approach can be used are called *normal* problems. In Appendix 18A we discuss how normal problems are characterized geometrically, why (18.30) is reasonable, and how the *multiplier* μ can be interpreted. For normal problems we can formulate the following theorem.

Theorem 18.3 The Maximum Principle (Normal Case) *Let the optimization problem (18.25) be normal and have a solution $u^*(t)$, $x^*(t)$. Then there are vectors $\lambda(t)$ and μ such that*

$$\min_{u \in U} H(x^*(t), u, \lambda(t)) = H(x^*(t), u^*(t), \lambda(t)), \quad 0 \le t \le t_f$$

(18.31)

where λ satisfies

$$\dot{\lambda}(t) = -H_x(x^*(t), u^*(t), \lambda(t))^T$$
$$\lambda(t_f)^T = \phi_x(t_f, x^*(t_f)) + \mu^T \psi_x(t_f, x^*(t_f))$$

(18.32)

If choosing t_f is part of the optimization problem, then there is also the condition

$$H(x^*(t_f), u^*(t_f), \lambda(t_f)) = -\phi_t(t_f, x^*(t_f)) - \mu^T \psi_t(t_f, x^*(t_f))$$

(18.33)

(The subscripts x and t denote differentiation with respect to the variable in question.)

Proof: See Appendix 18A.   □

This theorem is called the *Maximum Principle* because in the original formulation the variables were defined in such a way that (18.31) was a maximization instead of a minimization. It is sometimes called the Pontryagin maximum principle after one of the originators. Other researchers associated with it are Boltyanski, Gamkrelidze and Hestenes.

For optimization problems that are not normal, Theorem 18.3 cannot be used. Those problems are called *abnormal*. To get a formulation of the maximum principle that covers both normal and abnormal problems, one can define the Hamiltonian as

$$H(x, u, \lambda, n_0) = n_0 L(x, u) + \lambda^T f(x, u)$$

(18.34)

We then have

Theorem 18.4 The Maximum Principle *Let the optimal control problem (18.25) have a solution $u^*(t)$, $x^*(t)$. Then there is a vector $\lambda(t)$, a number $n_0 \geq 0$ and a vector μ such that $[n_0 \ \mu^T]$ is nonzero and*

$$\min_{u \in U} H(x^*(t), u, \lambda(t), n_0) = H(x^*(t), u^*(t), \lambda(t), n_0), \quad 0 \leq t \leq t_f$$
$$(18.35a)$$

where H is given by (18.34), and λ satisfies

$$\dot{\lambda}(t) = -H_x(x^*(t), u^*(t), \lambda(t), n_0)^T \qquad (18.35b)$$
$$\lambda(t_f)^T = n_0 \phi_x(t_f, x^*(t_f)) + \mu^T \psi_x(t_f, x^*(t_f)) \qquad (18.35c)$$

If t_f is not fixed, but has to be chosen as part of the optimization, then the condition

$$H(x^*(t_f), u^*(t_f), \lambda(t_f), n_0) = -n_0 \phi_t(t_f, x^*(t_f)) - \mu^T \psi_t(t_f, x^*(t_f))$$
$$(18.35d)$$

has to be added.

Proof: See Appendix 18A.   □

In the general formulation of the Maximum Principle given by Theorem 18.4, n_0, μ, and λ will not be unique. If the theorem is satisfied by one choice of them, it will also be satisfied if they are all multiplied by the same non-negative number. All cases where $n_0 > 0$ can (by multiplying n_0, μ, and λ with $1/n_0$) be normalized to the case $n_0 = 1$ and the formulation given by Theorem 18.3. There are really only two choices of n_0: $n_0 = 1$ (the normal case) and $n_0 = 0$ (the abnormal case). From Theorem 18.1 we see that optimal control problems with fixed final time, and no terminal constraints, are always normal. There are no general and simple criteria for normality of optimal control problems. In many cases one can only show normality by carrying out the calculations for $n_0 = 0$ and verifying that there is no solution.

There is an important corollary of Theorems 18.3 and 18.4.

Corollary 18.1 *Let the conditions of Theorem 18.3 or 18.4 be satisfied. Then the Hamiltonian is constant along an extremal.*

Proof: We show the corollary for the special case when the control signal lies in the interior of U. Since H is minimized we have $H_u = 0$ along an extremal. This gives the relation

$$\frac{d}{dt}H = H_x \dot{x} + H_\lambda \dot{\lambda} + H_u \dot{u} = H_x f - f^T H_x^T = 0$$

□

Example 18.2:   Optimal Heating

We consider an optimal heating problem. Let T be the temperature of an object that is heated with the power P. We assume that the environment has the temperature zero, and that the heat loss is proportional to the temperature difference. We then get

$$\dot{T} = P - T$$

where, for simplicity, we have assumed that all physical constants are equal to 1. Let the power be bounded by

$$0 \leq P \leq P_{max}$$

Consider the problem of raising the temperature one unit in one unit of time, i.e.,

$$T(0) = 0, \quad T(1) = 1$$

with minimal energy consumption, i.e.,

$$\text{minimize} \quad \int_0^1 P(t)dt$$

The Hamiltonian is

$$H = n_0 P + \lambda P - \lambda T$$

The equations (18.32) for λ are $\dot{\lambda} = \lambda$, $\lambda(1) = \mu_1$ with the solution $\lambda(t) = \mu_1 e^{t-1}$. Introducing

$$\sigma(t) = n_0 + \lambda(t) = n_0 + \mu_1 e^{t-1}$$

we can write $H = \sigma(t)P - \lambda T$, so that (18.31) gives

$$P(t) = \begin{cases} 0, & \sigma(t) > 0 \\ P_{max}, & \sigma(t) < 0 \end{cases}$$

Let us now consider the different possibilities.

1. $\mu_1 > 0$. This means that $\sigma(t) > 0$ for all t, giving $P(t) = 0$ during the whole time interval. Then $T(1) = 0$, which does not satisfy the terminal constraint. This case must thus be rejected.

2. $\mu_1 = 0$. This means that $\lambda(t) \equiv 0$. Since $[n_0 \ \mu^T]$ must not be zero, we get $n_0 > 0$, which means that $\sigma(t) > 0$ for all t. This case must then be rejected for the same reason as the previous one.

3. $\mu_1 < 0$. In this case $\sigma(t)$ is strictly decreasing. There is then at most one time instant t_1 such that $\sigma(t_1) = 0$.

Since case 3 above is the only possible one, we conclude that an extremal must be of the form

$$P(t) = \begin{cases} 0, & 0 \le t \le t_1 \\ P_{max}, & t_1 < t \le 1 \end{cases} \qquad (18.36)$$

Possibly t_1 is outside the time interval $0 \le t \le 1$ in which case $P \equiv 0$ or $P \equiv P_{max}$.

For P given by (18.36) we get

$$1 = T(1) = \int_{t_1}^1 e^{-(t-\tau)} P_{max} d\tau = (1 - e^{-1+t_1}) P_{max}$$

Solving for t_1 gives

$$t_1 = 1 + \ln(1 - P_{max}^{-1}) \qquad (18.37)$$

The condition $\sigma(t_1) = 0$ then gives

$$\mu_1 = -\frac{n_0}{1 - P_{max}^{-1}} \qquad (18.38)$$

Since P_{max} is positive, we automatically get $t_1 \le 1$. We see that $t_1 \ge 0$ if

$$P_{max} \ge \frac{1}{1 - e^{-1}} \qquad (18.39)$$

This condition also implies that $\mu_1 < 0$ in (18.38). If there is equality in (18.39), then $T(1) = 1$ is achieved by taking $P(t) = P_{max}$, $0 \le t \le 1$. This is the lowest value of P_{max} for which the optimal control problem is solvable. We conclude that there are two cases.

1. If P_{max} does not satisfy (18.39) there is no choice of $P(t)$ which satisfies the terminal constraints. There is thus no solution to the optimal control problem.

2. If P_{max} satisfies (18.39) there is an extremal given by (18.36) with t_1 defined by (18.37). From (18.38) we see that $n_0 \ne 0$ since $[n_0 \ \mu^T]$ must be nonzero. The extremal is thus normal.

Example 18.3: DC Motor with Power Constraint

Let us consider positioning with a current controlled DC motor. A simple model using the current as input signal u is

$$\dot{\omega} = u, \quad \dot{\theta} = \omega \tag{18.40}$$

where θ is the angle of rotation of the motor axis and ω the angular velocity. (Note that this model is different from the usual one where the input is the applied voltage. Instead of one time constant and one integrator we now have a double integrator.) Now suppose that the motor axis should be turned from one position to another in the shortest possible time. This gives the end point constraints

$$\omega(0) = 0, \quad \theta(0) = 0, \quad \omega(t_f) = 0, \quad \theta(t_f) = 1 \tag{18.41}$$

and the following criterion to minimize

$$\int_0^{t_f} 1 \, dt = t_f \tag{18.42}$$

Of course, there will not be a well defined problem if we do not have any restrictions on the choice of u. This is because the system is controllable, which means that it is possible to reach any state in an arbitrarily short time. Let us assume that heating of the motor is the limiting factor. Then it is natural to introduce a constraint of the form

$$\int_0^{t_f} u^2 \, dt \leq W_0 \tag{18.43}$$

since the amount of power being lost as heat is proportional to u^2. To get a problem of standard type we can introduce an extra state W, satisfying

$$\dot{W} = u^2, \quad W(0) = 0 \tag{18.44}$$

The condition (18.43) then becomes

$$W(t_f) \leq W_0 \tag{18.45}$$

Since it is clear, for physical reasons, that this constraint should be active, we can put

$$W(t_f) = W_0 \tag{18.46}$$

We now have an optimal control problem specified by (18.40), (18.44), (18.41), (18.46) and (18.42).

Using the general notation of (18.25) we get

$$x = \begin{bmatrix} \omega \\ \theta \\ W \end{bmatrix}, \quad f = \begin{bmatrix} u \\ \omega \\ u^2 \end{bmatrix}, \quad \psi = \begin{bmatrix} \omega \\ \theta - 1 \\ W - W_0 \end{bmatrix}$$

$$x_0 = 0, \quad L = 1, \quad \phi = 0$$

The Hamiltonian is

$$H = n_0 + \lambda_1 u + \lambda_2 \omega + \lambda_3 u^2 \tag{18.47}$$

and u has to be chosen to minimize this expression. From (18.32), λ has to satisfy

$$\dot{\lambda}_1 = -\lambda_2, \quad \dot{\lambda}_2 = 0, \quad \dot{\lambda}_3 = 0 \tag{18.48}$$

We see that λ_2 and λ_3 are constant. After solving for λ_1, we get

$$\lambda_1 = C_1 - C_2 t, \quad \lambda_2 = C_2, \quad \lambda_3 = C_3$$

for some constants C_1, C_2 and C_3.

The minimization of H in (18.47) is well posed if the coefficient of u^2 is positive, i.e., $\lambda_3 = C_3 > 0$. We will only consider this case. The minimization of H gives

$$\frac{\partial H}{\partial u} = C_1 - C_2 t + 2C_3 u = 0$$

i.e., a control that varies linearly with time. By renaming the constants we can write

$$u = A + Bt \tag{18.49}$$

where $A = -C_1/(2C_3)$, $B = C_2/(2C_3)$. The solution to the differential equations (18.40) then becomes

$$\omega = At + Bt^2/2, \quad \theta = At^2/2 + Bt^3/6$$

The terminal constraints give

$$A = 6/t_f^2, \quad B = -12/t_f^3 \tag{18.50}$$

The third state is given by

$$W = \frac{6}{t_f^3} \left(\left(\frac{2t}{t_f} - 1 \right)^3 + 1 \right)$$

The condition $W(t_f) = W_0$ gives the following value of t_f.

$$t_f = \left(\frac{12}{W_0} \right)^{1/3}$$

Substituting into (18.49) via (18.50) gives the extremal

$$u = W_0 \left(\frac{t_f}{2} - t \right), \quad t_f = \left(\frac{12}{W_0} \right)^{1/3}$$

In Figures 18.2a and 18.2b the control signal and state variables are shown for the case $W_0 = 12$.

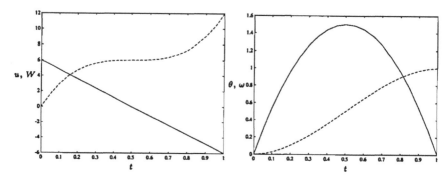

(a) The control signal u (solid) and the accumulated heat W (dashed).

(b) The angle θ (dashed) and the angular velocity ω (solid).

Figure 18.2: Time optimal control of a DC motor under power constraint.

18.3   Solution of the Goddard Rocket Problem

Let us now discuss the solution of the Goddard rocket problem, see Equations (18.1). The problem is not exactly of the type we have discussed, due to the inequality condition $m(t_f) \geq m_1$. Since it is fairly obvious that it is pointless to have fuel left at the final time, we can replace it with the equality constraint $m(t_f) = m_1$.

The Hamiltonian is

$$H = \lambda_1 \left(\frac{u - D(v, h)}{m} - g \right) + \lambda_2 v - \lambda_3 \gamma u$$

The λ-equations are

$$\dot{\lambda}_1 = -\lambda_2 + \lambda_1 D_v(v, h)/m, \quad \lambda_1(t_f) = 0$$
$$\dot{\lambda}_2 = \lambda_1 D_h(v, h)/m, \qquad\quad \lambda_2(t_f) = -n_0$$
$$\dot{\lambda}_3 = \lambda_1(u - D(v, h))/m^2, \quad \lambda_3(t_f) = \mu_1$$

where $D_v = \partial D/\partial v$, $D_h = \partial D/\partial h$. The coefficient of u in the Hamiltonian is

$$\sigma(t) = \frac{\lambda_1}{m} - \lambda_3 \gamma \tag{18.51}$$

The minimization of the Hamiltonian is achieved by the following choice of u.

$$u(t) = \begin{cases} 0 & \sigma(t) > 0 \\ u_{max} & \sigma(t) < 0 \end{cases}$$

The difficulties of solving the Goddard problem are related to the possibility of having $\sigma(t) = 0$. When this is the case, the Hamiltonian does not depend explicitly on u, and it is not obvious how u should be chosen. Problems where this phenomenon occur are called *singular*. That $\sigma(t)$ is zero at a single value of t does not matter, since the choice of u at a single value of t does not affect the solution of the differential equations. Complications arise if there is a time interval $[t_1, t_2]$ such that

$$\sigma(t) = 0, \quad t \in [t_1, t_2]$$

The optimal control problem is said to have a *singular trajectory* in the time interval $[t_1, t_2]$. Before discussing this possibility we will investigate the special case of no aerodynamic drag.

The Goddard Problem without Aerodynamic Drag

In the introductory remarks of Section 18.1 we said that the solution is intuitively clear when there is no drag. We will now confirm the intuitive reasoning by calculating an extremal given by the maximum principle. Since we have $D(v, h) \equiv 0$, the λ-equations become

$$\dot{\lambda}_1 = -\lambda_2, \qquad \lambda_1(t_f) = 0$$
$$\dot{\lambda}_2 = 0, \qquad \lambda_2(t_f) = -n_0$$
$$\dot{\lambda}_3 = \lambda_1 u/m^2, \qquad \lambda_3(t_f) = \mu_1$$

with the solution

$$\lambda_2 = -n_0, \qquad \lambda_1 = n_0(t - t_f)$$

We first investigate the possibility $n_0 = 0$, which gives

$$\lambda_1 = 0, \qquad \lambda_2 = 0, \qquad \lambda_3 = \text{konst.}$$

Since $H = 0$ for $t = t_f$ and H is constant, we have $H \equiv 0$. Since u has to be nonzero at some time (otherwise the rocket stays on the ground) it follows that $\lambda_3 = 0$. Then we get $\mu_1 = 0$, contradicting the condition that $[n_0 \ \mu^T]$ is nonzero. It follows that n_0 is nonzero, i.e., the problem is normal and we can take $n_0 = 1$. From (18.51) we get

$$\dot{\sigma}(t) = -\frac{\lambda_2}{m} + \frac{\lambda_1}{m^2}\gamma u - \gamma\frac{\lambda_1 u}{m^2} = \frac{1}{m} > 0$$

We see that an extremal must have a control signal of the form

$$u(t) = \begin{cases} u_{max} & 0 \le t < t_1 \\ 0 & t_1 \le t \le t_f \end{cases}$$

for some t_1, as we found by intuitive reasoning in Section 18.1. The value of t_f obviously is determined by the condition that the fuel is consumed.

The Goddard Problem with Aerodynamic Drag

To solve the general Goddard problem is difficult. It took about fifty years from the formulation of the problem until a fairly complete solution was available. The character of the solution of course depends on the properties of the function $D(v, h)$. For physically reasonable functions, the solution has the following character. The function σ of (18.51) is of the form

$$\sigma(t) = \begin{cases} < 0 & 0 \le t < t_1 \\ = 0 & t_1 \le t < t_2 \\ > 0 & t_2 \le t \le t_f \end{cases}$$

This means that

$$u(t) = \begin{cases} u_{max} & 0 \le t < t_1 \\ 0 & t_2 \le t \le t_f \end{cases}$$

What is the value of u in the intermediate interval $[t_1, t_2]$? During this interval we have $\sigma(t) \equiv 0$, which gives

$$\sigma(t) \equiv 0, \quad \dot{\sigma}(t) \equiv 0, \quad \ddot{\sigma}(t) \equiv 0, \quad \text{etc.}$$

We have

$$\sigma(t) = \frac{\lambda_1}{m} - \lambda_3\gamma \equiv 0 \tag{18.52}$$

Differentiation with respect to time gives

$$\dot{\sigma}(t) = \frac{1}{m^2}(\lambda_1 D_v - \lambda_2 m + \gamma D \lambda_1) \equiv 0$$

Substituting from (18.52) then gives

$$\lambda_3 \gamma (D_v + \gamma D) - \lambda_2 \equiv 0 \tag{18.53}$$

The time derivative of (18.53) is an expression which contains u explicitly. Solving for u gives

$$u = D + mg + m\frac{D_h - g\gamma \tilde{D} - \tilde{D}_h v}{\gamma \tilde{D} + \tilde{D}_v} \tag{18.54}$$

where

$$\tilde{D} = D_v + \gamma D$$

We have now an expression for u along a singular trajectory. Note that the control signal is in the form of a feedback. If (18.52) and (18.53) are used together with $H \equiv 0$ we get

$$\gamma \lambda_3 \left(-D - mg + (D_v + \gamma D)v \right) \equiv 0$$

Since $\lambda_3 \neq 0$ we get the condition

$$-D(v, h) - mg + (D_v(v, h) + \gamma D(v, h))v = 0 \tag{18.55}$$

This is the equation of a surface in the v-h-m-space in which a singular trajectory must lie.

It can now be shown that an extremal of Goddard's problem, for physically reasonable functions $D(v, h)$, takes the form $u(t) = u_{max}$ until equation (18.55) is satisfied. After that, the control signal is given by (18.54) until all fuel is spent. Finally, $u = 0$ until the speed is zero (this event specifies t_f). Possibly the surface given by (18.55) is not reached before the fuel is spent. In that case there is no singular trajectory and the control signal switches directly from u_{max} to zero, as in the case $D = 0$ above.

18.4   Minimum Time Problems

We now turn our attention to a special class of optimal control problems. In the general problem formulation (18.25), take $L(x(t), u(t)) = 1$, $\phi(t_f, x(t_f)) = 0$ with free final time t_f. The criterion is

$$\text{minimize} \int_0^{t_f} 1 \, dt = t_f$$

i.e., the time to satisfy the condition $\psi(t_f, x(t_f)) = 0$ is minimized. We had an optimization problem of this form in Example 18.3.

This type of problem, the *minimum time problem*, comes up in many situations, for instance when controlling industrial robots to perform certain tasks as fast as possible.

We will mainly consider the minimum time problem for linear systems with upper and lower bounds on the control signals. The problem formulation is then

$$\text{minimize} \quad t_f \tag{18.56a}$$
$$\dot{x} = Ax + Bu \tag{18.56b}$$
$$|u_i(t)| \le u_i^{max}, \quad i = 1, \ldots, m \tag{18.56c}$$
$$x(0) = x_0, \quad \psi(x(t_f)) = 0 \tag{18.56d}$$

For simplicity the control constraints are assumed to be symmetric around the origin.

The Hamiltonian of the problem is

$$H = n_0 + \lambda^T A x + \lambda^T B u \tag{18.57}$$

An extremal must then satisfy

$$\dot{\lambda} = -A^T \lambda, \quad \lambda(t_f) = \psi_x^T(t_f) \mu \tag{18.58}$$

$$H(x(t_f), u(t_f), \lambda(t_f)) = 0 \tag{18.59}$$

with u solving

$$\min_{|u_i(t)| \le u_i^{max}} \lambda(t)^T B u(t) \tag{18.60}$$

Introducing the vector $\sigma(t)$, defined by

$$\sigma(t)^T = \lambda(t)^T B \tag{18.61}$$

we can express the control signal as

$$u_i(t) = \begin{cases} u_i^{max} & \sigma_i(t) < 0 \\ -u_i^{max} & \sigma_i(t) > 0 \end{cases}$$

The control signal u_i thus achieves its maximal or minimal value as soon as σ_i is nonzero.

Definition 18.1 Bang-bang Control *A control signal which always takes its maximal or minimal value with finitely many switches between them is called a bang-bang control.*

Time-optimal control problems often have bang-bang solutions. We shall see that this property is coupled to the controllability of the system. Let the B-matrix have the columns b_i:

$$B = \begin{bmatrix} b_1 & b_2 & \dots & b_m \end{bmatrix}$$

Recall that the system is controllable from u_i, if the matrix

$$S_i = \begin{bmatrix} b_i & Ab_i & \dots & A^{n-1}b_i \end{bmatrix}$$

is nonsingular (Theorem 3.1). We now have the following result.

Theorem 18.5 *Assume that the system (18.56b) is controllable from u_i and that $\psi_x(t_f)$ has full rank. Then u_i is bang-bang along extremals of the minimum time problem.*

Proof: Assume that $\sigma_i(t)$ is zero for infinitely many values of t in the interval $[0, t_f]$. From the theorem of convergent subsequences follows the existence of some t_g and some sequence t_j, both in the interval $[0, t_f]$, such that

$$\sigma_i(t_j) = 0, \quad t_j \to t_g$$

Continuity implies that $\sigma_i(t_g) = 0$. Since $\sigma_i(t_j) = 0$ and $\sigma_i(t_{j+1}) = 0$, there is according to the Mean Value Theorem some intermediate point where $\dot{\sigma}_i$ is zero. Since this is true for any j, there is another sequence $\bar{t}_j$ of intermediate points such that

$$\dot{\sigma}_i(\bar{t}_j) = 0, \quad \bar{t}_j \to t_g$$

From continuity $\dot{\sigma}_i(t_g) = 0$. We can now repeat the reasoning about intermediate points for $\dot{\sigma}_i$ and conclude that $\ddot{\sigma}_i(t_g) = 0$. Continuing in this fashion we establish that the derivative of any order has to be zero:

$$\sigma_i^{(k)}(t_g) = 0, \quad k = 0, 1, 2, \dots$$

Since $\sigma_i(t) = \lambda(t)^T b_i$ and λ is the solution to (18.58) we get

$$\lambda(t) = e^{-A^T(t-t_f)}\psi_x^T(t_f)\mu$$

and

$$\sigma_i(t) = \mu^T \psi_x(t_f)e^{-A(t-t_f)}b_i$$

Differentiation gives

$$\sigma_i^{(k)}(t_g) = (-1)^k\mu^T\psi_x(t_f)e^{-A(t_g-t_f)}A^k b_i = 0, \quad k = 0,1,2,\ldots$$

Since the vectors $b_i,\ldots,A^{n-1}b_i$ are linearly independent (the controllability condition), we get

$$\mu^T\psi_x(t_f)e^{-A(t_g-t_f)} = 0$$

and, since ψ_x and $e^{-A(t_g-t_f)}$ have full rank, this implies $\mu = 0$. From (18.57) and (18.59) it follows that $n_0 = 0$. Then $[n_0 \ \mu^T]$ is zero, contradicting Theorem 18.4. It follows that $\sigma_i(t)$ can be zero only for finitely many values of t. The control u_i then has to be bang-bang.   □

Example 18.4:   A Mechanical System

Consider a mechanical system, where a unit mass moves in a straight line under influence of a bounded external force u, $|u| \le 1$. Let the position be x_1 and the velocity x_2. If the mass is also tied to a spring with spring constant 1, then the equations of motion are

$$\dot{x}_1 = x_2, \quad \dot{x}_2 = -x_1 + u, \quad |u| \le 1$$

Suppose we have the initial state

$$x(0) = [0 \ 6]^T$$

and that we want to reach the final state $x = 0$ as quickly as possible. Since the system is controllable and

$$\psi(x) = [x_1 \ x_2]^T, \quad \psi_x(x) = I$$

Theorem 18.5 can be used, and we conclude that an extremal is a signal switching between $+1$ and -1 finitely many times. To look at the control signal in detail, we form the Hamiltonian

$$H = n_0 + \lambda_1 x_2 + \lambda_2(-x_1 + u)$$

and conclude that (18.61) gives $\sigma = \lambda_2$ with λ satisfying

$$\dot{\lambda}_1 = \lambda_2, \quad \dot{\lambda}_2 = -\lambda_1$$

The function σ satisfies

$$\ddot{\sigma} + \sigma = 0$$

with a solution of the form

$$\sigma(t) = C_1 \sin(t + C_2)$$

We conclude that the time span between two changes of sign is π. The solution of the system equations for $u = \pm 1$ over a time interval $t_a \le t \le t_a + T$ is

$$x_1(t_a + T) = x_1(t_a)\cos T + x_2(t_a)\sin T \pm (1 - \cos T) \tag{18.62a}$$
$$x_2(t_a + T) = -x_1(t_a)\sin T + x_2(t_a)\cos T \pm \sin T \tag{18.62b}$$

In particular, we have for $T = \pi$

$$x_1(t_a + \pi) = -x_1(t_a) \pm 2 \tag{18.63a}$$
$$x_2(t_a + \pi) = -x_2(t_a) \tag{18.63b}$$

After some calculations one finds that the final state can be reached using three changes of sign, starting with $u = -1$. The result is

$$u(t) = \begin{cases} -1 & 0 \le t < T_1 \\ 1 & T_1 \le t < T_1 + \pi \\ -1 & T_1 + \pi \le t < T_1 + 2\pi \\ 1 & T_1 + 2\pi \le t < T_1 + 2\pi + T_2 \end{cases}$$

From (18.62) we get

$$x_1(T_1) = 6\sin T_1 + \cos T_1 - 1$$
$$x_2(T_1) = 6\cos T_1 - \sin T_1$$

and using (18.63) twice we get

$$x_1(T_1 + 2\pi) = 6\sin T_1 + \cos T_1 - 5$$
$$x_2(T_1 + 2\pi) = 6\cos T_1 - \sin T_1$$

and, finally, using (18.62)

$$x_1(T_1 + 2\pi + T_2) = (6\sin T_1 + \cos T_1 - 5)\cos T_2$$
$$+ (6\cos T_1 - \sin T_1)\sin T_2 + 1 - \cos T_2$$
$$x_2(T_1 + 2\pi + T_2) = -(6\sin T_1 + \cos T_1 - 5)\sin T_2$$
$$+ (6\cos T_1 - \sin T_1)\cos T_2 + \sin T_2$$

The condition that this final state is zero gives the following system of equations after some simplifications.

$$6\cos T_1 - \sin T_1 + \sin T_2 = 0$$
$$6\sin T_1 + \cos T_1 + \cos T_2 = 6$$

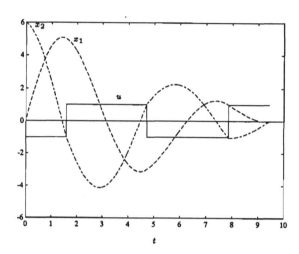

Figure 18.3: u, x_1 and x_2 as functions of time, using minimum time control of a mechanical system.

A solution is $T_1 = T_2 = \pi/2$. We get the extremal

$$u(t) = \begin{cases} -1 & 0 \leq t < \pi/2 \\ 1 & \pi/2 \leq t < 3\pi/2 \\ -1 & 3\pi/2 \leq t < 5\pi/2 \\ 1 & 5\pi/2 \leq t < 3\pi \end{cases}$$

shown in Figure 18.3.

In the mechanics example above, we can get arbitrarily many switches by placing the initial x far enough from the final x. For systems with real eigenvalues, there turns out to be an upper bound on the number of switches of a bang-bang control.

Theorem 18.6 *Let the conditions of Theorem 18.5 be satisfied and let the $n \times n$ matrix A have real eigenvalues. Define σ by (18.61). Then $\sigma_i(t)$ has at most $n - 1$ zeros, and consequently u_i changes sign at most $n - 1$ times.*

Proof: We consider first the situation when A has distinct eigenvalues $k_1, \ldots, k_n$. Using diagonalization it is easy to show that each element of e^{At} is a linear combination of $e^{k_1 t}, \ldots, e^{k_n t}$. This will also hold for the functions σ_i

$$\sigma_i(t) = c_1 e^{k_1 t} + \cdots + c_n e^{k_n t}$$

where not all the constants $c_1, \ldots, c_n$ are zero. By suitable ordering we can make one of the constants $c_2, \ldots, c_n$ nonzero. We see immediately that the theorem is true for $n = 1$. Suppose the theorem holds for n-values up to and including $n_0 - 1$. Then assume that

$$\sigma_i(t) = c_1 e^{k_1 t} + \cdots + c_{n_0} e^{k_{n_0} t}$$

bar n_0 zeros. Then the expression

$$c_1 + c_2 e^{(k_2 - k_1)t} + \cdots + c_{n_0} e^{(k_{n_0} - k_1)t}$$

also has n_0 zeros. But then its derivative must have $n_0 - 1$ zeros according to the Mean Value Theorem. This means that the expression

$$c_2(k_2 - k_1)e^{(k_2 - k_1)t} + \cdots + c_{n_0}(k_{n_0} - k_1)e^{(k_{n_0} - k_1)t}$$

has $n_0 - 1$ zeros contradicting the assumption that the theorem holds for $n = n_0 - 1$. We have thus shown the theorem for the case of all eigenvalues being distinct.

If all eigenvalues are equal $(= k)$ we have

$$\sigma_i(t) = (c_1 t^{n-1} + \cdots + c_{n-1} t + c_n)e^{kt}$$

and we see directly, since a polynomial of degree $n - 1$ cannot have more than $n - 1$ zeros, that the theorem holds.

The general case, where some, but not all, eigenvalues are multiple, can be shown by a combination of the techniques above.   □

Using Theorems 18.5 and 18.6, it is often possible to see directly what minimum time extremals look like.

Example 18.5: DC Motor

Consider the DC motor of Example 4.2

$$\dot{x}_1 = x_2, \quad \dot{x}_2 = -x_2 + u, \quad |u| \le 1$$

where we have added a bound on u. In matrix form we have

$$\dot{x} = \begin{bmatrix} 0 & 1 \\ 0 & -1 \end{bmatrix} x + \begin{bmatrix} 0 \\ 1 \end{bmatrix} u$$

Let the optimal control problem be to move from an arbitrary initial state to $x = 0$ in the shortest possible time. Since

$$[B \quad AB] = \begin{bmatrix} 0 & 1 \\ 1 & -1 \end{bmatrix}$$

is nonsingular, and the eigenvalues of A are real (0 and -1), we conclude from Theorem 18.6 that an extremal must be of the form

$$u(t) = \begin{cases} 1 & 0 \leq t < t_1 \\ -1 & t_1 \leq t \leq t_f \end{cases}, \quad \text{or} \quad u(t) = \begin{cases} -1 & 0 \leq t < t_1 \\ 1 & t_1 \leq t \leq t_f \end{cases}$$

There are two unknowns (t_1 and t_f) which can be determined from the two conditions $x_1(t_f) = 0$ and $x_2(t_f) = 0$. In this example it is also interesting to regard the problem from a geometric point of view in the x_1-x_2-plane. For $u = 1$ the state equations are

$$\dot{x}_1 = x_2, \quad \dot{x}_2 = -x_2 + 1$$

Eliminating the time variable by forming $dx_2/dx_1 = \dot{x}_2/\dot{x}_1$ gives

$$\frac{x_2}{1 - x_2} \frac{dx_2}{dx_1} = 1$$

with the solution

$$-x_2 - \log(|1 - x_2|) = x_1 + C \tag{18.64}$$

where C is an arbitrary constant. In the same way $u = -1$ gives

$$x_2 - \log(|1 + x_2|) = -x_1 + C \tag{18.65}$$

These curves are shown in Figures 18.4a, 18.4b. An extremal must thus be composed of a piece of curve from each figure in such a way that it ends up at the origin. Figure 18.4c shows the curve segments which end at the origin. Obviously an extremal must finish by following one of these curve segments to the origin. We realize that it is possible to reach the curve of Figure 18.4c by following curves corresponding to $u = -1$ in the upper part of the figure and curves corresponding to $u = 1$ in the lower part. The curve of Figure 18.4c summarizes (18.64) (the left part) and (18.65) (the right part) for $C = 0$ and can be written

$$x_1 = -x_2 + \text{sgn}(x_2) \log(1 + |x_2|) \tag{18.66}$$

Since we choose $u = -1$ above and $u = 1$ below the curve, the control signal can be written

$$u = -\text{sgn}\left(x_1 + x_2 - \text{sgn}(x_2) \log(1 + |x_2|)\right) \tag{18.67}$$

We have thus expressed the control signal as a *feedback* from the state variables. In Figure 18.4d the phase plane is shown with this feedback. The solid curve is the boundary between the sets where $u = 1$ and $u = -1$, the so called *switching curve*. The extremals for some different starting points are also shown. Note that they follow the switching curve towards the origin.

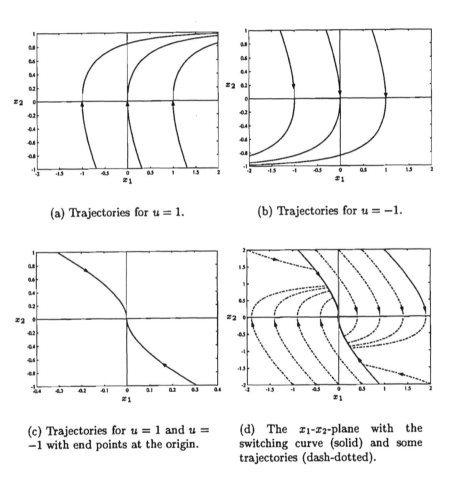

(a) Trajectories for $u = 1$.

(b) Trajectories for $u = -1$.

(c) Trajectories for $u = 1$ and $u = -1$ with end points at the origin.

(d) The x_1-x_2-plane with the switching curve (solid) and some trajectories (dash-dotted).

Figure 18.4: Minimum time control of a DC motor.

18.5   Optimal Feedback

In Example 18.5 we saw that it is sometimes possible to express the optimal control as a feedback from the state:

$$u = \ell(x)$$

where ℓ is some function, typically nonlinear. We will now discuss the calculation of an optimal feedback in more general cases. To do that we will consider a more general version of the optimal control problem, where both starting time and starting point vary. Denoting the starting time τ and the starting point z we have

$$\text{minimize} \int_{\tau}^{t_f} L(x(t), u(t))\, dt + \phi(t_f, x(t_f)) \qquad (18.68\text{a})$$

$$\dot{x}(t) = f(x(t), u(t)) \qquad (18.68\text{b})$$

$$u(t) \in U, \quad 0 \leq t \leq t_f \qquad (18.68\text{c})$$

$$x(\tau) = z, \quad \psi(t_f, x(t_f)) = 0 \qquad (18.68\text{d})$$

Assume that we find functions $u^*(t)$ and $x^*(t)$ which solve this optimization problem. We then define

$$V(\tau, z) = \int_{\tau}^{t_f} L(x^*(t), u^*(t))\, dt + \phi(t_f, x^*(t_f))$$

The function V, which we call *value function* or *optimal return function*, represents the optimal value of the criterion when starting at time τ and state z. From the definition we see that V has to satisfy

$$V(t_f, x) = \phi(t_f, x) \quad \text{if} \quad \psi(t_f, x) = 0 \qquad (18.69)$$

An important property of V is the following

Lemma 18.1 *Let u be arbitrary. If the optimal return $V(x(t))$ is differentiable with respect to time, where $x(t)$ is a solution to (18.68b), then*

$$\frac{d}{dt} V(t, x(t)) + L(x(t), u(t)) \geq 0 \qquad (18.70)$$

If $x(t), u(t)$ are optimal, then equality holds.

Proof: Suppose the control $u(t)$ is used during the time interval $[t, t+h]$ and that the control is optimal for $t \geq t + h$. The criterion then takes the value

$$\int_t^{t+h} L(x(r), u(r))\, dr + V(t + h, x(t + h))$$

from the definition of V. If the control is optimal all the time, the criterion takes the value $V(t, x(t))$. Since this value is optimal we must have

$$V(t, x(t)) \leq \int_t^{t+h} L(x(r), u(r))\, dr + V(t + h, x(t + h)) \qquad (18.71)$$

or, rearranging the terms and dividing by h

$$\frac{V(t + h, x(t + h)) - V(t, x(t))}{h} + \frac{1}{h} \int_t^{t+h} L(x(r), u(r))\, dr \geq 0$$

Letting $h \to 0$ gives (18.70). For an optimal control there is by definition equality in (18.71) which gives equality in (18.70). $\qquad\square$

Let us now assume that the optimal return function is differentiable and introduce the notation

$$V_t = \frac{\partial V}{\partial t}, \quad V_x = \begin{bmatrix} \frac{\partial V}{\partial x_1} & \cdots & \frac{\partial V}{\partial x_n} \end{bmatrix}$$

We now have the following basic result

Theorem 18.7 *If the value function V is differentiable, then it satisfies the relation*

$$-V_t(t, x) = \min_{u \in U} (V_x(t, x) f(x, u) + L(x, u)) \qquad (18.72)$$

Proof: A differentiation using the chain rule results in the relation

$$\frac{d}{dt} V(t, x(t)) = V_t(t, x(t)) + V_x(t, x(t)) f(x(t), u(t))$$

Substituting from (18.70) then gives

$$-V_t(t, x(t)) \leq V_x(t, x(t)) f(x(t), u(t)) + L(x(t), u(t))$$

with equality for an optimal u. The result is (18.72). $\qquad\square$

The partial differential equation (18.72) is called the *Hamilton-Jacobi equation* or the *Hamilton-Jacobi-Bellman equation*. We have shown in Theorem 18.7 that it is satisfied for an optimal return function which is differentiable. This is actually a severe restriction. If we look at, for instance, the minimum time problems considered in the previous section, we find that their optimal return functions, in general, have corners. They are therefore not differentiable at all points. It turns out that the Hamilton-Jacobi equation is satisfied also in these cases, provided V_x is suitably interpreted. We will not pursue this matter further here.

There are two important contributions from the Hamilton-Jacobi equation, from a control point of view. The first one is the possibility of generating an optimal feedback. When evaluating the right hand side of (18.72) we get

$$u = \ell(t,x), \quad \ell(t,x) = \mathrm{argmin}_{u \in U} \left(V_x(t,x)f(x,u) + L(x,u) \right)$$

where argmin denotes the minimizing value (for simplicity we assume that it exists and is unique). Since the control is a function of the state (possibly time-varying) this is a feedback form of the optimal control.

The second contribution of the Hamilton-Jacobi equation is its ability to show that an extremal really is optimal, that is to give *sufficient conditions* for optimality. All our results up to now have been necessary conditions.

Theorem 18.8 *Let $V(t,x)$ be a differentiable function satisfying the partial differential equation (18.72) and the end point condition (18.69). Let $u^*(t)$ be a control signal on the interval $[t_0, t_f]$ and let $x^*(t)$ be the corresponding state with $x^*(t_0) = x_0$. Assume that the constraints $u^*(t) \in U$ and $\psi(t_f, x^*(t_f)) = 0$ are satisfied and that the right hand side of (18.72) is minimized, i.e.,*

$$-V_t(t, x^*(t)) = V_x(t, x^*(t))f(x^*(t), u^*(t)) + L(x^*(t), u^*(t))$$

Then $u^*(t)$ is optimal and the corresponding value of the criterion is $V(t_0, x_0)$.

Proof: Let $u(t)$ be an arbitrary control signal satisfying $u(t) \in U$ and let $x(t)$ be the corresponding state, satisfying $x(t_0) = x_0$ and $\psi(t_f, x(t_f)) = 0$. Then the value of the optimization criterion is

$$\int_{t_0}^{t_f} L(x(t), u(t))dt + \phi(t_f, x(t_f)) = \int_{t_0}^{t_f} L(x(t), u(t))dt + V(t_f, x(t_f))$$

$$= V(t_0, x_0) + \int_{t_0}^{t_f} \left(L(x(t), u(t)) + \frac{d}{dt} V(t, x(t)) \right) dt = V(t_0, x_0)$$

$$+ \int_{t_0}^{t_f} (L(x(t), u(t)) + V_t(t, x(t)) + V_x(t, x(t)) f(x(t), u(t))) \, dt$$

$$\geq V(t_0, x_0)$$

where the inequality follows from the fact that V satisfies the Hamilton-Jacobi equation. If we replace u, x with u^*, x^*, then the last inequality is replaced by an equality. It follows that the value of the optimization criterion is $V(t_0, x_0)$ for u^*, x^* and larger for any other choice of u satisfying the constraints.   □

A problem with the Hamilton-Jacobi equation is the difficulty of computing a solution, since it is a nonlinear partial differential equation. We will give some examples where a solution can be found.

Example 18.6: Optimal Control of a Scalar System

Consider the optimal control problem

$$\min \frac{1}{2} \int_0^\infty (u^2 + x^4) dt, \quad \dot{x} = u \tag{18.73}$$

where there are no constraints. Since the control takes place on an infinite interval and all functions are time invariant, the value of the criterion must be the same, regardless of when the control begins. This means that the optimal return function has to be time independent:

$$V(t, x) = V(x), \quad \Rightarrow \quad V_t = 0$$

The Hamilton-Jacobi equation takes the form

$$0 = \min_u (V_x u + \frac{1}{2} u^2 + \frac{1}{2} x^4)$$

The minimization with respect to u gives

$$u = -V_x$$

which gives

$$V_x^2 = x^4$$

when substituted into the Hamilton-Jacobi equation. From the form of the criterion (18.73) it follows that $V(x) > 0$ when $x \neq 0$ and $V(0) = 0$. Therefore we choose the sign of V_x in the following way.

$$V_x = \begin{cases} x^2 & x \geq 0 \\ -x^2 & x < 0 \end{cases}$$

We then get

$$V(x) = \frac{1}{3}|x|^3$$

with the optimal feedback $u = -x|x|$.

Example 18.7: Linear Quadratic Problem

Let us now consider the problem

$$\min \frac{1}{2} \int_0^{t_f} \left(x^T(t)Q_1 x(t) + u^T(t)Q_2 u(t) \right) \, dt + \frac{1}{2} x^T(t_f) Q_0 x(t_f)$$

$$\dot{x} = Ax + Bu$$

where Q_2 is a positive definite matrix. This is a variant of the LQ-problem we considered in Chapter 9. There are no constraints and the final time t_f is fixed. The Hamilton-Jacobi equation becomes

$$- V_t(t,x)$$

$$= \min_u \left(\frac{1}{2} x^T Q_1 x + \frac{1}{2} u^T Q_2 u + V_x(t,x) Ax + V_x(t,x) Bu \right) \quad (18.74)$$

The condition (18.69) gives

$$V(t_f, x) = \frac{1}{2} x^T Q_0 x \tag{18.75}$$

Rearranging the right hand side gives

$$- V_t(t,x) = \min_u \left(\frac{1}{2} x^T Q_1 x + V_x(t,x) Ax - \frac{1}{2} V_x B Q_2^{-1} B^T V_x^T \right.$$

$$\left. + \frac{1}{2} (u + Q_2^{-1} B^T V_x^T)^T Q_2 (u + Q_2^{-1} B^T V_x^T) \right)$$

which is minimized by the choice

$$u = -Q_2^{-1} B^T V_x^T \tag{18.76}$$

The resulting V-equation is

$$-V_t(t,x) = \frac{1}{2} x^T Q_1 x + V_x(t,x) Ax - \frac{1}{2} V_x B Q_2^{-1} B^T V_x^T \tag{18.77}$$

Since the criterion is quadratic and V is quadratic at t_f according to (18.75), it is reasonable to guess that V is quadratic in x for every t. We will therefore try to satisfy the Hamilton-Jacobi equation with the function

$$V(t,x) = \frac{1}{2} x^T S(t) x$$

where $S(t)$ is a symmetric matrix to be determined. We get

$$V_t(t,x) = \frac{1}{2}x^T \dot{S}(t)x, \quad V_x(t,x) = x^T S(t)$$

Substituting into (18.77) gives

$$- x^T \dot{S}(t)x$$
$$= x^T Q_1 x + x^T S(t)Ax + x^T A^T S(t)x - x^T S(t)BQ_2^{-1}B^T S(t)x$$

where we have used the symmetrization $2x^T S(t)Ax = x^T(S(t)A + A^T S(t))x$. This equation is satisfied identically in x, if and only if $S(t)$ satisfies

$$-\dot{S}(t) = Q_1 + S(t)A + A^T S(t) - S(t)BQ_2^{-1}B^T S(t) \tag{18.78}$$

From (18.75), S has to satisfy the boundary condition

$$S(t_f) = Q_0 \tag{18.79}$$

and from (18.76) the control signal is given by

$$u = -L(t)x, \quad L(t) = Q_2^{-1}B^T S(t) \tag{18.80}$$

We have thus shown that we can solve the Hamilton-Jacobi equation, provided the *Riccati equation* (18.78), (18.79) can be solved. Note that $S(t)$ and $L(t)$ normally vary with time. The optimal control is thus a linear but *time-varying* feedback.

What happens when $t_f = \infty$? We can reason as we did in Example 18.6 and conclude that V must be a function of x only. It is then natural to test a solution of the form $V(t,x) = V(x) = \frac{1}{2}x^T Sx$ where S is constant. Equation (18.78) is replaced by

$$0 = Q_1 + SA + A^T - SBQ_2^{-1}B^T S \tag{18.81}$$

which is precisely Equation (9.7b) of Section 9.2. The regulator that we discussed there can be seen as a special case of a more general optimal control problem. One can also show that (18.81) can be obtained as a limit of (18.78) when $t_f \to \infty$ by showing that $S(t)$ tends to a stationary value (under certain regularity conditions).

Example 18.8: Time-Varying Feedback

Let us consider the following special case of the previous example.

$$\min \int_0^1 \frac{1}{2}u^2 dt + \frac{1}{2}x(1)^2, \quad \dot{x} = u$$

The Riccati equation (18.78), (18.79) is

$$-\dot{S}(t) = -S(t)^2, \quad S(1) = 1$$

with the solution

$$S(t) = \frac{1}{2-t}$$

and the optimal control

$$u(t) = -\frac{x(t)}{2-t}$$

Note that the gain of the feedback increases as the final time comes closer. This is logical, since the value of x is penalized by the criterion only at the final time.

18.6   Numerical Methods

We have seen that solving optimal control problems can be quite complicated. For many problems it is not possible to get closed form analytic expressions for solutions. It is then natural to look for numerical solution methods.

Numerical Methods for Open Loop Control

When looking for a solution in the form of an open loop control $u^*(t)$, it is natural to use the maximum principle, Theorem 18.4. We will only discuss normal extremals, i.e., the case $n_0 = 1$. Let us also specialize to the case when the minimization of the Hamiltonian

$$\min_{u \in U} H(x, u, \lambda(t), n_0)$$

has a unique solution u for every choice of x and λ:

$$u = h(x, \lambda)$$

If the vector x has dimension n, we get a system of $2n$ differential equations

$$\dot{x} = f(x, h(x, \lambda)) \tag{18.82}$$

$$\dot{\lambda} = -f_x^T(x, h(x, \lambda))\lambda - L_x^T(x, h(x, \lambda)) \tag{18.83}$$

These differential equations have conditions both at $t = 0$ and the final time $t = t_f$. It is thus a *two point boundary value problem* for ordinary differential equations. This is an important point. In principle, it is not difficult to solve differential equations numerically if all variables are given at the initial point. In this case, however, we must find a solution which fits at both ends. A natural approach would be to guess $\lambda(0)$ and then solve the differential equations up to time t_f. Then the solution could be adjusted depending on the deviations in $x(t_f)$ and $\lambda(t_f)$ from the boundary conditions. There is, however, a fundamental difficulty, which is most easily described for the case of a linear f:

$$f(x, u) = Ax + Bu$$

The λ-equation becomes

$$\dot{\lambda} = -A^T \lambda - L_x^T(x, h(x, \lambda))$$

If A has its eigenvalues in the left half plane, so that the $\dot{x}$-equation is stable in the forward direction, the $\dot{\lambda}$-equation will be unstable in the forward direction. This will often lead to numerical difficulties. Many numerical methods therefore solve the $\dot{x}$-equations in the forward direction and the $\dot{\lambda}$-equations in the backward direction.

It is possible to show that the solution of the $\dot{\lambda}$-equations backwards in time, in principle, corresponds to the calculation of a gradient of the optimization problem. Some numerical methods can therefore be regarded as "steepest descent" methods for optimal control. If the λ-equations are augmented with a Riccati equation, second order effects can be estimated and it is possible to design Newton-type methods.

Numerical Calculation of a State Feedback

Numerical computation of a feedback control law could be based on the Hamilton-Jacobi equation (18.72):

$$-V_t(t, x) = \min_{u \in U} \left(V_x(t, x) f(x, u) + L(x, u) \right) \tag{18.84}$$

$$u = \ell(t, x), \quad \ell(x) = \operatorname{argmin}_{u \in U} \left(V_x(t, x) f(x, u) + L(x, u) \right) \tag{18.85}$$

If f and L are real analytic (that is, given by convergent power series) close to the origin, it is possible to calculate a feedback by specifying power series expansions for $V(t,x)$ and $\ell(t,x)$. If these series are substituted into (18.84) and (18.85), a number of equations are obtained for the coefficients of the series. Solving these equations determines the coefficients of the power series. The coefficients of the lowest order terms are obtained from a Riccati equation, while the higher order terms lead to linear equations. In principle, one can determine arbitrarily many terms in the series expansions of V and ℓ. It is possible to show that these series converge (under mild conditions) so that V and ℓ are real analytic functions close to the origin. The computational complexity is however very high for this methodology. There is furthermore no guarantee that the series expansions converge, except close to the origin.

18.7   Comments

Main Points of the Chapter

- The optimal control problem is

$$\text{minimize} \int_0^{t_f} L(x(t), u(t))\, dt + \phi(x(t_f))$$
$$\dot{x}(t) = f(x(t), u(t))$$
$$u(t) \in U, \quad 0 \le t \le t_f$$
$$x(0) = x_0, \quad \psi(x(t_f)) = 0$$

- The Hamiltonian of a normal optimal control problem is

$$H(x, u, \lambda) = L(x, u) + \lambda^T f(x, u)$$

- Necessary conditions for x^*, u^* to be optimal ("the Maximum principle") for a normal problem are

$$\min_{u \in U} H(x^*(t), u, \lambda(t)) = H(x^*(t), u^*(t), \lambda(t)), \quad 0 \le t \le t_f$$

where λ satisfies

$$\dot{\lambda}(t) = -H_x(x^*(t), u^*(t), \lambda(t))^T$$
$$\lambda(t_f)^T = \phi_x(t_f, x^*(t_f)) + \mu^T \psi_x(t_f, x^*(t_f))$$

If the choice of t_f is included in the optimization problem, then there is the additional condition

$$H(x^*(t_f), u^*(t_f), \lambda(t_f)) = -\phi_t(t_f, x^*(t_f)) - \mu^T \psi_t(t_f, x^*(t_f))$$

- An optimal feedback can be calculated by solving the Hamilton-Jacobi equation

$$-V_t(t, x) = \min_{u \in U} \left(V_x(t, x) f(x, u) + L(x, u) \right)$$

The minimizing u is a function of x (and possibly t) and constitutes the optimal feedback.

Literature

Optimal control has its roots in the branch of mathematics called calculus of variations, with contributions by Euler, Lagrange, Hamilton, Weierstrass and others. In the 1940s, the American mathematician Hestenes showed how optimal control problems arising in aeronautical applications could be solved, and formulated a first version of the maximum principle. A more general version was presented 1956 by Pontryagin, Boltyanski, and Gamkrelidze. Work on optimal feedback and the Hamilton-Jacobi equation was done by Bellman, Boltyanski, and others around 1960. The work on aerospace problems led to computer based algorithms formulated by Bryson, Jacobson, Mayne and others around the same time.

Two classical textbooks on optimal control, with numerous examples from aerospace applications are Bryson & Ho (1975) and Leitmann (1981). The proof of the maximum principle in Appendix 18A is based on Fleming & Rishel (1975).

Software

There is a toolbox – RIOTS95 – for MATLAB which can be used to solve optimal control problems.

Appendix 18A: Proof of the Maximum Principle

In this appendix we discuss a geometric interpretation of normal and abnormal problems and sketch a proof of the Maximum Principle. We consider the optimal control problem with terminal constraints. To get necessary conditions for optimality, we can use the same tools as in Section 18.2. We use the same type of variation of the control signal, (18.4) and get as in (18.10) the equation

$$\phi(x(t_f, \epsilon)) = \phi(x^*(t_f)) + \epsilon\phi_x(x^*(t_f))\eta(t_f) + \rho_1(\epsilon) \tag{18.86}$$

where ρ_1 satisfies $\lim_{\epsilon \to 0} \rho_1(\epsilon)/\epsilon = 0$. In the same way, the terminal constraints ψ satisfy

$$\psi(x(t_f, \epsilon)) = \psi(x^*(t_f)) + \epsilon\psi_x(x^*(t_f))\eta(t_f) + \rho_2(\epsilon) \tag{18.87}$$

where ρ_2 is subject to the same conditions as ρ_1. Assuming that the terminal condition $\psi(x^*(t_f)) = 0$ is satisfied, we get

$$\psi(x(t_f, \epsilon)) = \epsilon\psi_x(x^*(t_f))\eta(t_f) + \rho_2(\epsilon) \tag{18.88}$$

Let us consider the problem geometrically in a coordinate system, where we plot $\phi(x(t_f, \epsilon)) - \phi(x^*(t_f))$ along the first axis and $\psi(x(t_f, \epsilon))$ along the other ones. The first order (in ϵ) change in these quantities made by the control (18.4) is then proportional to the vector

$$v = \begin{bmatrix} \phi_x(x^*(t_f))\eta(t_f) \\ \psi_x(x^*(t_f))\eta(t_f) \end{bmatrix} \tag{18.89}$$

See Figure 18.5. Now suppose that two changes of the type given by (18.4)

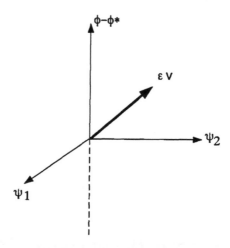

Figure 18.5: First order change of the criterion ϕ and the constraint ψ when changing u. The vector v is given by (18.89).

are made:

$$u(t) = \begin{cases} u & t_1 - \epsilon_1 \leq t \leq t_1 \\ \hat{u} & t_2 - \epsilon_2 \leq t \leq t_2 \\ u^*(t) & \text{otherwise} \end{cases} \qquad (18.90)$$

If the change of u at t_1 alone gives the vector

$$\epsilon_1 v_1 = \epsilon_1(\phi_x(x^*(t_f))\eta_1(t_f), \ \psi_x(x^*(t_f))\eta_1(t_f))$$

and the change at t_2 alone gives the vector

$$\epsilon_2 v_2 = \epsilon_2(\phi_x(x^*(t_f))\eta_2(t_f), \ \psi_x(x^*(t_f))\eta_2(t_f))$$

then the total change gives the first order change

$$\epsilon_1 v_1 + \epsilon_2 v_2, \quad \epsilon_1 > 0, \ \epsilon_2 > 0$$

We thus get linear combinations with positive coefficients of vectors that have the form (18.89). Plotting all those linear combinations for all choices of u of the form (18.4), we get a cone, see Figure 18.6. Intuitively it is clear that this

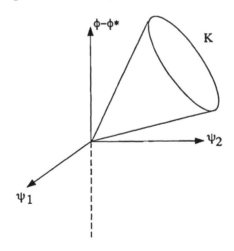

Figure 18.6: Changes of criterion and constraints when u is changed. The first order changes are in the cone K.

cone cannot contain the dashed ray of the figure, because in that case it would be possible to change the control signal to get a better value of the criterion while still satisfying the constraints (at least to first order). It is then also intuitively clear that one can find a hyperplane separating the cone and the dashed ray, see Figure 18.7. Letting $n^T = [n_0 \ \mu^T]$ (where μ is a vector of the

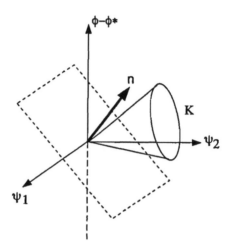

Figure 18.7: The cone K with the normal n of a separating plane.

same dimension as ψ) be the upward-pointing normal of this hyperplane, we have

$$n_0 \geq 0, \quad n^T k \geq 0$$

for all vectors k of the cone. We have thus given an intuitive derivation of the following lemma.

Lemma 18.2 *Let u^*, x^* solve the optimal control problem (18.25) and let K be the cone generated by all positive linear combinations of all vectors of the form (18.89) when all possible variations of type (18.4) are made. Then there is a nonzero vector n, whose first coordinate is zero or positive, such that*

$$n^T k \geq 0, \quad \textit{for all } k \textit{ in } K \tag{18.91}$$

Proof: Even if our geometric intuition tells us that the lemma should be true, according to the reasoning above, a strict mathematical proof is difficult. The difficulties are related to the presence of the nonlinear terms ρ_1 and ρ_2 in (18.86) and (18.88). A formal proof can be based on the Brouwer fixed point theorem.   □

The normal n of the plane can be chosen to have the first component n_0 strictly greater than zero, except in the case where the dashed ray of Figure 18.6 is tangent to the cone. In the latter case a vertical plane has to be chosen, giving $n_0 = 0$. The optimal control problem is then said to be *abnormal*. In the more typical case of $n_0 > 0$ the problem is said to be *normal*. We will see below that this is in agreement with the definitions given in Section 18.2.

Since the length of the vector n is unimportant in (18.91), we can choose $n_0 = 1$, so that n takes the form $n^T = [1 \ \mu^T]$. Since the cone K in particular contains vectors of the form (18.89), we get from (18.91) the following relation

$$0 \leq [1 \ \ \mu^T)] \begin{bmatrix} \phi_x(x^*(t_f))\eta(t_f) \\ \psi_x(x^*(t_f))\eta(t_f) \end{bmatrix}$$
$$= (\phi_x(x^*(t_f)) + \mu^T \psi_x(x^*(t_f)))\eta(t_f) \quad (18.92)$$

We can compare this with (18.13) in the derivation of Theorem 18.1. We see that the only difference is the replacement of ϕ with $\phi + \mu^T \psi$. With this change, we continue the calculations as we did in the derivations of Theorems 18.1 and 18.2. This gives directly (18.31) and (18.32) of Theorem 18.3. To get (18.33) we have to consider what happens if t_f is varied. We will do this for the case $L = 0$. Since x at the final time has a velocity of $\dot{x}^* = f(x^*(t_f), u^*(t_f))$, a change of t_f with the amount $\pm \epsilon$ gives the change $\pm \epsilon f(x^*(t_f), u^*(t_f))$ in $x^*(t_f)$ (to first order). The first order changes in ϕ and ψ are then given by $\pm \epsilon v$, where

$$v = \begin{bmatrix} \phi_t(t_f, x^*(t_f)) + \phi_x(t_f, x^*(t_f))f(x^*(t_f), u^*(t_f)) \\ \psi_t(t_f, x^*(t_f)) + \psi_x(t_f, x^*(t_f))f(x^*(t_f), u^*(t_f)) \end{bmatrix}$$

This vector has to be taken into the cone K of Figures 18.6 and 18.7. Equation (18.91) then gives $\pm n^T v \geq 0 \Rightarrow n^T v = 0$. For the normal case this relation can be rewritten

$$\phi_t(t_f, x^*(t_f)) + \mu^T \psi_t(t_f, x^*(t_f))$$
$$+ (\phi_x(t_f, x^*(t_f)) + \mu^T \psi_x(t_f, x^*(t_f)))f(x^*(t_f), u^*(t_f)) = 0$$

Using the final value of λ from (18.32) gives

$$\lambda(t_f)^T f(x^*(t_f), u^*(t_f)) = -\phi_t(t_f, x^*(t_f)) - \mu^T \psi_t(t_f, x^*(t_f))$$

which is (18.33) for the case $L = 0$. The generalization to nonzero L-functions is done in analogy with Theorem 18.2.

We have thus shown the Maximum Principle for normal problems, Theorem 18.3. The general formulation, Theorem 18.4, is shown by doing the calculations above for general n_0 and the general definition of H:

$$H(x, u, \lambda, n_0) = n_0 L(x, u) + \lambda^T f(x, u)$$

Chapter 19

Some Final Words

The past 450 pages have dealt with quite a bit of theory and methodology. Which are the central results? What should one remember of all this? How to approach a practical case? Let us here summarize the core material of the book.

Limitations

Capacity Limitations. An important message is that there are inherent performance limitations in all control systems, that do not depend on the controller or on how it is calculated. These limitations might be of practical nature, and are then related to bounds on the control signal and its rate of change. This may prevent us from making the system as fast as desired, or obtaining reasonable disturbance rejection. Scaling the variables and making back-of-an-envelope calculations as in Chapter 7 is the natural start of any controller design in practice.

Fundamental Limitations. The limitations may also be of principal character, like that the sensitivity may not be arbitrarily good, no matter what input amplitudes are allowed. In particular if the system is unstable and/or non-minimum-phase, there are strong basic performance limitations. The non-minimum phase property is particularly treacherous in the multivariable case, since it is not visible from the individual transfer functions. The discussion and results of Chapter 7 are therefore of central importance. There are cases where theoretically impossible performance specifications have been given, and where considerable time has been spent on trying to find an acceptable controller.

Disturbances and Noise. Another fundamental problem in the control design is related to the basic difference between system disturbances and measurement disturbances. The impact of the former on the controlled variable z should be damped or eliminated by the controller, while the latter's influence on the measured output y should be neglected, if possible, when u is formed. (If not, the measured output y would follow r, which means that the control error e would be equal to the measurement error n.) It is thus necessary to strike a balance in the treatment of measurement and system disturbances. For a linear system we have

$$y = z + n = z_u + z_w + n$$

where z_u is that component of z that originates from the control, and z_w originates from the system disturbance. Since w is not measured, there is no possibility to distinguish between these two contributions to y, other than by some *a priori* information. This may be of the form that "the measurement noise is primarily of high frequency, while the system disturbances are dominated by low frequency signals" or "there is a constant error is the sensors, due to misalignment of the gyros". The observer and the Kalman filter of Chapter 5 allow a good distinction between the disturbances, provided the model picks up their essential properties.

When is the Theory Required?

Multivariable Theory. A multivariable system that is almost decoupled, so that the different loops do not interact very much, can of course be well controlled by simple regulators, designing one SISO controller loop at a time. With strong cross couplings present, we have seen that it may be quite difficult to understand why such a controller does not work. It is then necessary to treat the multivariable system as one unit, studying the zeros and the singular values for the entire system, and not only for the individual loops.

Nonlinear Theory. All real-life systems are nonlinear, but this may be manifested in different ways. Suppose the system can be linearized around an equilibrium that varies much slower than the linear dynamics. Then using linear design methodology will often work well. Aircraft control is a typical example. The linearized dynamics (as in

Example 2.3) strongly depends on velocity and altitude, but these vary quite slowly compared to the time constants of the system. In other cases, the nonlinearity is essential, and then the nonlinear theory has to be used. See below.

Sampled Data Theory. It is important to realize that the sampling and time discretization give phase losses in the loop. This may mean that a controller that has been designed using continuous time models, may give worse stability margins than expected when it is implemented digitally. If the implementation shows some problems, the sampled data, closed loop system, according to Section 6.7 should be analyzed. The result may be that it is necessary to directly synthesize the sampled data controller using a discrete time description of the plant. Today continuous time design dominates, together with fast sampling, except for model predictive control, which by nature is sampled.

How to Choose the Design Method?

We have treated a number of different design methods in this book, and it is not easy to know which one to apply in a given case. It is also true that there is no "best" method for a given problem.

The basic principle, as in all engineering work, is "to try simple things first". If a simple PID-regulator gives satisfactory behavior, there is no need to bring in the $\mathcal{H}_\infty$-machinery.

Linear Systems. For a multivariable system the first task is to find out the strength of the cross couplings between the signals. This can simply be done by studying how steps in the different control inputs affect the outputs. For better understanding it may be necessary to set up and analyze a model of the system, as described in Part II. If simple PID-regulators are believed to be sufficient for the control, decoupled or decentralized control will be a good starting point. For better performance, the systematic design methods of Chapters 9 and 10 may be easier to use.

Perhaps, the most important message about linear controllers is to draw and scrutinize (and understand what is seen!) the singular values of the different closed loop transfer functions, regardless of how the controller was designed. The resulting system must also be simulated with all real-life constraints like input saturation and rate limitations, etc.

Nonlinear Systems. For nonlinear systems there are several approaches. When the system can be well described by a linearized system, with an operating point (equilibrium) that slowly varies, it is a normal approach to use linear design techniques for a number of typical operating points. Then the controller that corresponds to the closest operating point is used. This is the standard solution in, for example, aircraft control. Linear interpolation between adjacent controllers can also be used. This method is known as *Gain scheduling*.

If the nonlinearity primarily is caused by the actuator, like saturation, rate limitation, hysteresis, etc., it is important to analyze what this may lead to. The circle criterion and describing functions can be used for that. If the nonlinearity is essential in the sense that the limitation is active most of the time, the theory of Chapter 18 may give guidance for the controller construction. So called *fuzzy control* may be a "soft" way of implementing such nonlinear features in the controller. Otherwise, model predictive control is an increasingly popular method to handle input optimization in the presence of essential constraints.

The third and most difficult case is that the nonlinearity is part of the dynamics, and cannot be simplified by linearization, since the operating points change too fast. If a good model is available, like, e.g., for industrial robots, aircraft etc., it is natural to check if exact linearization can be used. A reasonable model can also be the basis of model predictive control.

In general, there is also industrial tradition that affects which design method is used. In process industry, for example, PID-thinking dominates, now more and more complemented by model predictive control. In the aerospace industry, the methods of Chapters 9 and 10, as well as optimal control have seen a breakthrough with the latest generation of products.

Bibliography

Anderson, B. D. O. (1967), 'An algebraic solution to the spectral factorization problem', *IEEE Trans. Autom. Control* **12**, 410–414.

Anderson, B. D. O. & Moore, J. B. (1979), *Optimal Filtering*, Prentice-Hall, Englewood Cliffs, NJ.

Anderson, B. D. O. & Moore, J. B. (1989), *Optimal Control: Linear Quadratic Methods*, Prentice-Hall.

Åström, K. J. & Wittenmark, B. (1997), *Computer-Controlled Systems, Theory and Design*, 3rd edn, Prentice-Hall, Englewood Cliffs, NJ.

Åström, K. J., Hagander, P. & Sternby, J. (1984), 'Zeros of sampled systems', *Automatica* **20**, 31–38.

Atherton, D. P. (1982), *Nonlinear Control Engineering*, Van Nostrand Reinhold, New York.

Bode, H. W. (1945), *Network Analysis and Feedback Amplifier Design*, Van Norstrand, New York.

Bracewell, R. N. (1978), *The Fourier Transform and Its Applications*, McGraw-Hill, New York.

Bristol, E. H. (1966), 'On a new measure of interactions for multivariable process control', *IEEE Trans. Autom. Control* **11**, 133–134.

Bryson, Jr., A. E. & Ho, Y. C. (1975), *Applied Optimal Control*, John Wiley, New York.

Camacho, E. F. & Bordons, C. (1995), *Model Predictive Control in the Process Industry*, Springer Verlag, New York.

Churchill, R. V. & Brown, J. W. (1984), *Complex Variables and Applications*, 4:th edn, McGraw-Hill, New York.

Cook, P. A. (1994), *Nonlinear Dynamical Systems*, Prentice Hall, Englewood Cliffs, NJ.

Dorf, R. C. & Bishop, R. H. (1995), *Modern Control Systems*, Addison-Wesley, Reading, MA.

Doyle, J. C. & Stein, G. (1981), 'Multivariable feedback design concepts for a classical modern syntheis', *IEEE Trans. Autom. Control* **26**, 4–16.

Fleming, W. H. & Rishel, R. W. (1975), *Deterministic and Stochastic Optimal Control*, Springer-Verlag, New York.

Franklin, G. F., Powell, J. D. & Amami-Naeini, A. (1994), *Feedback Control of Dynamic Systems*, 3rd edn, Addison-Wesley, Reading, MA.

Franklin, G. F., Powell, J. D. & Workman, M. L. (1990), *Digital Control of Dynamic Systems*, 2nd edn, Addison-Wesley, Reading, MA.

Freudenberg, J. S. & Looze, D. P. (1988), *Frequency Domain Properties of Scalar and Multivariable Feedback Systems*, Springer-Verlag, New York.

Friedland, B. (1988), *Control System Design*, McGraw-Hill, New York.

Glover, K. & McFarlane, D. (1989), 'Robust stabilization of normalized coprime factor plant descriptions with H_∞ bounded uncertainty', *IEEE Trans. Autom. Control* **34**, 821–830.

Golub, G. H. & Loan, C. F. V. (1996), *Matrix Computations*, 3rd edn, Johns Hopkins Univ. Press, Baltimore.

Heymann, M. (1968), 'Comments on pole assignment in multi-input linear systems', *IEEE Trans. Autom. Control* **13**, 748–749.

Horn, R. A. & Johnson, C. R. (1985), *Matrix Analysis*, Cambridge University Press.

Horowitz, I. M. (1993), *Quantative Feedback Design Theory*, QFT Publications, Boulder, Colorado.

Isidori, A. (1995), *Nonlinear Control Systems*, Springer-Verlag, New York.

James, H. M., Nichols, N. B. & Phillips, R. S. (1947), *Theory of Servomechanisms*, McGraw-Hill, New York.

Jury, E. I. (1958), *Sampled-Data Control Systems*, John Wiley, New York.

Kailath, T. (1980), *Linear Systems*, Prentice-Hall, Englewood Cliffs, NJ.

Kalman, R. E. (1960a), 'Contributions to the theory of optimal control', *Bol. Soc. Matem. Mex.* pp. 109–119.

Kalman, R. E. (1960b), 'A new approach to linear filtering and prediction problems', *Trans ASME, Ser. D. Journal Basic Eng.* **82**, 24–45.

Kalman, R. E. (1960c), On the general theory of control systems, *in* 'Proc. First IFAC Congress', Vol. 1, Butterworths, Moscow, pp. 481–492.

Kalman, R. E. & Bucy, R. S. (1961), 'New results in linear filtering and prediction theory', *Trans ASME, Ser. D. Journal Basic Eng.* **83**, 95–107.

Khalil, H. K. (1996), *Nonlinear Systems*, McMillan, New York.

Kwakernaak, H. & Sivan, R. (1972), *Linear Optimal Control Systems*, Wiley Interscience, New York.

Leigh, J. R. (1983), *Essentials of Nonlinear Control Theory*, Peter Peregrinus, London.

Leitmann, G. (1981), *The Calculus of Variations and Optimal Control*, Plenum Press, New York.

Ljung, L. (1999), *System Identification – Theory for the User*, 2nd edn, Prentice Hall, Upper Saddle River, NJ.

Ljung, L. & Glad, T. (1994), *Modeling of Dynamic Systems*, Prentice Hall, Upper Saddle River, NJ.

Lyapunov, A. M. (1992), *Stability of Motion*, Taylor and Francis, London.

Maciejowski, J. M. (1989), *Multivariable Feedback Design*, Addison-Wesley Publishing Company, Reading, MA.

Minkler, G. & Minkler, J. (1993), *Theory and Application of Kalman Filtering*, Magellan Book Company, Palm Bay, FL.

Moore, B. (1981), 'Principal component analysis in linear systems: Controllability, observability and model reduction', *IEEE Trans. Autom. Control* **26**, 17–31.

Morari, M. & Zafiriou, E. (1989), *Robust Process Control*, Prentice-Hall, Englewood Cliffs, NJ.

Newton, G. C., Gould, L. A. & Kaiser, J. F. (1957), *Analytical Design of Linear Feedback Control*, Wiley, New York.

Nijmeijer, H. & van der Schaft, A. (1990), *Nonlinear Dynamical Control Systems*, Springer-Verlag, New York.

Oppenheim, A. V. & Schafer, R. W. (1989), *Discrete-Time Signal Processing*, Pentice-Hall, Englewood Cliffs, NJ.

Papoulis, A. (1977), *Signal Analysis*, McGraw-Hill, New York.

Prett, D. M. & Garcia, C. E. (1988), *Fundamental Process Control*, Butterworths, Stonehan, MA.

Ragazzini, J. R. & Franklin, G. F. (1958), *Sampled-Data Control Systems*, McGraw-Hill, New York.

Safonov, M. G. (1980), *Stability and Robustness of Multivariable Feedback Systems*, MIT Press, Cambridge, MA.

Seborg, D. E., Edgar, T. F. & Mellichamp, D. A. (1989), *Process Dynamics and Control*, John Wiley, New York.

Seron, M. M., Braslavsky, J. H. & Goodwin, G. C. (1997), *Fundamental Limitations in Filtering and Control*, Springer-Verlag, New York.

Skogestad, S. & Postlethwaite, I. (1996), *Multivariable Feedback Control*, John Wiley, New York.

Slotine, J.-J. E. & Li, W. (1991), *Applied Nonlinear Control*, Prentice-Hall, Englewood Cliffs, NJ.

Stein, G. & Athans, M. (1987), 'The LQG/LTR procedure for multivariable feedback control design', *IEEE Trans. Autom. Control* **32**, 105–114.

Thompson, J. M. T. & Stewart, H. B. (1986), *Nonlinear Dynamics and Chaos*, Wiley.

Tsypkin, Y. Z. (1958), *Theory of Impulse Systems*, State Publisher for Physical Mathematical Sciences.

Van Dooren, P. (1981), 'A generalized eigenvalue approach for solving Riccati equations', *SIAM Journal Sci. Stat. Comp.* **2**, 121–135.

Verhulst, F. (1990), *Nonlinear Differential Equations and Dynamical Systems*, Springer-Verlag, New York.

Vidyasagar, M. (1993), *Nonlinear Systems Analysis*, Prentice-Hall, Englewood Cliffs, NJ.

Wiener, N. (1949), *The Extrapolation, Interpolation and Smoothing of Stationary Time Series*, John Wiley, New York.

Wong, E. (1983), *Introduction to Random Processes*, Springer-Verlag, New York.

Youla, D. C., Jabr, H. A. & Bongiorno, J. J. (1976), 'Modern Wiener-Hopf design of optimal controllers. Part II: The multivariable case.', *IEEE Trans. Autom. Control* **21**, 319–338.

Zhou, K., Doyle, J. C. & Glover, K. (1996), *Robust and Optimal Control*, Prentice-Hall, Englewood Cliffs, NJ.

Index

Index of Examples

Milton Keynes UK
Ingram Content Group UK Ltd.
UKHW031536071024
449327UK00023B/1844